Experimental Brain Research Supplementum 10

Hand Function and the Neocortex

Edited by
A. W. Goodwin and I. Darian-Smith

With 114 Figures and 5 Tables

Springer-Verlag
Berlin Heidelberg GmbH 1985

Dr. A. W. Goodwin
Dr. I. Darian-Smith

University of Melbourne, Department of Anatomy
Parkville, Victoria 3052, Australia

ISBN 978-3-642-70107-8   ISBN 978-3-642-70105-4 (eBook)
DOI 10.1007/978-3-642-70105-4

Library of Congress Cataloging in Publication Data. Main entry under title:
Hand function and the neocortex. (Experimental brain research. Supplemen-
tum; 10) Summary of a satellite meeting of the 29th Congress of the International Union of Physiological Sciences, held in Melbourne, Aug. 24–26,
1983. Includes bibliographies and index. 1. Cerebral cortex-Congresses.
2. Hand-Congresses. 3. Sensory-motor integration-Congresses. I. Goodwin,
A. W. II. Darian-Smith, I. (Ian) III. International Union Physiological Sciences.
Congress (29th: 1983: University of Sydney) IV. Series. [DNLM: 1. Cerebral   Cortex-physiology-congresses.   2.   Hand-physiology-congresses.
W1 EX485B v. 10 1983/WE 830 H2324 1983]
QP383.H36 1985 599.8'041852 84-23581

Offsetprinting: Beltz Offsetdruck, Hemsbach/Bergstr.
Binding: J. Schäffer OHG, Grünstadt

2125/3140-543210

Preface

To watch a curious young macaque explore and manipulate his immediate surroundings must surely delight and amaze us all. The monkey selects a particular object for scrutiny, reaches out and grasps it, handles it delicately with the fingers or more forcibly with the hand, scans the various surfaces with the finger pads, and finally identifies it. If we compare this monkey's adroitness and discriminative capacities with our own, at once we have a concise statement of important biological similarities and differences among primates. While there are species differences in the functional anatomy of the hand, in tactile sensibility, and in the control of hand movements, these are relatively minor. The real difference in the dexterity of the two species relates to the complexity of the task that is executed: no monkey, and for that matter only one man in a century, paints a Guernica, although an astute investigator can readily teach the monkey to flourish a brush and paint a few strokes.

The goal of the symposium, of which this book is a summary, was to examine current knowledge of those cortical mechanisms that determine the sensorimotor functions of the hand that are common to man and the monkey – mechanisms accessible to analysis by recording the responses of single cortical neurons while the monkey explores and manipulates his surroundings. The issue we hoped to highlight at this meeting was the remarkable interdependence of the sensory and motor functions of the hand and the neural processes mediating them. Dextrous hand movements depend not only on their appropriate planning and "programming" in neural terms but also on a continuous feedback of visual and somatic sensory information about the moment-to-moment position and movement of the hand. Similarly, identifying an object or surface by touch depends on the subject directing exploratory movements of the fingers that ensure a rapid and efficient acquisition of sensory information, movements that are continuously being adjusted to optimize this search.

This symposium, held in Melbourne August 24–26, 1983, was a satellite meeting of the 29th Congress of the International Union of Physiological Sciences, in Sydney. It was hosted by the Department of Anatomy, University of Melbourne. The Department of Anatomy of this University has a long tradition relating to hand function. Frederick Wood Jones, who wrote extensively on the evolution and functional anatomy of the hand, was professor of anatomy from 1930 to 1937. Later, Sir Sydney Sunderland occupied the same chair, during which time he wrote his greatly respected *Nerves and Nerve Injuries.* It was a great pleasure that Sir Sydney, now emeritus professor, participated in our meeting and was chairman of its first session.

Hand Function and the Neocortex was generously supported by the University of Melbourne, The Van Cleef Foundation, and ICI Australia Operations Pty Ltd., each of whom we thank warmly. The lecture given by Dr. E. V. Evarts was one of a series of public lectures organized by the University.

Finally, we wish to thank Heather Jessell for her attention to every detail of the meeting, ensuring that it ran smoothly.

Melbourne, Australia
December, 1984

A. W. Goodwin
I. Darian-Smith

Contents

List of Contributors*

Arbib, M. A.   *111*[1]
Caminiti, R.   *175*
Carlson, M.   *1*
Chapman, C. E.   *196*
Cheney, P. D.   *211*
Cicirata, F.   *259*
Darian-Smith, I.   *17*
Evarts, E. V.   *130*
Fetz, E. E.   *211*
Geffen, G.   *232*
Gemba, H.   *275*
Georgopoulos, A. P.   *175*
Goodwin, A.   *17*
Heywood, J.   *17*
Hikosaka, O.   *44*
Iberall, T.   *111*
Iwamura, Y.   *44*
Johnson, J. I.   *294*
Kalaska, J. F.   *175, 248*
Kasser, R. J.   *211*

Kubota, K.   *288*
Lamarre, Y.   *196*
Lederman, S. J.   *77*
Lyons, D.   *111*
Muir, R. B.   *155*
Nelson, R. J.   *59*
Nilsson, J.   *232*
Pons, C.   *259*
Quinn, K.   *232*
Rispal-Padel, L.   *259*
Roland, P. E.   *93*
Sakamoto, M.   *44*
Sasaki, K.   *275*
Smith, A. M.   *248*
Spidalieri, G.   *196*
Sugitani, M.   *17*
Tanaka, M.   *44*
Tanji, J.   *184*
Teng, E. L.   *232*
Wetts, R.   *248*

---

* The adress of the authors is given on the first page of each contribution

[1] Page, on which contribution commences

Significance of Single or Multiple Cortical Areas for Tactile Discrimination in Primates

M. Carlson

Departments of Psychiatry, Anatomy and Neurobiology, School of Medicine and McDonell Center for Higher Brain Function, Washington University, St. Louis, MO 63110, USA

As a student of primate biology, very early in my career I became interested in the changes that may have occurred in the hand and brain
of our ancestors during the course of primate evolution. Although
manual sensitivity and dexterity are notable achievements of primates,
the general structure of the hand is "primitive," relative to most
other mammalian species (Wood Jones 1916). To comprehend those evolutionary changes responsible for the greater behavioral capacity of the
human hand, it is necessary to look at the progressive developments in
the sensorimotor regions of the cerebral neocortex in a series of
representative primate species. Some of the living primate species,
the Old World prosimians and anthropoids, can be selected to represent
such an approximation of these developments (Clark 1959). Over the
last 10 years, in collaboration with other colleagues, I have examined
the organization of the primary somatic sensory cortex (SmI) and tactile discrimination capacity in a variety of primates in an attempt to
determine what features of cortical organization relate to increased
tactile capacity.

SmI in Prosimian Primates

Our first study of the SmI hand area was of the prosimian, *Galago
crassicaudatus*, a small, nocturnal, African species with a primitive
pentadactyl hand, similar in structure to fossil primates and the
human hand. In contrast with the earlier studies of *Macaca* by Paul
and colleagues (1972), in which two separately organized cutaneous

projections of the hand were described, in *Galago* we found only a single glabrous (G) projection pattern (Carlson and Welt 1980). As illustrated in Fig. 1, the anterior half of this single area receives
input from the glabrous digits and palm and the posterior half from
the dorsal hairy digits and dorsal hairy hand. Neurons with glabrous
receptive fields are associated with a dense granular architectonic
field, area 3, and those with receptive fields on the hairy digits and
hand with a less granular field, area 1. This appears to be the general pattern of SmI organization in primates: the separation of the
cortical projections from the glabrous and hairy surfaces of the hand,
each associated with a distinct architectonic field.

Next we examined another African species, *Perodicticus*, a slow-moving,
quadrupedal climber, whose second digit (D2) has been lost in the
adaptation of its hand for clinging and grasping (Fitzpatrick et al.
1982; see Fig. 2). The organization of the SmI hand area in this
species is similar to that in *Galago* in several ways: (a) there is
separation of glabrous and hairy input to anterior and posterior SmI
respectively and (b) each projection is associated with a distinctive
cytoarchitectural field. In addition, two special features are seen
in the SmI map of *Perodicticus* as presented in Fig. 1. The projection
areas for the second palm pad (P2, between D1 and D2) and the dorsal
hairy surface of the D2 stub are increased in size relative to those
in *Galago*. Only a small projection from the glabrous surface of the
stub D2 is found in SmI, among palm fields P1-P2. The mediolateral
width of the remaining digit projections are similar to those in
Galago, but the anteroposterior dimension of the glabrous digit projection is extended by twofold. Concomitant with that expansion, the
palm fields are compressed medially, posteriorly, and laterally into
the projection area for the hairy hand and digits. Interestingly, the
cytoarchitectural features of that portion of the palm projection pro-
truding into the hairy hand region are characteristic of area 1, the
hairy hand region. Although the glabrous area expands into the hairy
hand region, the architectonic features remain unchanged. The expansion of the remaining digit fields, just like the elimination of D2,
may have been related to the requirement for greater hand sensitivity
and control in this behaviorally specialized species.

An Asian species we studied, *Nycticebus*, has made a behavioral adapta-
tion quite similar to *Perodicticus* (Carlson and Fitzpatrick 1982).
The second digit in this species is present but reduced in size
(Fig. 2). The same general pattern seen in *Galago*, an anterior gla-

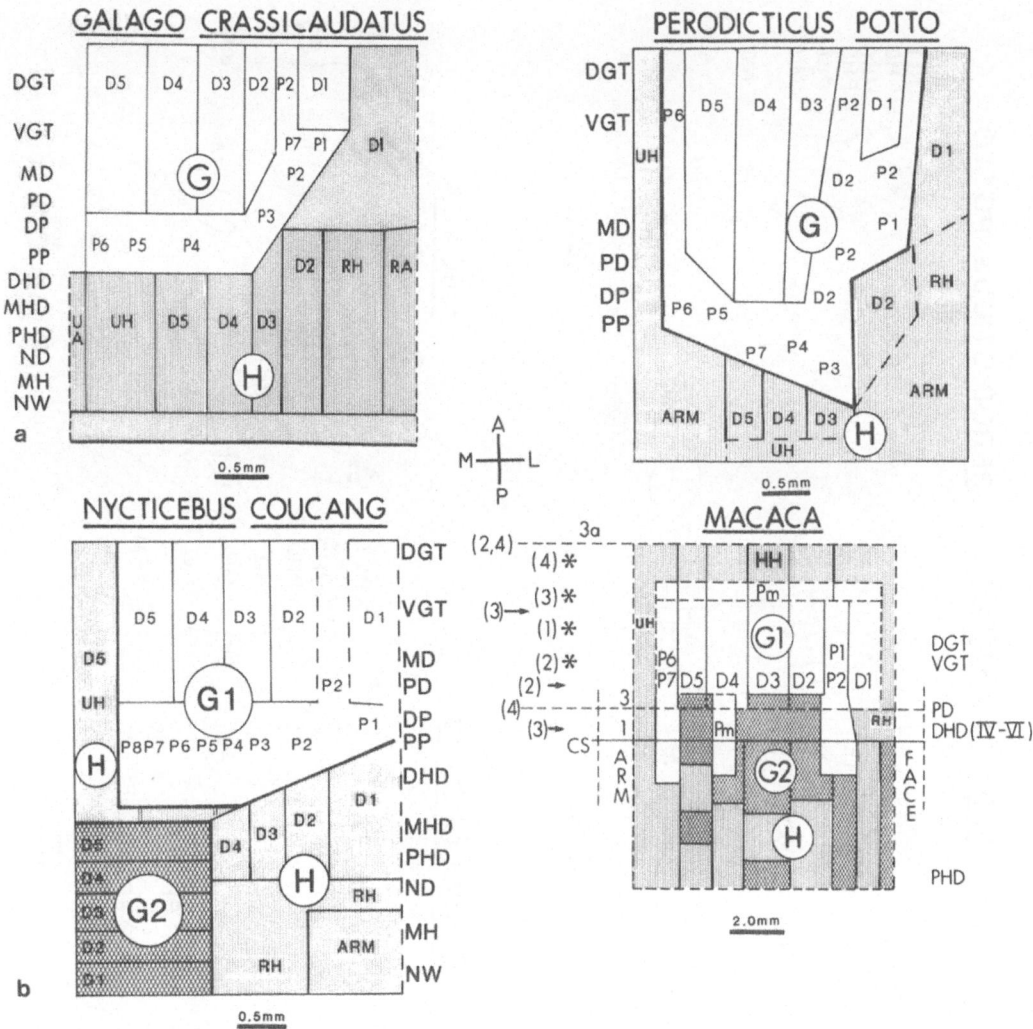

Fig. 1a,b. Summary diagrams of topographic relationships between gla-brous *(G)* and hairy *(H)* input from the hand to SmI. **a** Species with single glabrous projection pattern *(Galago* and *Perodicticus).* **b** Spe-cies with multiple glabrous projection *(Nycticebus* and *Macaca).* Cortical projection sites are labeled according to the abbreviations used in Fig. 2. In the *Macaca* diagram, numbers with *asterisks* indi-cate the extent of the area 3 hand area mapped in different studies: *(1)* Paul et al. 1972, *(2)* Kaas et al. 1979, *(3)* McKenna et al. 1982, and *(4)* Carlson et al. 1982. *Upper dashed line* (2,4) and highest *arrow* (3) indicate location of area 3-3a border in those same studies. The *lower dashed line* (4) and *lower arrows* (2) and (3) indi-cate the placement of area 3-1 border in studies referenced above. The posterior bank of the CS in *Macaca* is "folded up" into an antero-posterior plane for illustration. In all diagrams the *white areas* represent glabrous projections *(G), striped areas* represent hairy pro-jections *(H)*, and *crosshatched areas* represent the secondary glabrous *(G2)* projections to SmI. *HH* refers to large receptive fields on the hairy hand and *Pm* to fields on the palm

4

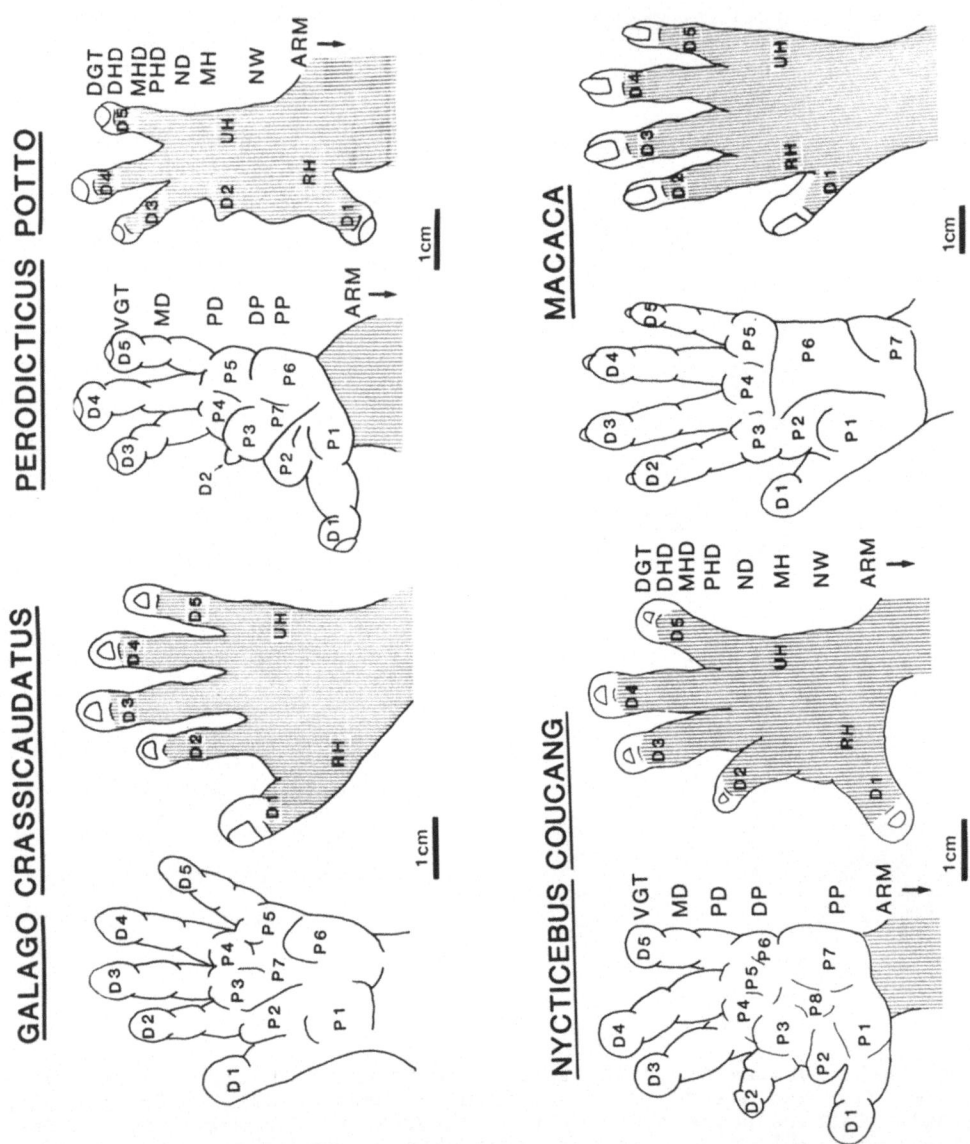

Fig. 2. Ventral and dorsal surfaces of the hands of three prosimian (*Galago*, *Perodicticus*, *Nycticebus*) and one anthropoid (*Macaca*) primate species. Glabrous skin is illustrated in *white* and hairy skin is *shaded*. The identification of the digits *(D1-D5)*, the palmar pads *(P1-P7 or P8)*, and radial *(RH)* and ulnar *(UH)* hand is given. The distal-to-proximal segments of the ventral digits are ventral glabrous tips *(VGT)*, middle digit *(MD)*, and proximal digit *(PD)*. The palm fields are distal pads *(DP)* and proximal pads *(PP)*. The segments of the dorsal surface of the digits are: dorsal glabrous tips *(DGT)*, distal hairy digits *(DHD)*, middle hairy digit *(MHD)*, and proximal hairy digit *(PHD)*. Fields on the dorsal hand are: near digits *(ND)*, middle hand *(MH)* and near wrist *(NW)*. Note the normal length of D2 in *Galago* and *Macaca*, the diminuation of D2 in *Nycticebus*, and the absence of that digit in *Perodicticus*

brous projection (G) and posterior hairy projection (H), is seen in this species - but with one novel addition. In the medial half of the posterior region, a second projection, primarily from the glabrous digits, is found. This second projection (G2) is not organized as a mirror image reversal of the first glabrous region (Kaas et al. 1979); rather, most fields are on the glabrous tips of the digits and the topographic pattern is at a 90° angle relative to the anterior pattern.

The primary glabrous projection in *Nycticebus* is associated with a separate architectonic field, like in *Galago*. In an earlier study of *Nycticebus* (Krishnamurti et al. 1976), only the glabrous input was mapped, and it was contained within a distinctive architectonic field, koniocortex (Ks), a homologue to area 3 in *Macaca*. This was similar to our findings. The projection of the hairy hand and digits that we described was co-extensive with their field, parakoniocortex (ParK), suggested as a homologue to area 1 or areas 1 and 2 (Sanides and Krishnamurti 1967). There was no distinction between the cytoarchitectonic features of the medially situated posterior glabrous field (G2) and the laterally placed hairy hand (H) projection in the posterior half of SmI. In spite of glabrous input to the area, the features were similar to those of the field receiving input from the hairy hand and digits in the other two species.

The appearance of a second glabrous projection in *Nycticebus* may, as suggested for *Perodicticus*, be another evolutionary mechanism for increasing sensory capacity by increasing the SmI hand area in this species, which has become more specialized for grasping than *Galago*. *Perodicticus* and *Nycticebus* appear to have independently made the same behavioral adaptation, but they have increased tactile sensitivity by increasing the cortical projection of glabrous receptors to cortex by very different mechanisms. In the first, the single digit area has expanded and, in the second, a novel projection appeared in a preexisting area.

SmI in the Anthropoid Primate, *Macaca*

Our original interest in prosimian species was to look for a simple SmI organization to compare with the complex topography described by Paul and colleagues (1972) for *Macaca*. However, a continuing controversy surrounds the specific details of SmI organization, particularly in area 1, in that genus. Conflicting studies prompted our own SmI mapping studies in anesthetized and in awake, moving *Macaca*. By use of these two approaches, we can compare results with our previous studies of prosimians and to all previous studies of *Macaca*.

In the anterior half of the postcentral gyrus, multineuronal responses were recorded near layer IV in two species of *Macaca*: *M. mulatta* and *M. nemestrina* (Carlson, Welt, and Fitzpatrick, unpublished observations). In area 1, large patches of input from the hairy digits were found among other patches of input from the glabrous digits (see Fig. 1). We estimated that about 25% of area 1 receives input from the hairy surface of the digits; this is similar to the findings of Paul et al. (1972).

In tangential penetrations through the crown of the gyrus near the 3-1 border, fields on the distal hairy digits (DHD) are found in the infragranular layers (IV-VI) in both anesthetized and awake *Macaca* (Carlson et al. 1982). Fields on the glabrous digits or palmar pads are located in the supragranular layers. Among the glabrous fields in this area, proximodistal receptive field sequences are seen in the dorsoventral (or anteroposterior) direction. Some of these progressions begin near the area 3-1 border and continue 1-2 mm into the sulcus. Other sequences begin in area 1 and continue *without reversals* across the 3-1 border. In some dorsoventral penetrations through the supragranular layers, fields on the digit tips may be found from area 1 continuously across the 3-1 border into area 3.

In the *Macaca* summary in Fig. 1, several features of the general plan proposed for *Galago* can be seen:
1. Area 3 contains projections primarily from glabrous digits and palm, with medial and lateral margins receiving input from the hairy hand and hairy surface of D1 respectively
2. Fields on the hairy surfaces of remaining digits (D2-D4) are in area 1, with distal hairy digits (DHD) at the 3-1 border
3. Areas G and H are associated with separate architectonic fields, namely areas 3 and 1.
In spite of the G2 projection to area H in *Macaca* and *Nycticebus*, these areas retain the architectonic features of area 1.

Differences from the *Galago* plan include an expansion of both palm (Pm) and proximal digit fields across the 3-1 border in some locations, similar to *Perodicticus*. In *Macaca*, the fields on hairy digits are not displaced medially, posteriorly, and laterally, as in *Nycticebus* and *Perodicticus*, but rather occupy the infragranular layers at the 3-1 border and remaining portions of area 1. In area 1 of *Macaca*, as in *Nycticebus*, a separate projection of the *glabrous digits* (mainly distal tips) occurs. In *Nycticebus*, the G2 area is concentrated in the medial corner of area 1; in *Macaca*, it appears distributed across the central region of area H. The intermittent patches of glabrous and hairy fields (also shown by Paul and colleagues 1972) may be due to slight variations in depth at which the receptive fields of neuron clusters are defined. Finally, if recordings are extended into the fundus of the CS, fields of the palm (see McKenna et al. 1982) and on the dorsal hairy hand can be found. Whether these fields are within area 3 or 3a depends on where borders are placed by individual investigators. In our studies, such fields were found above the area 3-3a border, the dashed line labeled (2,4) in Fig. 1.

These studies in *Macaca* were an effort to reconcile the debate concerning the numbers and organization of topographic patterns in areas 3 and 1 and their possible evolutionary origins. The general designation of cytoarchitectonic boundaries (indicated by dashed lines and arrows) and extent of area 3 mapped (see asterisks), shown in Fig. 1, appear as important differences among the various SmI maps. The area 3-1 border in our studies frequently fell at a point or transition from fields on DHD to fields on the proximal glabrous digits or palm. Kaas and colleagues (1979) place the 3-1 border slightly more ventrally, dividing the proximal glabrous digit or palm projection into two halves such that areas 3 and 1 appear as "roughly, but not precisely, mirror images of one another." However, if this area 3-1 border is moved less than 0.5 mm dorsally, the pattern described by Paul et al. (1972) as "serially related" is seen as a transition from DGT (or DHD, depending on cortical layer) to PD or palm fields. Neurons with receptive field centers on the digit tips are found abundantly at the area 2-1 border, at the area 3-1 border, and within area 1, primarily in the supragranular layers. Minor proximodistal variations in receptive field location may occur repeatedly in anteroposterior recording sequences throughout area 1.

Our studies in both anesthetized and awake, moving *Macaca* suggest that the real distinctions between findings of these studies are not due to methodology as much as to theoretical orientation. There is considerable variation in receptive field organization within area 1 across the area 3-1 border, wherever it may be placed, and between supra- and infragranular layers at different mediolateral locations in the hand area. Whitsel's group (McKenna et al. 1982) recorded much further down the bank (3) (see Fig. 2) than the Kaas group (2), where the vast majority of cells have fields on the distal digits, and they chose to emphasize the homogeneity of receptive field locations in the summary of their data. Kaas et al. (1979) recorded predominantly from the area 1-2 to a few millimeters past the 3-1 border, though they placed the 3-3a border near the fundus of the CS. They chose to emphasize the variations in receptive field sequences. Neither group has stressed the tremendous variabitity that occurs in the distoproximal sequences of receptive fields within area 1 and at the 3-1 border. Our interpretation of the SmI map or maps of SmI in *Macaca* is that the glabrous input in area 1, as in *Nycticebus*, represents an ectopic projection primarily from the digit tips, which is not somatotopically organized, but possibly organized by some other criteria. This new projection results in the displacement of the hairy input to deep cortical layers relative to that in the earlier evolved primate species. The functional significance of this second glabrous projection area in area 1 of *Macaca* will be discussed in the next sections describing tactile discrimination capacity in *Macaca* and *Galago*.

Tactile Discrimination Deficits in SmI-Lesioned Adult *Macaca*

My interest in the behavioral significance of multiple topographic patterns in the SmI hand area of primates was prompted by the innovative studies of Paul et al. (1972) showing separate topographies for the hand in area 1 and 3 in *Macaca*. Based on his preliminary studies, we (Randolph and Semmes 1974) ablated separately the two predominantly glabrous areas, 3 and 1, and the remaining SmI subdivision, area 2, which receives input primarily from noncutaneous tissues (Powell and Mountcastle 1959; Iwamura and Tanaka 1978). Following the unilateral removal of area 3, the acquisition of a variety of tactile discrimination tasks was prevented. When the two posterior subdivisions, areas 1 or 2, were separately removed, specific patterns of deficits were found. Removal of area 2 interfered selectively with the acquisition of shape and form discriminations (see Fig. 3). Removal of the second glabrous projection in area 1 interfered with acquisition of tasks involving texture differences (see Fig. 4). Interpretations

SIZE

◄──────── CONTRALATERAL   IPSILATERAL ────────►

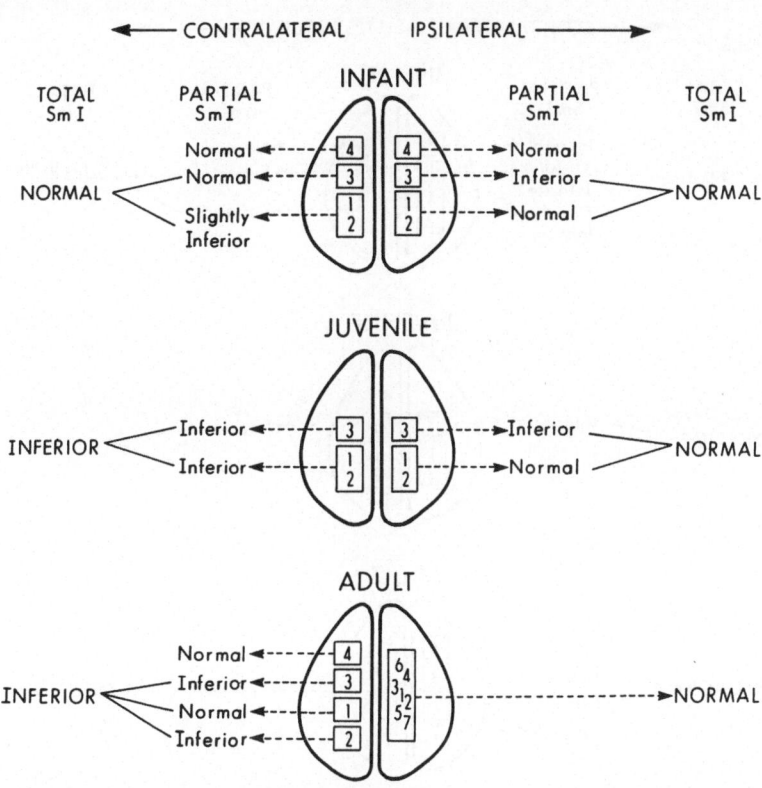

Fig. 3. Summary of size discrimination performance in infant, adult, and juvenile *Macaca* following contralateral or ipsilateral sensorimotor lesions. The *numbers* on the surfaces of the contralateral and ipsilateral hemispheres indicate the cytoarchitectural region or regions removed. In the case of partial SmI lesions in the infant, if area 4 (motor cortex) or area 3 hand region is removed unilaterally, size thresholds on the contralateral hand are normal, whereas if area 1 and 2 are removed in combination, thresholds are slightly inferior. In contrast, thresholds on the ipsilateral hand are normal following an area 4 and area 1-2 combined lesion but are inferior after the area 3 lesion. However, if areas 3, 1, and 2 are removed together (i.e., total SmI lesion), size thresholds on both the contralateral and ipsilateral hands are normal. In juvenile animals, both partial SmI lesions (i.e., areas 3 or 1-2 combined) and total SmI lesions produce inferior size threshold performance on the contralateral hand. On the ipsilateral hand, the area 3, but not the area 1-2 or total SmI, lesion influences size thresholds. In the adult animal, area 3, area 2, and total SmI lesions cause deficient size thresholds in the contralateral hand. However, removal of motor (areas 4, 6) and parietal regions (areas 3, 1, 2, 5, 7) combined does not influence size thresholds on the ipsilateral hand. Data from SmI-lesioned infant and juvenile *Macaca* are from Carlson (1984a, b, c), comparing performance on ALL threshold discrimination tasks (see text) with that of normal controls. Data on area 4 lesions in adults are from Semmes and Porter 1972; on areas 3, 1, and 2 from Randolph and Semmes 1974; on areas 1 and 2 from Carlson 1981; and on unilateral total SmI lesions from Semmes et al. 1974. Data on extensive unilateral sensorimotor removal of areas 6, 4, 3, 1, 2, 5, and 7 are from Semmes and Mishkin 1965

10

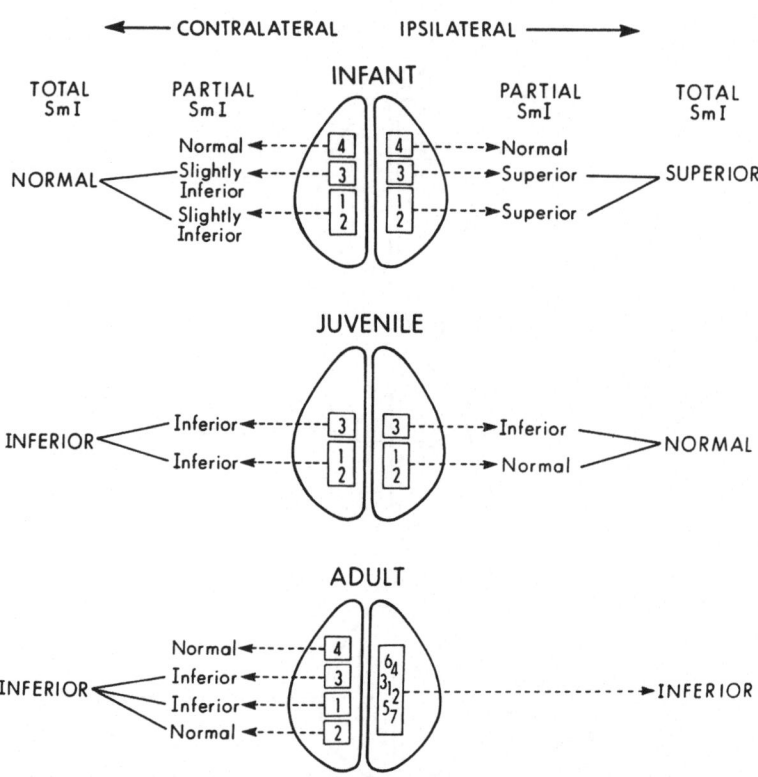

TEXTURE

Fig. 4. Summary of texture discrimination performance in infant, adult, and juvenile *Macaca* following contralateral or ipsilateral sensorimotor lesions. The conventions in this figure are the same as in Fig. 3

of the special consequences for lesions in areas 1 and 2 were based
upon the different distributions of cutaneous and noncutaneous classes
of input to the separate cytoarchitectonic areas in SmI (Powell and
Mountcastle 1959). Along with the findings of Paul et al. (1972),
these studies have led to current notions about the functional spe-
cialization with SmI subdivisions. Our original behavioral results
and their subsequent replication (Carlson 1981) were consistent with
the assumption that texture discrimination is dependent upon cutaneous
input to area 1 and form and size discrimination is dependent upon
noncutaneous input to area 2. Area 3, with its predominance of cutaneous input, is critical for all forms of tactile discrimination.
Though the specific contributions of areas 1 and 2 were demonstrated
in these studies, it seemed paradoxical that after partial SmI lesions, the two remaining areas (each with some redundancy of cutaneous
and noncutaneous input) (Powell and Mountcastle 1959; Hyvärinen and
Poranen 1978) could not mediate normal discrimination capacity.

**Behavioral Differences Between *Macaca* and *Galago* on Tactile Discrimi-
nation Tasks**

With a demonstration of the relationship between cutaneous input to
area 1 and texture discrimination, and noncutaneous input to area 2
and size discrimination, it is possible to compare the tactile discrimination capacity of species with different patterns of submodal
input to SmI and demonstrate this same brain-behavior relationship
without surgical intervention. We have recently completed such a comparative analysis of the complexity of SmI organization and behavioral
capacity of *Galago* and *Macaca* (Carlson and Charlton, unpublished observations). We tested *Galago* on the same graded series of size and
texture tasks used in previous studies of *Macaca* and found that although they made more errors during acquisition than *Macaca*, they
could master the same series of size tasks. On the complicated ALL
task (in which they were required to make relative discriminations
among six different pairs of stimuli ranging from 7 to 18 mm in diameter), they performed at 89% correct over all comparisons, as compared
with 92% for *Macaca*. Similarly on a texture acquisition task, *Galago*
made more errors during acquisition than *Macaca* prior to reaching criterion, but masterad all levels of difficulty of texture discriminations. Unlike performance on the size ALL tasks, they continued to
perform 10% or more below *Macaca* on the texture ALL task (in which six

pairs of sandpapers from 320 to 40 grains per linear inch were compared). Over all texture comparisons, *Macaca* performed at 89% correct compared with 77% for *Galago*. Just as the surgical removal of area 1 in adult macaques leads to inferior texture, but not size, performance, it is suggested that the difference in discrimination performance, seen only on texture comparisons, relates to the absence of the additional source of glabrous input to SmI in *Galago*.

Tactile Discrimination Deficits in Infant and Juvenile *Macaca* with SmI Lesions

In another series of studies, the relationship between SmI subdivisions and tactile capacity was examined again, this time from a developmental perspective. Beginning with studies of normal infants, it was discovered that infants as young as 8-10 weeks (when they first had the motor control to perform the task) could master size and texture discriminations at the level of adult thresholds (Carlson 1984a). When partial SmI lesions (area 3 or areas 1 and 2 combined) were made at 3-7 weeks of age, slight deficits in size acquisition and severe deficits in texture acquisition were observed several weeks postoperatively at the onset of testing (Carlson 1984b). Infants with a motor cortex or area 3 lesion were not different from normal infants on size discriminations with the contralateral hand, but infants with a lesion in areas 1 and 2 were slightly inferior to normals. Compared with adults from previous studies or animals operated as juveniles (18-24 mon, having preoperative training on the same tasks), infants performed remarkably well (see Fig. 3). However, with the hand ipsilateral to the lesion, infants and juveniles with area 3 lesions each showed inferior performance on the size ALL task. For the juveniles it may have resulted from the inferior performance on the contralateral hand providing less transfer to the second-trained, ipsilateral hand. For the infants, however, the result seems real, as they performed normally on the first-trained, contralateral hand. These ipsilateral deficits on size tasks in infants, but not in adults, with massive sensorimotor lesions (Semmes and Mishkin 1965) suggest that laterality of noncutaneous input or connections between areas of noncutaneous input may be different in the immature and adult macaque. Unfortunately, at this time data on SmI input and connections in primates is available only on adult animals.

On texture tasks, the results for the contralateral hand were similar
to those for size but considerably more striking. Severe impairment
was seen during texture acquisition, requiring from 2 to 4 months of
training to reach 80% criterion on all levels of difficulty. Lesions
in area 3 and in areas 1 and 2 produced severe deficits, implying that
each contributed to texture discrimination capacity at this age. When
tested on texture ALL tasks, infants with partial SmI lesions were
slightly but significantly inferior to normal infants with the contralateral hand (Fig. 4). An unexpected consequence of the partial
SmI lesions in infants was the finding of superior performance with
the ipsilateral hand by both groups of infants, even though performance on the first-trained, contralateral hand was slightly inferior
to normal infants. Among the animals operated as juveniles, both groups were inferior with the contralateral hand, and animals with area 3
lesions were inferior with the ipsilateral hand as well. The level of
recovered function in infants contrasts with that of the juveniles
(who received extensive preoperative training on the same tasks) and
the drastic impairment seen in adults with partial SmI lesions (Carlson 1981).

The superior performance on the texture ALL task and the inferior performance on the size ALL task with the ipsilateral hand suggest
differences in laterality of cutaneous as well as noncutaneous input
in the infant macaque as compared with the adult. Recent studies in
infant rodents have shown "exuberant" interhemispheric connections
between SmI and SmII, which retract as the infant matures. These connections have been shown to remain in the infant with an SmI lesion
(Caminiti and Innocenti 1981) providing a possible source of ipsilateral input to SmII from the contralateral SmI. Such an aberrant connection in infant macaques could explain the superior behavior with
the ipsilateral hand on the texture ALL task, but not the deficits on
the size ALL task or the recovery of tactile function found in the
contralateral hand. In an effort to determine if the remaining SmI
subdivisions in the lesioned hemisphere were mediating recovery in the
contralateral hand, infants with unilateral removals of areas 3, 1,
and 2 (total SmI lesion) were prepared.

The results of these studies not only leave open the question of what
areas contribute to recovered function in the infants with partial SmI
lesions but raise many additional questions as well. Partial SmI lesions in infants resulted in recovered tactile function, but total SmI
lesions resulted in "spared" function (Carlson 1984c). Infants with

total SmI lesions were not retarded on the acquisition of either size or texture tasks compared with normal infants. When tested on size and texture ALL tasks with the contralateral hand, their performance is also normal, although infants with partial SmI lesions are slightly below normal on these same tasks (see Figs. 3 and 4). As in the group with partial SmI lesions, performance on the texture ALL task with the ipsilateral hand is superior to that of normal infants. In juveniles with total SmI lesions, as in adults, severe deficits are seen in size and texture acquisition and inferior performance on size and texture ALL tasks (Carlson 1984c; Semmes et al. 1974). If studies of infants with partial SmI lesions had not preceded these studies of total SmI lesions, it might have appeared that SmI does not contribute to tactile capacity in infants. At the very least these studies raise the question of the projection of cutaneous and noncutaneous input to contralateral and ipsilateral SmI, SmII, and other sensorimotor areas in the immature primate. The results in the infants with total SmI lesions imply that those classes of input restricted to SmI in adult macaques are more widely distributed in the immature animal. Without some data on the normal physiology and anatomy of infant macaques, it is difficult to propose a unique mechanism for spared function after total SmI lesions in infants. We are currently examining the contribution of SmII to spared tactile function following neonatal SmI lesions (Carlson and Burton, research in progress).

Conclusions

The novel projection of glabrous input to area 1 in *Macaca* compared with *Galago* appears to be a remarkable evolutionary achievement and seems to confer an advantage upon this species for tactile discrimination capacity. Over the course of primate evolution, a multiplication of glabrous projection areas has occurred in diverse groups of primates (Carlson et al. 1982; Kaas et al. 1979; Paul et al. 1972). In the light of the recent data on development of tactile function in normal and *Macaca* with SmI lesions, it is evident that the significance of multiple cortical areas must be viewed from an ontogenetic as well as phylogenetic perspective. It is frequently the case in comparative studies that distinctions are made on the basis of mature members of a species. Yet, it is important to emphasize that ontogeny plays an important role in the evolutionary process (Gould 1977). If, as has been demonstrated in anatomical studies and suggested by the lesion experiments, somatic sensory projections and connections are more "exuberant" in the infant than in the adult primate, questions concerning the origins and significance of multiple somatic sensory areas in SmI of primates might be rephrased: Why does the adult *Galago* retain only a single area of glabrous input to the SmI hand area?

Acknowledgements. This work was supported by the National Institutes of Health (NIMH, NINCDS), the National Science Foundation (BNS), and the Milton Fund of Harvard University.

References

CAMINITI R, INNOCENTI GM (1981) The postnatal development of somato-sensory callosal connections after partial lesions of somatosensory cortex. Exp Brain Res 42: 53-62

CARLSON M (1981) Characteristics of sensory deficits following lesions of Brodmann's areas 1 and 2 in the postcentral gyrus of *Macaca mulatta*. Brain Res 204: 424-430

CARLSON M (1984a) Development of tactile discrimination capacity in *Macaca mulatta*. I. Normal infants. Developmental Brain Research 16: 69-82

CARLSON M (1984b) Development of tactile discrimination capacity in *Macaca mulatta*. II. Effects of partial removal of primary somatic sensory cortex (SmI) in infants and juveniles. Developmental Brain Research 16: 83-101

CARLSON M (1984b) Development of tactile discrimination capacity in *Macaca mulatta*. III. Effects of total removal of primary somatic sensory cortex (SmI) in infants and juveniles. Developmental Brain Research 16: 103-117

CARLSON M, FITZPATRICK, K (1982) Organization of the hand area in the primary somato sensory cortex (SmI) of the prosimian primate, *Nycticebus coucang*. J Comp Neurol 204: 280-295

CARLSON M, WELT C (1980) Somatic sensory cortex (SmI) of the prosimian primate *Galago crassicaudatus*: organization of mechanoreceptive input from the hand in relation to cytoarchitecture. J Comp Neurol 189: 249-271

CARLSON M, SHAO DH, XÜ BX (1982) Significance of topography and submo-dality in the organization of Brodmann's area 1 of SmI of *Macaca*. Soc Neurosci Abstr 8: 15.3

CLARK WE (1959) The antecedents of man. Edinburgh University Press, Edinburgh

FITZPATRICK K, CARLSON M, CHARLTON J (1982) Topography, cytoarchitecture, and sulcal patterns in primary somatic sensory cortex (SmI) of the prosimian primate, *Perodicticus potto*. J Comp Neurol 201: 296-310

GOULD SJ (1977) Ontogeny and phylogeny. Belknap, Cambridge

HYVÄRINEN J, PORANEN A (1978) Receptive field integration and submodality convergence in the hand area of the post-central gyrus of the alert monkey. J Physiol (Lond) 283: 539-556

IWAMURA Y, TANAKA M (1978) Postcentral neurons in hand region of area 2: their possible role in the form discrimination of tactile objects. Brain Res 150: 662-666

KAAS JH, NELSON RJ, SUR M, LIN C-S, MERZENICH MM (1979) Multiple representations of the body within the primary somatosensory cortex of primates. Science 204: 521-523

KRISHNAMURTI A, SANIDES F, WELKER WI (1976) Microelectrode mapping of modality-specific somatic sensory cerebral neocortex in slow loris. Brain Behav Evol 13: 267-283

McKENNA TM, WHITSEL BL, DREYER DA (1982) Anterior parietal cortical topographic organization in macaque monkey: a re-evaluation. J Neurophysiol 48: 289-314

PAUL RL, MERZENICH M, GOODMAN H (1972) Representation of slowly and rapidly adapting cutaneous mechanoreceptors of the hand in Brodmann's area 3 and 1 of *Macaca mulatta*. Brain Res 36: 229-249

POWELL TPS, MOUNTCASTLE VB (1959) Some aspects of the functional organization of the cortex of the postcentral gyrus of the monkey: a correlation of findings obtained in a single unit analysis with cytoarchitecture. Bull Johns Hopkins Hosp 105: 133-162

RANDOLPH M, SEMMES J (1974) Behavioral consequence of selective subtotal ablations in the postcentral gyrus of *Macaca mulatta*. Brain Res 70: 55-70

SANIDES F, KRISHNAMURTI A (1967) Cytoarchitectonic subdivision of sensorimotor and prefrontal regions and of bordering insular and limbic fields in slow loris (*Nycticebus coucang*). J Hirnforsch 9: 225-252

SEMMES J, MISHKIN M (1965) Somatosensory loss in monkeys after ipsilateral cortical ablation. J Neurophysiol 23: 473-486

SEMMES J, PORTER L (1972) A comparison of precentral and postcentral cortical lesions on somatosensory discrimination in the monkey. Cortex 8: 264-294

SEMMES J, PORTER L, RANDOLPH MC (1974) Further studies of anterior postcentral lesions in monkeys. Cortex 10: 55-68

WOOD JONES F (1916) Arboreal man. Arnold, London

Scanning a Textured Surface with the Fingers: Events in Sensorimotor Cortex

I. Darian-Smith, A. Goodwin, M. Sugitani, and J. Heywood

University of Melbourne, Department of Anatomy, Parkville, Victoria 3052, Australia

We grasp objects and explore their surfaces with the fingers every day of our lives. Wood Jones (1926), a former professor of anatomy in this University, would have quoted this uniquely primate action as illustrative of the "emancipation of the forelimb", an inheritance of our arboreal ancestry. Regardless of how this capacity to manipulate and explore objects with the hand evolved among primates, it is certainly true that identifying textured surfaces is the essence of the sense of touch. In order to examine the full capacities of the central nervous system to register and use tactile information, it is important to analyze the central neural events that occur while a subject scans various surfaces with the fingers and identifies them. We have been examining some of these events in the cerebral cortex of the monkey.

Consider first how one touches a surface to identify it. With most familiar surfaces it is sufficient to rub the finger pads to and fro across them a few times. The actual pattern of the scanning movement of the fingers is not critical, since this may be substantially changed without degrading the subject's capacity to identify the surface. Similarly, the contact force between the finger pads and the surface may be varied considerably without altering the subject's performance (Morley et al. 1983; Lederman 1974). Finally, with many, but not all, textured surfaces identification is as readily accomplished by moving the surface across the stationary finger as by actively scanning the stationary surface with the fingers (Lamb 1983). What is essential is that the surface moves tangentially relative to the skin. Vertical indentation of the skin alone without this lateral movement limits the subject to identifying only quite coarsely patterned surfaces.

Experimental Brain Research, Suppl. 10
© Springer-Verlag Berlin · Heidelberg 1985

These features of the subject's tactile performance imply certain matching features of the central neural events on which identification of the surface is based. Firstly, the tangential skin movement is essential to generate this activity. However, although sustained by this tangential movement, the parameters of the neural activity which define the spatial characteristics of the surface must be largely independent of the actual pattern of movement of the fingers.

In the experiments to be described, the surfaces scanned were rigid, plastic, regular gratings. The fingers scanned the grating at right angles to the ridges. Gratings, rather than more commonplace surfaces such as grained wood, sandpaper, or velvet, were used because they can be fully specified by two measures: the spatial frequency and the profile of the individual cycle. Furthermore, series of these gratings can be readily manufactured in which either the frequency or the cycle profile is changed by increments (Darian-Smith et al. 1982, 1984).

A human subject, by comparing pairs of periodic gratings, can reliably differentiate incremental changes in the spatial period of 3% - 5% for spatial periods in the range 750-1500 μm (Morley et al. 1983). If the subject freely selects the scanning movement, an oscillatory, near-sinusoidal pattern is commonly used. Lamb (1983), in our laboratory, has made similar, more extended observations on a subject's discrimination of geometric dot patterns and confirmed that resolution of these simple, periodic surfaces is largely independent of the pattern of finger movement that is used. This study also demonstrated that with these surfaces comparable resolution of the spatial pattern of the surface is achieved either by actively scanning the surface with the fingers or by moving the surface across the stationary finger pad.

Since a subject can resolve small incremental changes in the spatial frequency of a grating, this parameter of the surface must be represented with considerable fidelity in the discharge of the digital nerve fiber populations engaged during the scanning of the surface. Darian-Smith and Oke (1980) showed that this scanning action, in fact, engages all the low-threshold mechanoreceptive fibers that innervate the skin touching the grating. The Meissner, the pacinian, and the slowly adapting mechanoreceptive afferents each respond somewhat differently during the passage of the skin across the surface, but the discharge in each fiber always depends both on the characteristics of the grating and on the pattern of finger movement. This means that no single fiber provides unambiguous information about the spatial features of the surface: that information is confounded with information concerning the pattern of finger movement.

In order to specify the surface, the brain must have additional, independent information about the pattern of finger movement. Some possible sources of this information about the movement of the fingers are (a) the temporal order of the discharge in an assembly of mechanoreceptive fibers that innervate adjacent regions of the skin touching the surface (Darian-Smith and Oke 1980), (b) the discharge in the various groups of proprioceptive fibers activated by the scanning movement, and (c) the neuronal activity within the central nervous system which specifies the final motor discharge responsible for the scanning movement. When the fingers actively scan the surface, all of these sources may contribute to defining the movement of the skin relative to the surface. However, with the surface being moved across the stationary finger pad, only the responding digital nerve fiber population can signal this information to the central nervous system. This issue is considered further, later in this chapter.

Correlating Events in the Sensorimotor Cortex with the Tactile Scanning of Gratings

The present experiments were done using monkeys (*Macaca nemestrina*) trained to a visually guided manual tracking task (Fig. 1). The visual target was a vertical band displayed on an oscilloscope screen. The monkey inserted the fingers into a carriage overhanging the grating, so that the extended fingers, when slightly flexed at the metacarpophalangeal joints, made light contact with the surface. The carriage moved freely to and fro a few millimeters above the surface, being pivoted at a point below the elbow. The position of the carriage (and hence of the fingers) was signaled on the screen by a vertical cursor line, which brightened when the monkey pressed on the grating with a force between 30 and 80 g. The monkey was rewarded with water when he simultaneously positioned the cursor over the target and pressed on the grating with the appropriate contact force. The trained monkey could track sinusoidal oscillations with frequencies of 0.1 - 1.2 Hz, and a peak-to-peak amplitude of 80 mm with considerable precision for up to 2 h, although at the high frequencies he needed a short rest at intervals of 5 - 10 min.

Brodmann Area 2: Neuron with direction selective response
Fingers rubbed to and fro across grating (spatial period 3000 μm)

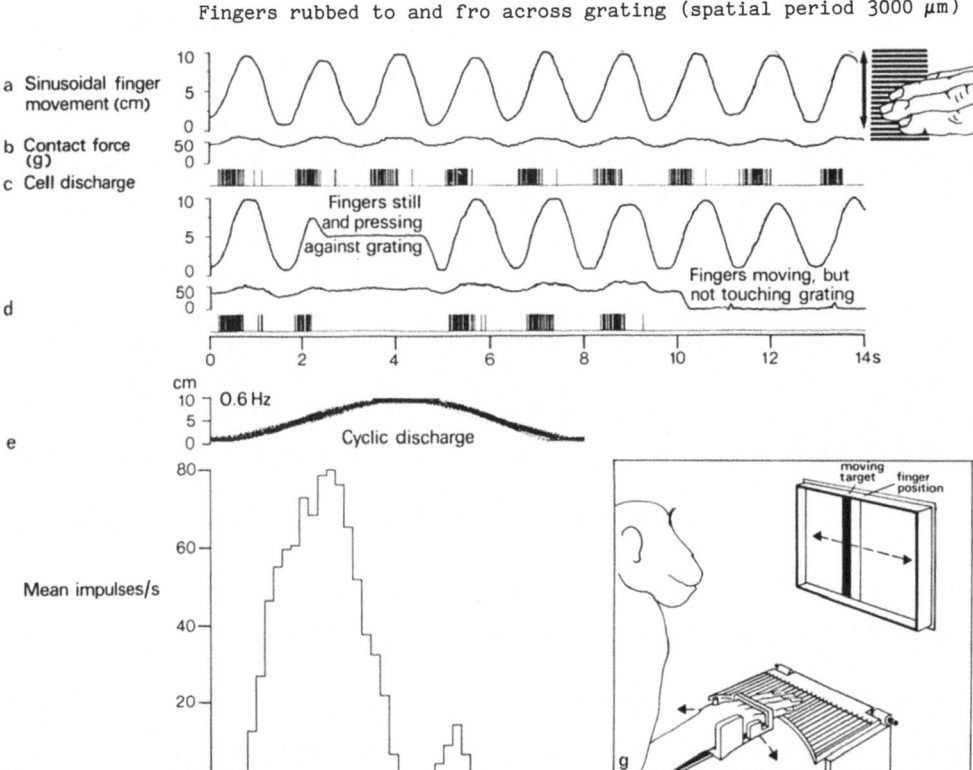

Fig. 1. Responses of neuron in Brodmann's area 2 elicited by to-and-fro movement of contralateral finger pads across a grating (spatial period, 3000 μm). The neuron had a discontinuous cutaneous receptive field on the finger pads of the 3rd, 4th, and 5th digits **(g)**. The cell was silent except when pads rubbed across the grating from left to right **(c)**; it was direction selective. **(d)** Lack of response, either to steady contact of the pads with the grating or while fingers were moved to and fro without touching the grating. **(a)** The sinusoidal finger movement and the contact force between the pads and the surface. Histogram **(f)** illustrates the mean discharge rate throughout a single cycle of finger movement: **(e)** record of cyclic finger movement. *Insert* **(g)** illustrates the experimental arrangement in which the monkey manually tracked a horizontally moving visual target. The target movement was sinusoidal, in which the frequency and amplitude could be varied. Finger position indicated by vertical cursor

Once the monkey's tracking performance was satisfactory, the responses of single cortical neurons in the hemisphere contralateral to the tracking arm were recorded, using a chamber similar to those used in other laboratories (Evarts 1966; Mountcastle 1975), and glass-coated platinum-iridium microelectrodes. The present data were obtained from three adult male macaques (6-8 kg), in which five hemispheres were studied extensively. Head fixation was found to be unnecessary with the last monkey used. Figure 1 illustrates typical data recorded during the scanning of a surface; the neuron discharged cyclically during the sinusoidal scanning of the surface and was silent when the finger pads rested on the grating or when the sinusoidal movement of the fingers was sustained but with the pads not touching the surface. Using these maneuvers it could be determined whether the cortical neuron had a mechanoreceptive input from the skin touching the grating, and if so, whether it adapted quickly or slowly to steady indentation, and finally, whether the cell responded to the movement alone of the arm, independently of cutaneous input. In addition, the temporal relationship between the neuron's discharge and the onset of oscillatory movement of the arm could also be measured (Fig. 10 and 11).

The most immediate processes in the identification of the textured surface by touch are the execution of the scanning movement and the inflow of sensory information to the forebrain that results from this movement. In daily life these key events do not occur in isolation but rather depend on other neural processes preceding them, occurring simultaneously, and following them. Firstly, the subject must plan the actions required for exploring and identifying the surface. This plan must then be translated or "programmed" in terms of the movements and sequences of muscle contractions involved. Once the fingers touch and begin scanning the surface, the resulting sensory inflow is "viewed" continuously until sufficient information accumulates to allow reliable identification. The effectiveness of this last task depends not only on the sufficiency of the sensory information relayed to the forebrain but also on the effective use of this information. If, for example, the subject is not attending to this sensory inflow, the process of identifying the surface becomes inefficient, takes longer, or may not occur.

The inflow of sensory information from the hand is used not only to identify the surface but also to provide feedback on the scanning action of the fingers: are the pattern of movement and the contact force between the fingers and the surface appropriate, as planned, and optimal for identifying the particular surface? Some of the resulting adjustments will depend not only on the inflow from the finger pads touching the surface but also on that from proprioceptive afferents. They may also be determined by the subject's previous experience with similar tasks.

In the experiments to be described only part of the whole goal-oriented task of identifying a grating was carried out. The monkey certainly used scanning movements like those used by a human subject when identifying a surface, and rubbing the finger pads over the surface generated a continuing inflow to the brain of sensory information about the surface. It will be recalled, however, that the monkey's scanning movement was part of a tracking task and that the monkey attended to the surface only to complete this task satisfactorily, and not in order to identify the surface.

Regions of cerebral cortex expected to be active during the manual scanning of a textured surface include areas 3b, 1, 2, and probably 3a of the postcentral gyrus; areas 4, 6, and the supplementary motor cortex; somatosensory area II; and some neuron populations in areas 5 and 7 of the posterior parietal cortex. The purpose of our study has been to attempt to work out the characteristics of the neuronal activity in each of these areas during the digital scanning of a surface and to determine how these regional responses interrelate. Since there is substantial evidence both in man and in the monkey (Carlson 1981, 1984; Randolph and Semmes 1974; Semmes et al. 1974) that the identification of textured surfaces depends on the integrity of areas 3b and 1, the postcentral cortex was the first studied, and more recently the precentral cortex has been examined.

Responses in Neuron Populations in Postcentral Gyrus

All the neurons to be described were initially identified by the discharge evoked when the monkey scanned a grating with the contralateral fingers (see Powell and Mountcastle 1959).

Areas 3b and 1

The cells isolated within areas 3b and 1 (Powell and Mountcastle 1959) were similar in their response properties (cf Kaas et al. 1981; McKenna et al. 1982; Hyva~rinen and Poranen 1978a, b). Most cells had small, sharply defined receptive fields, 2-10 mm in diameter, on a single finger pad; were silent at rest; and adapted rapidly to sustained indentation of the finger pad. Figure 2 illustrates the distribution of these cells within areas 3b and 1. There was a separate, orderly representation of each finger pad, that of the index finger being lateral and of the 5th digit medial, in accord with other studies (Kaas et al. 1981). Neurons with receptive fields on a single finger pad were not evenly distributed within the projection of that digit: rather, in area 3b they were concentrated deep in the posterior lip of the central sulcus and in the caudal part of area 1 (Fig. 2).

All neurons in areas 3b and 1 that had receptive fields on the finger pad skin touching the grating discharged when the monkey moved the fingers across the surface. Figure 3 illustrates the discharge in an area 1 neuron elicited when the monkey scanned a particular grating using different patterns of finger movement. The grating had a spatial period of 3000 μm (ridge to groove ratio 1:7). The oscillatory finger movement was sinusoidal (frequency 0.3, 0.6, and 1.2 Hz; amplitude 80 mm; contact force 30-60 g). This cell discharged only when the distal pad of the 3rd digit rubbed across the grating; then its response was cyclic and in synchrony with the sinusoidal finger movement. Within each cycle of finger movement this cell fired vigorously at the onset and again on slowing of the finger movement, that is, when finger movement was near its peak acceleration. This cyclic pattern was similar in profile at different frequencies of finger movement. This was the commonest pattern of cyclic discharge in responding cells in areas 3b and 1; with others the cyclic discharge reflected more closely the velocity profile of finger movement, like in Fig. 4, rather than the acceleration. Often, some asymmetry of the discharge during the two halves of the cyclic finger movement was observed. This could be regularly explained by the position of the cell's receptive field on the finger pad. Movement of the finger to, say, the left brought that zone of skin firmly into contact with the grating, whereas movement in the reverse direction lifted that skin away from the surface. Skin mechanics rather than neurally de-

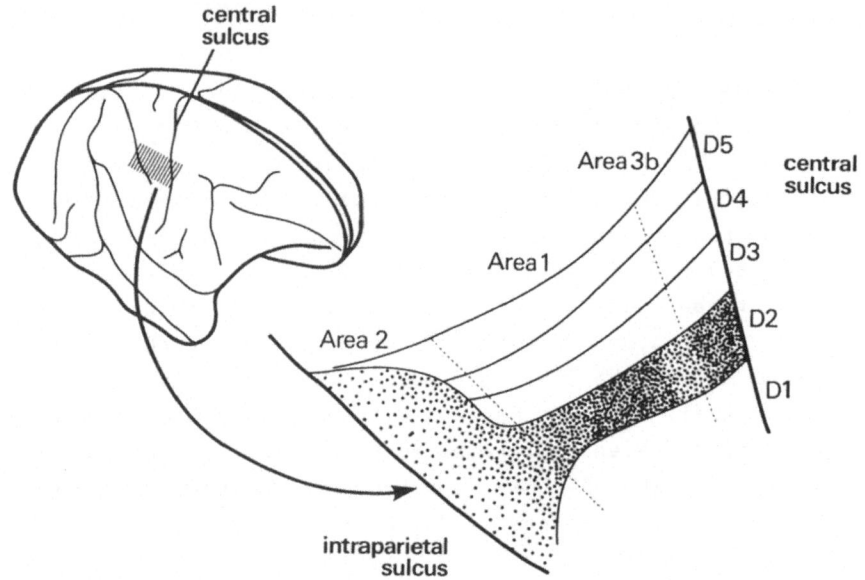

Fig. 2. Distribution of cortical neurons within areas 3b, 1, and 2 when monkey scans a grating with the contralateral fingers. Each strip, designated *D2, 3, 4,* and *5,* indicates the distribution of cells with cutaneous receptive fields on a particular finger pad (*D2,* second digit; *D3,* third digit, etc.). The *dot pattern* within D2 indicates approximately the density of the distribution of these cells. Note that within area 2, because most cells had receptive fields that included two or more finger pads, the representation of each finger pad expanded mediolaterally. (Darian-Smith et al. 1984)

Fig. 3. Discharge in a single neuron in Brodmann's area 1 (receptive field 5 mm in diameter on distal pad of third digit) recorded while monkey moved the finger to and fro with frequencies of 0.3 **(a)**, 0.6 **(b)** and 1.2 Hz **(c)**, and a fixed peak-to-peak amplitude of 80 mm. Within each data block, upper superimposed traces indicate the instantaneous position of the finger pad on the grating, the next sequence of traces indicates the instantaneous contact force, and the following plot of vertical bars indicates the occurrence of single action potentials recorded while the monkey performed these finger movements. The correlation between the finger movement and the cell's discharge is apparent. Surface was a nylon periodic grating with a spatial period of 3000 μm and ridge-to-groove ratio of 1:7. (Darian-Smith et al. 1982)

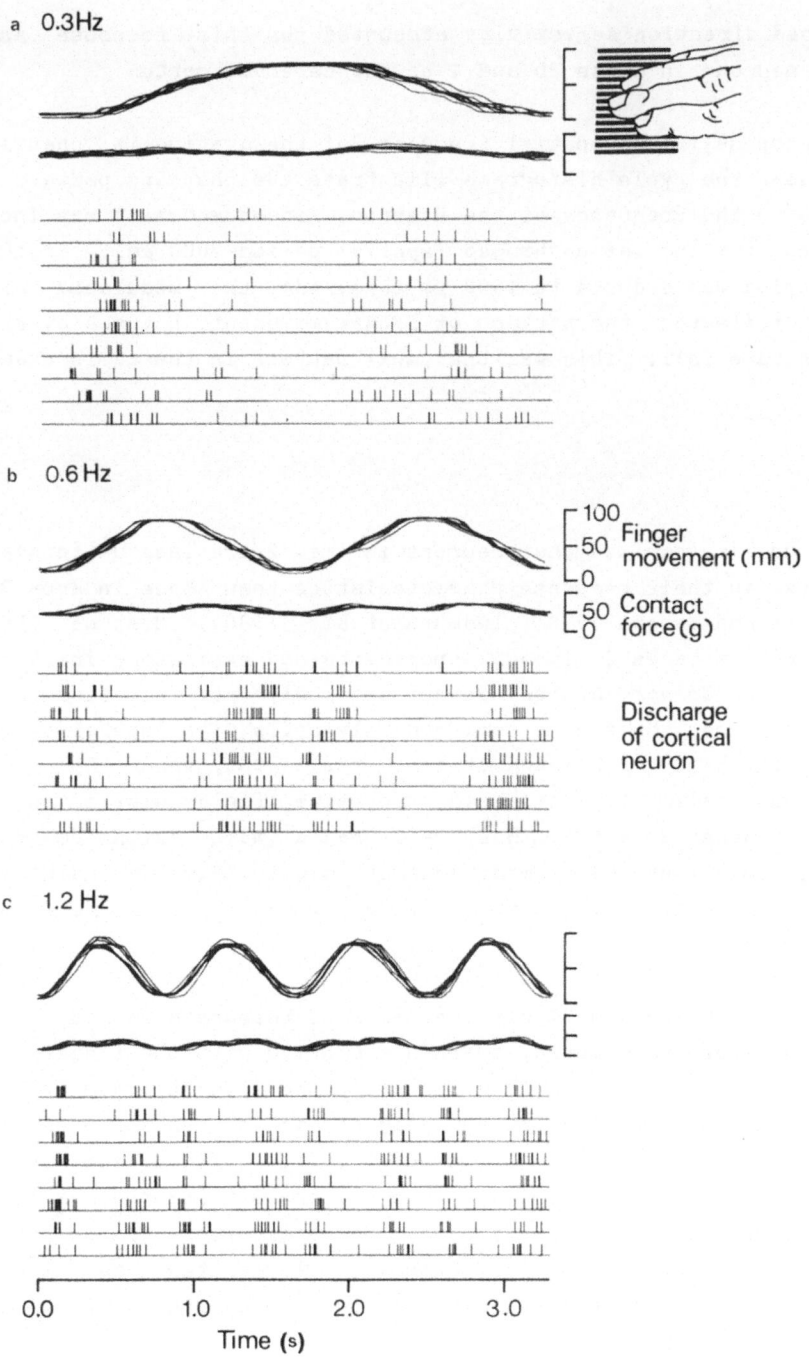

a 0.3 Hz

b 0.6 Hz

┌ 100   Finger
├ 50   movement (mm)
└ 0

┌ 50   Contact
└ 0   force (g)

Discharge
of cortical
neuron

c 1.2 Hz

0.0   1.0   2.0   3.0

Time (s)

termined direction selectivity accounted for this response asymmetry among neurons in areas 3b and 1 of the cerebral cortex.

What happens if the spatial frequency of the grating is changed? In Fig. 4a-c the cycle histograms illustrate the changing pattern of discharge as the frequency of the scanning finger movement was increased, but the grating was unchanged (spatial period 3000 μm). If the grating period was reduced to 1500 μm (Fig. 4d), the modulated discharge still reflected the pattern of finger movement, but the overall discharge rate fell. This was the usual pattern in the cells examined.

Area 2

It is well recognized that neurons in area 2 are less uniform and more complex in their response characteristics than those in area 3b and 1 (Iwamura and Tanaka 1978; Iwamura et al. 1980). Most of the cells isolated in area 2 in our experiments had excitatory input from the skin of one or more digits but not obviously from deep tissues, probably because the sample included only cells with input from the contralateral hand (cf McKenna et al. 1982). Only about 10% of our sample had small, low-threshold receptive fields similar to those of cells in areas 3b and 1. Most cells had a more extensive receptive field, involving two or more adjacent digits (Fig. 5); discontinuous receptive fields, limited to the finger pads of two or more fingers, were common.

For all their individual differences, the responses evoked in area 2 neurons when the monkey scanned a grating with the fingers could be subdivided into a few major groups. In the largest group in our sample, although the receptive field of each neuron was extensive, the responses on scanning a grating were similar to those seen in area 3b and 1 cells (Fig. 5). Both the pattern of finger movement and the spatial characteristics of the grating were represented in the cell's discharge. Figure 5, for example, illustrates the differential response of a cell to three different gratings (spatial periods 3000, 1500, and 750 μm).

Figure 1 illustrates the responses of one of a second group of neurons found in area 2. This cell had a discontinuous receptive field on the pads of the second, third, and fourth digits, was silent at rest, and adapted quickly to indentation of the skin. When the monkey scanned

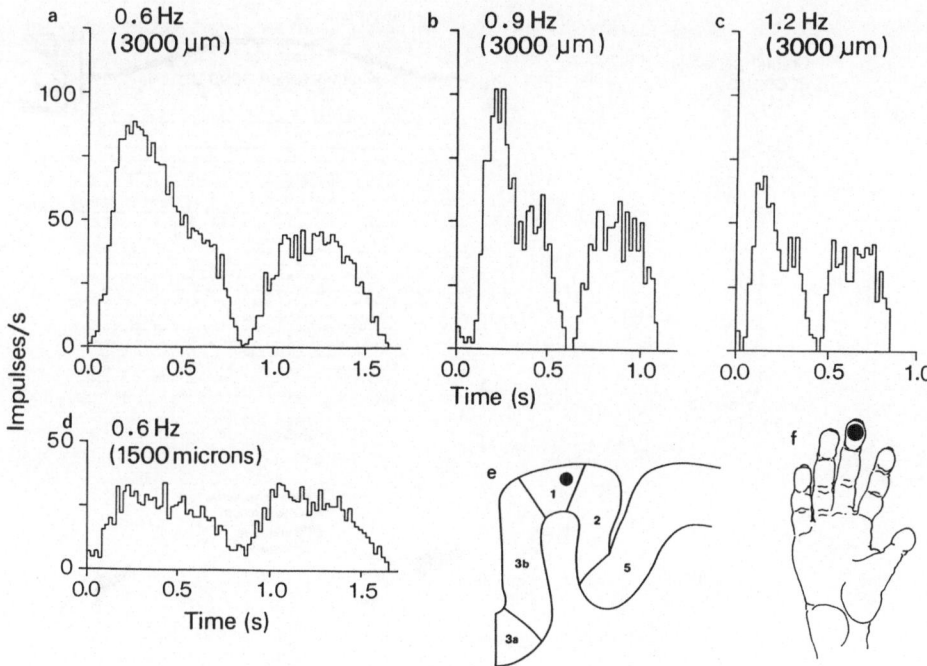

Fig. 4. Patterns of cyclic discharge in an area 1 neuron **(e)** elicited by sinusoidal scanning of a grating. In **a, b,** and **c** the spatial fre-quency of the grating was fixed (spatial period 3000 μm; ridge-to-groove ratio 1:7) but the frequency of the sinusoidal finger movement was changed from 0.6 Hz through 0.9 Hz to 1.2 Hz. In **a** and **d** the finger movement was fixed, with a frequency of 0.6 Hz, but the grating period was changed from 3000 μm to 1500 μm. There was an overall reduction in discharge with the scanning of the finer grating. The cell had a small cutaneous receptive field on the terminal pad of the third digit **(f)**. Amplitude of oscillatory finger movement was 80 mm, and contact force was 50-80 g. (Darian-Smith et al. 1984)

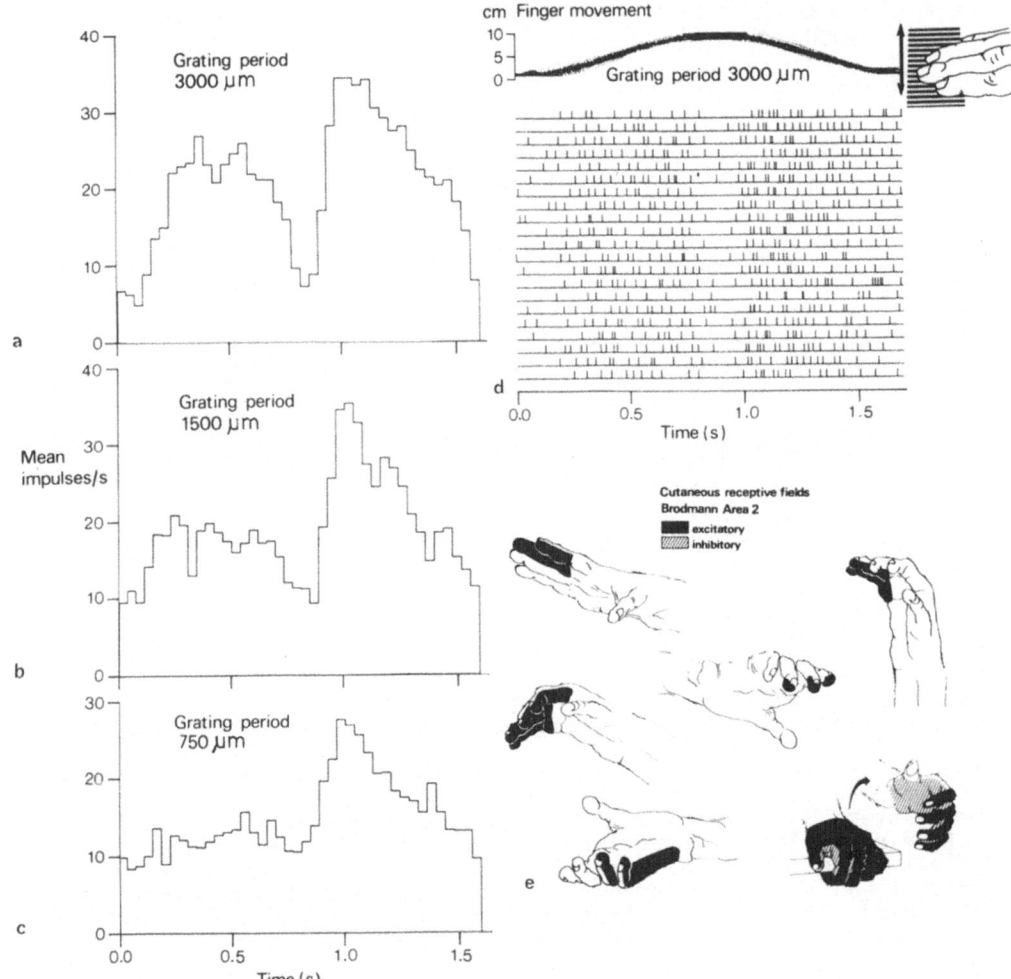

Fig. 5. Effect of changing the spatial period of a grating on the cyclic discharge elicited in an area 2 neuron when the monkey scanned the surface with a fixed sinusoidal movement of the fingers (frequency 0.6 Hz; amplitude 80 mm; contact force 40-70 g). This cell had a continuing discharge of 12 impulses/s when the hand was stationary and not touching the surface. The modulation of the discharge elicited by the gratings was graded, being largest with the coarse grating **(a)** and less with finer gratings **(b, c)**. This neuron's cutaneous receptive field was on the glabrous skin of the third and fourth digits. Successive cyclic responses **(d)**, with each vertical bar indicating the occurrence of an action potential. **e** Typical cutaneous receptive fields of the other cells in area 2. (Darian-Smith et al. 1984)

the surface, the neuron discharged only when the fingers moved to the
right but not when the movement was to the left. About 20% of the
cells sampled in area 2 showed a similar direction selectivity which,
unlike that seen in cells in areas 3b and 1, could not be accounted
for by the lateral position of the receptive field on the finger pad.
The extensive contact between the cell's receptive field and the grating made this unlikely, and as suggested by Hyvärinen and Poranen
(1978b), Gardner and Costanzo (1980), and Costanzo and Gardner (1980),
a central inhibitory mechanism is more probable.

In a third group of area 2 neurons, a vigorous discharge was elicited
on brushing the finger pads, but when the monkey scanned the grating
there was little or no response. With some of these neurons it was
possible to demonstrate a distinctive sensitivity to the orientation
of the grating relative to the direction of the scanning movement.
Figure 6 illustrates the responses of a neuron in area 2 which
responded vigorously only when the orientation of the grating was
within a narrow range.

About one-third of the area 2 neurons examined responded differentially to gratings with different spatial periods. Cells with this differential response were without direction selectivity. Direction- and
orientation-selective neurons need to be more fully characterized to
determine whether they signal information to the brain about the spatial period of the grating being scanned.

Comment

Our observations have shown that while the monkey scans a surface with
the fingers (a) both pattern of finger movement and the spatial
features of the surface are represented in the discharge of most
responding neurons in areas 3b and 1 and (b) a significant fraction of
cells responding in area 2 also signal information about these two
parameters. However, this representation of the finger movement and
the surface is quite limited in the responses of individual neurons.
Since those cells that responded to changes in the surface also
responded to changes in the finger movement, their discharge could
never uniquely specify the surface itself. Similarly, the pattern of
finger movement was never completely specified in the responses of
single neurons, not only because of the confounding of information

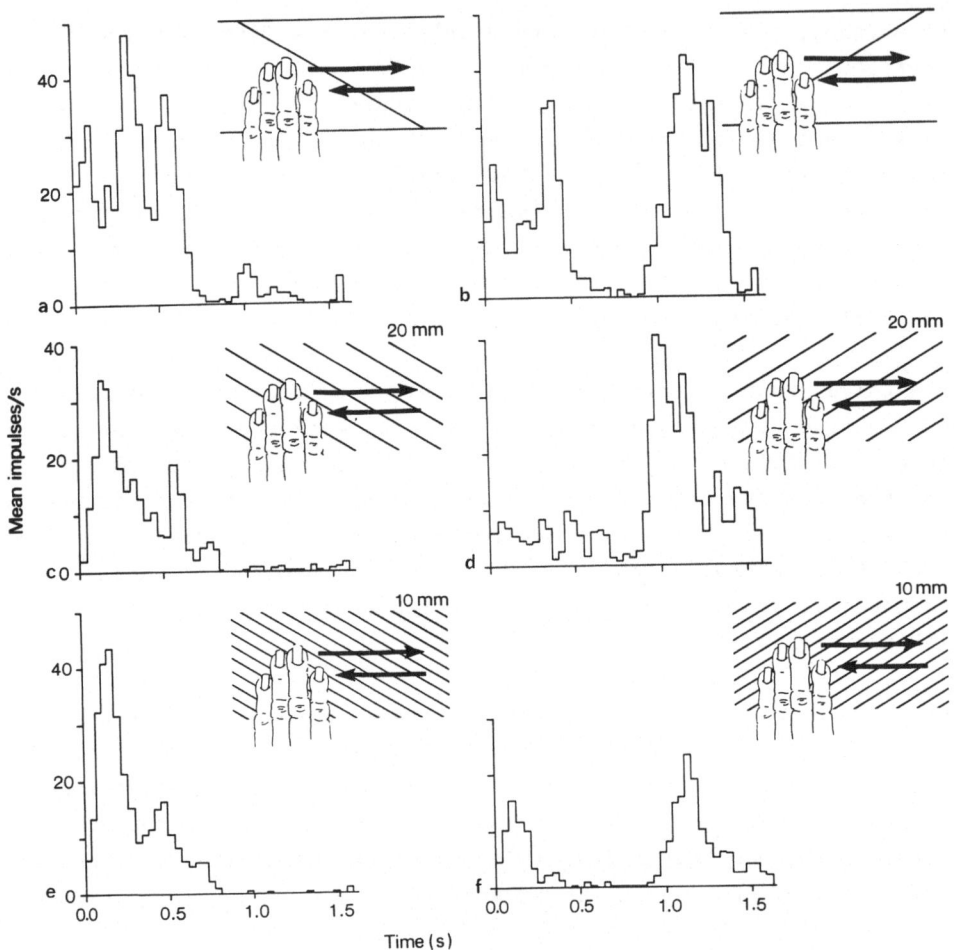

Fig. 6. Responses of neuron located in Brodmann's area 2 elicited by sinusoidal scanning of different gratings with the contralateral fingers. This cell responded vigorously to brushing the finger pads of the third and fourth digits, but only when the direction of brushing was toward the palm; it did not respond on rubbing the fingers to and fro across a standard grating. However, if the orientation of the ridges of the grating, relative to the axis of finger movement, was changed, scanning movements elicited a sharp response. The pairs of cycle histograms illustrate the neuron's responsiveness to the passage of the fingers to and fro across a single ridge (**a, b**), and across different coarse gratings (20-mm (**c, d**) and 10-mm (**e, f**) spatial period) in which the orientation of the ridges was as shown. The cell responded most vigorously when the point of contact of the skin with each ridge was moving toward the palm. Each histogram illustrates the mean discharge rate throughout a single sinusoidal sweep of the fingers across the grating. Frequency of finger movement 0.6 Hz; amplitude 80 mm; contact force 40-60 g. (Darian-Smith et al. 1984)

about finger movement and the surface features but also because most
cells discharged only through part of the cycle of to-and-fro finger
movement (e.g., Figs. 1,3,4,6). At best, the simultaneous discharge
of populations of these cells could define the pattern of finger movement and, in turn, the spatial features of the surface.

Our observations on *single* cortical neurons do not answer the important question: how well do the *populations* of neurons in the postcentral gyrus define the spatial features of a grating or other textured
surfaces? Indirect evidence suggests that for simply patterned surfaces, such as a grating, their representation in the postcentral
gyrus is necessary for their identification and discrimination.
Lesions in the hand projection zone in areas 3b and 1 impair texture
discrimination both in man and in the macaque (Carlson 1981, 1984;
Randolph and Semmes 1974; Semmes et al. 1974). The functional integrity of this cortex is necessary for texture discrimination.
Further, simply textured surfaces may be differentiated using "passive" touch, in which the surface is moved tangentially across the
stationary finger pad (Morley et al. 1983). During such a task the
regional cortical activity that is evoked is limited mainly to the
hand projection zones of the postcentral gyrus and somatosensory
area II. This implies that within the primary somatosensory cortex
alone, the independent representation of both the movement of the skin
relative to the surface and the spatial features of the grating are
sufficient to allow identification of the surface.

Contrasting with the identification of simply patterned surfaces, it
is common experience to find that with many complex, irregular surfaces, such as that of the skin of an orange, identification is possible only when the subject actively explores the surface with the
finger pads; rubbing the orange skin across the fingers is a much
less efficient way of examining it. Presumably, in the active mode,
on the basis of the immediate inflow of sensory information to the
brain, the subject directs and modifies the finger movements to optimize the further collection of information about the surface. This implies that the pattern of finger movement will be continuously
represented in the responding populations of neurons with a precision
sufficient to allow the extraction of separate information about the
spatial features of the surface. With active movement of the hand and
fingers, much more information about the relative movement between
skin and surface is, in fact, represented in the neural responses than
is so with the surface rubbing across the stationary fingers. At the

periphery, for example, kinesthetic information is relayed not only in the responding cutaneous mechanoreceptive fiber population in the digital nerve but also in the mechanoreceptive fiber populations innervating muscle, tendons, and joints relevant to the movement (for review, see Darian-Smith 1984a, b). Similarly, at the level of the cerebral cortex, not only is this multiple input from the peripheral tissues processed but, in addition, those neuron populations concerned with the planning, programming, and execution of the scanning movements can contribute to their specification at the cortical level (Fetz et al. 1980; Lemon et al. 1976; Lemon and Porter 1976; Soso and Fetz 1980; Evarts 1981; Evarts and Tanji 1976).

With this role of the precentral cortex in mind, we explored the responses of neurons in areas 4 and 6 that were correlated with the monkey's digital scanning of gratings.

Responses of Neurons in Precentral Cortex

Figure 7 illustrates the distribution in both the precentral and postcentral cerebral cortex of neurons that responded cyclically during the scanning of a grating with the contralateral fingers. The map is of a single hemisphere in which a total of 484 such cells were identified. The topographical projection that was described earlier (of cells with small mechanoreceptive fields on the terminal pads of the digits) is plotted in the postcentral gyrus. In the precentral cortex, the neuron responses elicited while the monkey scanned a surface differed substiantially from those of postcentral neurons. Firstly, nearly all these cells continued to discharge cyclically with the to-and-fro movement of the fingers even when they were not touching the grating. Hence, a component of the response was related to the pattern of finger movement. With many cells in and adjacent to the anterior lip of the central sulcus (A in Fig. 7), however, the response was less when the fingers were not touching the surface. Most of these cells had cutaneous receptive fields that included one or more finger pads. The fields were typically larger than those of cells in areas 3b and 1 in the postcentral gyrus, with less well-defined boundaries. With adjacent, more medially located cells (B in Fig.7), the cutaneous receptive fields were mostly on the forearm rather than on the fingers.

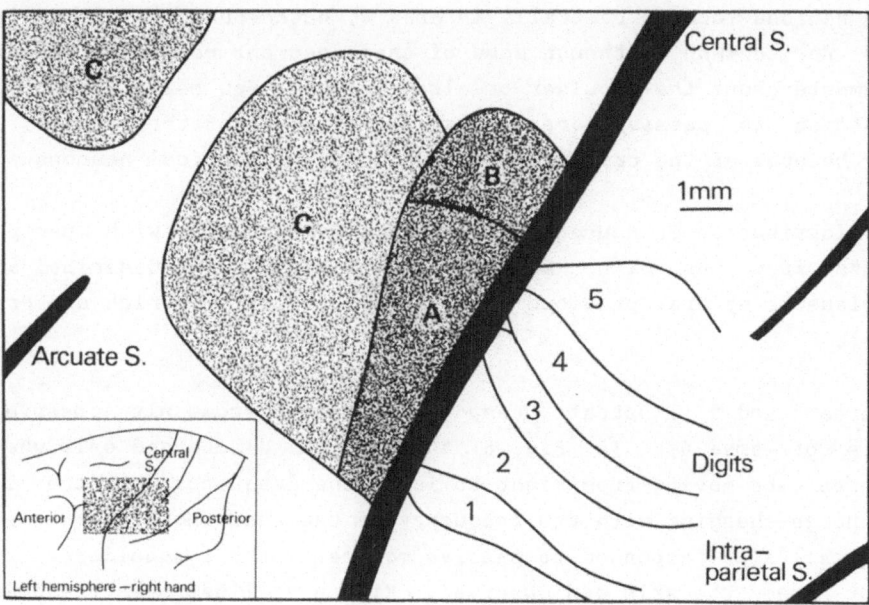

Fig. 7. Map of left sensorimotor cortex illustrating distribution of neurons that responded cyclically in synchrony with sinusoidal scanning movement of the right hand. *1, 2, 3, 4,* and *5* indicate the projections of terminal finger pads of the five digits. *A,* distribution of precentral neurons with receptive fields on finger pads; *B,* distribution of neurons with cutaneous receptive fields on forearm; *C,* distribution of cells without cutaneous input, but in which discharge was modulated by scanning movement of fingers (not responding to eye movements)

With neurons located rostrally in area 4, cutaneous receptive fields were not common, although some of these neurons responded to passive movements about the shoulder or elbow joints. Few cells in our sample responded to passive movements at the wrist or of the fingers, probably because of the criteria used for sampling cortical neurons.

The distribution of neurons within the motor cortex, with peripheral inputs from the skin, muscle, tendons, and joints described above, complements several previous reports (Evarts 1981; Strick and Preston 1979).

Figures 8 and 9 illustrate response patterns commonly observed in cells of area 4. In Fig. 8 the neuron discharged only when the fingers were moving from right to left, the frequency of the cyclic discharge changing with the frequency of the scanning finger movement. This cell also responded to passive movement of the shoulder, but no input from the skin was observed. Figure 9 illustrates the changing cyclic discharge in an area 4 neuron when the fingers were lifted away from the grating, but the sinusoidal finger movement was maintained. This cell had a cutaneous mechanoreceptive field on the terminal pads of the second, third, and fourth digits and was located in the anterior lip of the central sulcus.

In the present experiments, the dominant input to neurons in areas 3b, 1 and 2 was from the digital skin touching the surface and from deep tissues of the hand and forearm. Neurons in the precentral cortex, in addition to receiving this feedback of cutaneous and proprioceptive information, have an essential role in the initiation and continuing execution of the hand movements. This cell population receives a "programmed" input from widely distributed parts of the cerebral cortex and from the cerebellum and basal ganglia. Included in this continuous inflow to precentral cells is information about the error adjustment required to sustain the visually guided tracking movement of the hand.

Fig. 8. Responses of precentral neuron (*insert* indicates location) to sinusoidal scanning movement of contralateral fingers. Finger touched grating with a contact force of 30-50 g. Finger movement, contact force, and discharge when frequency of oscillating finger movement was **(a)** 0.6 Hz and then **(b)** 0.3 Hz. Cell discharged mainly with movement from right to left and was not influenced by contact of finger pads with surface. It did, however, respond to passive movement about shoulder joint. c Cyclic histogram of cell discharge with frequency of finger movement of 0.6 Hz (amplitude 80 mm)

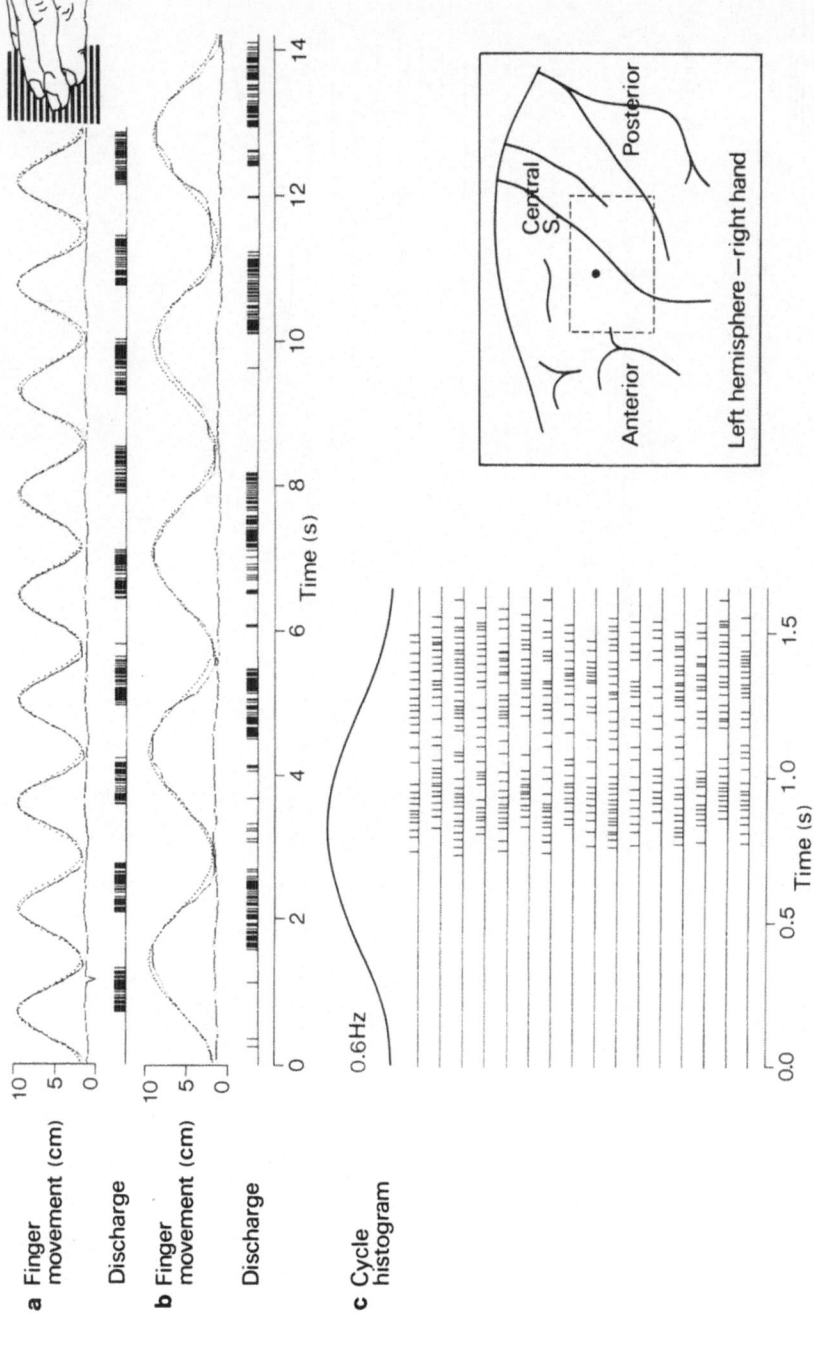

a Finger movement (cm)

Discharge

b Finger movement (cm)

Discharge

c Cycle histogram

0.6 Hz

Time (s)

Left hemisphere – right hand

36

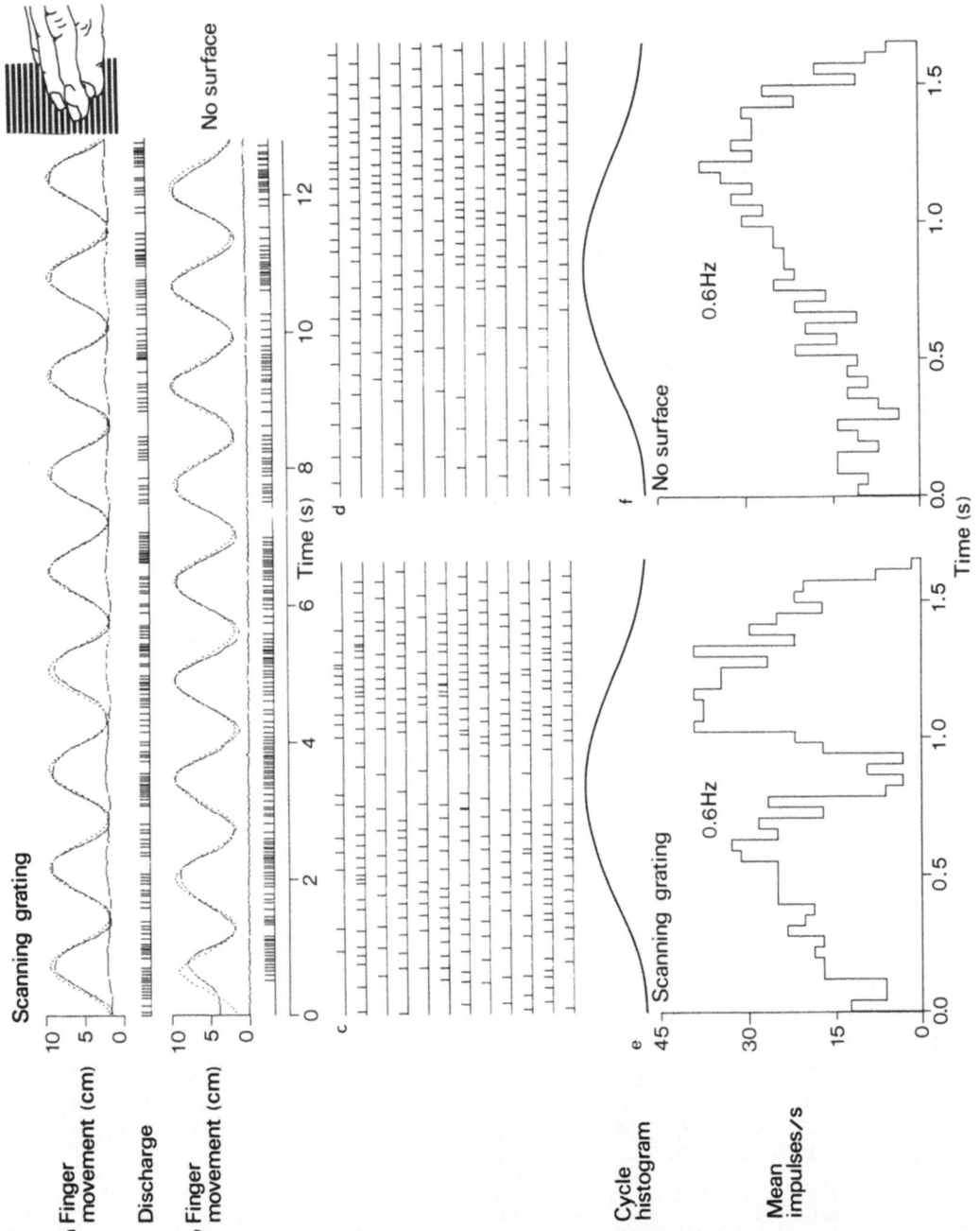

To what extent does the cyclic discharge of individual precentral neurons, such as those shown in Figs. 8-11, depend on (a) the centrally
generated programmed input that determines the hand movement and (b)
the feedback of sensory information that is generated by this
movement?

Input from the finger pads to a particular precentral neuron could be
assessed simply by removing the grating (as illustrated in Figs. 1 and
8). However, differentiating centrally determined activity in a neuron from the response to kinesthetic input to the same cell was difficult. With sudden ballistic movements, the latency of onset of the
response relative to that of the movement may provide information on
the central or peripheral origin of the input if the sensory response
begins after the movement (see Evarts 1981). However, with slow, continuously controlled movements, as in our experiments, latency measurements are of limited value in assessing the origin of the input
evoking the cortical response. Figure 10 illustrates the correlation
of the cyclic discharge in a neuron in area 4 with the onset of sinusoidal oscillatory involvement of the contralateral fingers. This
monkey always commenced tracking the visual target only after half of
the first oscillatory cycle had been completed, and his performance
was independent of the initial direction of movement of the target.
The main discharge in the cell was correlated with the fingers moving
from left to right, but whether this discharge contributed to generating the movement or resulted from input from receptors in muscles,
tendons, and joints of the moving arm or from a combination of these
central and peripheral inputs to the motor cortex cannot be determined
from these observations.

Figure 11 illustrates one other discharge pattern commonly observed in
neurons in the anterior part of area 4. With these cells the onset of
activity matched with the onset of movement in either direction of the

◁ **Fig. 9.** Responses of precentral cortical neuron located within area
designated *A* in Fig. 7 during scanning movement of contralateral hand.
a Finger movement, contact force, and discharge with finger pads
touching a grating and then **(b)** with fingers no longer touching it.
Finger movement was sinusoidal (frequency 0.6 Hz; amplitude 80 mm)
and contact force used with upper sequence was 30-40 g. Discharge
during a succession of cycles of finger movement **(c)** with fingers
touching surface and **(d)** without this contact. **e, f** Cycle histograms
of the cell's response during the cyclic finger movement (average of
30 cycles in each histogram). Grating scanned had a spatial period of
3000 μm with ridge-to-groove ratio of 1:7

38

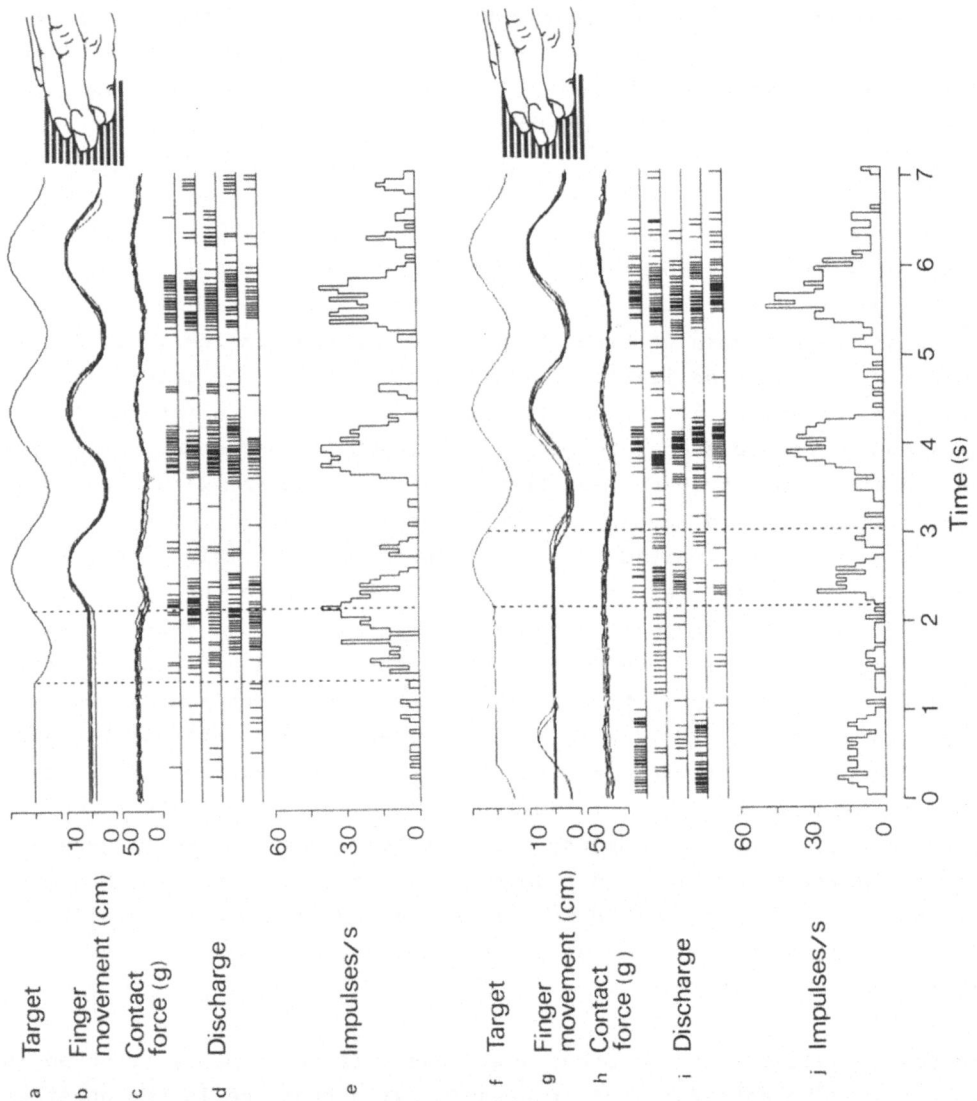

Fig. 10. Responses of precentral neuron (in area designated *C* in Fig. 7) with onset of scanning movement of contralateral fingers across grating. a-d Target movement, finger movement, contact force, and discharge pattern as in previous figures. e, j Histograms are the average of the 5 responses displayed. f-i Onset of movement of visual target was delayed by one-half cycle. Cell's discharge was tightly locked to finger movement and occurred mainly when the finger was moving from left to right. Grating scanned had spatial period of 1500 μm (ridge-to-groove ratio 1:7). Note that monkey always delayed tracking movement to commence one-half cycle after onset of movement of target

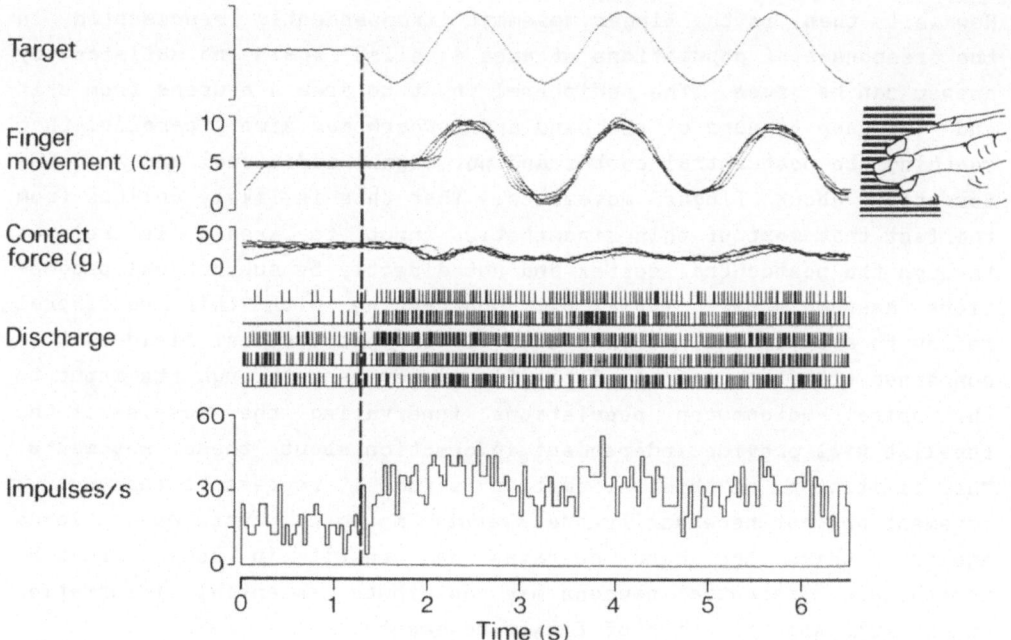

Fig. 11. Response of precentral neuron (in area designated *C* in Fig. 7) with onset of scanning movement of contralateral fingers. Display as in Fig. 10. With this neuron response occurred with onset of movement of *visual target* and not of hand. Also the continuing discharge was not cyclic and synchronized with finger movement. This was possibly an anticipatory response in the neuron, starting almost 1 s before onset of finger movement

visual target, well before the beginning of finger movement. The responses in these cells were not cyclic but continued throughout the period of movement. Some cells of this type discharged continuously when the fingers were stationary; with the onset of finger movement this discharge was often inhibited rather than enhanced. Similar anticipatory neurons have been commonly observed in the monkey's motor cortex (Tanji and Evarts 1981; Weinrich and Wise 1982).

Comment

The present observations have demonstrated a representation of the monkey's scanning movement in the responses of single precentral neu-rons. With many such cells, unlike those in the hand projection in areas 3b and 1, this response was independent of the characteristics of the textured surface that was scanned.

How well, then, is the finger movement independently represented in the responses of populations of area 4 cells? Again, no satisfactory answer can be given. The peripheral input to area 4 neurons from skin and the deep tissues of the hand and forearm may simply parallel that reaching the postcentral cortex and not signal additional and new information about finger movements. That this is likely follows from the fact that most of this kinesthetic input to area 4 is relayed through the postcentral cortex and not directly by subcortical projections (Asanuma et al. 1983a, b). Contrasting with this peripheral inflow to area 4, the neuronal activity in this cortical field that is concerned with the control of the finger movement through its input to the spinal motoneuron populations innervating the muscles of the forelimb will provide independent information about these movements. The limitation of this information is that it represents the *planned* movement and not necessarily the *executed* movement; unforeseen loads against which the hand operates may result in some mismatch. Nonetheless, precentral neurons may contribute essential information in the full specification of finger movements.

Conclusion

Much study of sensorimotor interaction at the cortical level has been directed to the input of kinesthetic information to the precentral cortex and the part it plays in the control of hand movements. Equally important is the role of exploratory finger movements used in identifying surfaces and objects by touch and the respective contributions of "sensory" and "motor" areas of cortex in these sensory tasks. We conclude from the present experiments that:

1. During the digital scanning of different gratings, single neurons in postcentral areas 3b, 1, and 2 in the macaque respond differently to different gratings but that these responses also reflect the pattern of finger movement relative to the surface.

2. Accurate, independent specification of the scanning finger movement in the cortical neuronal response is necessary if the spatial features of the patterned surface are to be represented in the response of postcentral neuron populations.

3. The pattern of finger movement is represented in the response of many neurons within the "hand" zone of area 4, and at least some of this representation is independent of that in the postcentral gyrus. To this extend the precentral representation may contribute

to the cortical neuronal representation of the spatial pattern of the surface explored with the fingers.

4. Cortical area 5, somatosensory area II, and possibly area 6 may also contribute to the tactile identification of textured surface and need further study.

Acknowledgements. The author's laboratory has been generously support-ed by the University of Melbourne, National Health and Medical Research Council of Australia, the Ian Potter Foundation, Jack Brock-hoff Foundation, Van Cleef Foundation, and Australian Brain Founda-tion.

References

ASANUMA C, THACH WT, JONES EG (1983a) Distribution of cerebellar ter-minations and their relation to other afferent terminations in the ventral lateral thalamic region of the monkey. Brain Res Rev 5: 237-265

ASANUMA C, THACH WT, JONES EG (1983b) Anatomical evidence for segre-gated focal groupings of efferent cells and their terminal ramifica-tions in the cerebellothalamic pathway of the monkey. Brain Res Rev 5: 267-297

CARLSON M (1981) Characteristics of sensory deficits following lesions of Brodmann's areas 1 and 2 in the postcentral gyrus of *Macaca mulatta*. Brain Res 204: 424-430

CARLSON M (1984) The significance of single or multiple cortical areas for tactile discrimination in primates. In: GOODWIN AW, DARIAN-SMITH I (eds) Hand function and the neocortex. Springer, Berlin Heidelberg New York, pp 1-16

COSTANZO RM, GARDNER EP (1980) A quantitative analysis of responses of direction-sensitive neurons in somatosensory cortex of awake monkeys. J Neurophysiol 43: 1319-1341

DARIAN-SMITH I (1984a) The sense of touch: performance and peripheral neural processes. In: DARIAN-SMITH I (ed) Sensory processes. Handbook of physiology, section I, vol III. Am Physiol Soc, Bethesda, pp 739-788

DARIAN-SMITH I (1984b) Touch and thermal sensibility. In: LUCE RD, HERRNSTEIN RJ, ATKINSON RC, LINDZEY G (eds) Stevens' handbook of ex-perimental psychology. New York (in press)

DARIAN-SMITH I, OKE LE (1980) Peripheral neural representation of the spatial frequency of a grating moving across the monkey's finger pad. J Physiol 309: 117-133

DARIAN-SMITH I, SUGITANI M, HEYWOOD J, KARITA K, GOODWIN A (1982) Touching textures surfaces: cells in somatosensory cortex respond both to finger movement and to surface features. Science 218: 906-909

DARIAN-SMITH I, GOODWIN AW, SUGITANI M, HEYWOOD J (1984) The tangible features of textured surfaces: their representation in the monkey's somatosensory cortex. In: EDELMAN GM, COWAN WM, GALL WE (eds) Dynamic aspects of neocortical function. Wiley, New York

EVARTS EV (1966) Methods for recording activity in individual neurons in moving animals. In: RUSHMER RF (ed) Methods in medical research. Year Book Chicago, pp 241-250

EVARTS EV (1981) Role of motor cortex in voluntary movements in primates. In: BROOKS VB (ed) The nervous system, vol II; Motor control. American Physiological Society, Bethesda, pp 1083-1120 (Handbook of physiology, section I)

EVARTS EV, TANJI J (1976) Reflex and intended responses in motor cortex pyramidal tract neurons of monkey. J Neurophysiol 39: 1096-1108

FETZ EE, FINOCCHIO DV, BAKER MA, SOSO M (1980) Sensory and motor responses of precentral cortex cells during comparable passive and active joint movements. J Neurophysiol 43: 1070-1089

GARDNER EP, COSTANZO RM (1980) Neuronal mechanisms underlying direction sensitivity of somatosensory cortical neurons in awake monkeys. J Neurophysiol 43: 1342-1354

HYVARINEN J, PORANEN A (1978a) Movement-sensitive and direction and orientation-selective cutaneous receptive fields in the hand of the post-central gyrus in monkeys. J Physiol 283: 523-537

HYVARINEN J, PORANEN A (1978b) Receptive field integration and submodality convergence in the hand area of the post-central gyrus of the alert monkey. J Physiol 283: 539-556

IWAMURA Y, TANAKA M (1978) Postcentral neurons in hand region of area 2: their possible role in the form discrimination of tactile objects. Brain Res 150: 662-666

IWAMURA Y, TANAKA M, HIKOSAKA O (1980) Overlapping representation of fingers in the somatosensory cortex (area 2) of the conscious monkey. Brain Res 197: 516-520

KAAS JH, NELSON RJ, SUR M, MERZENRICH MM (1981) Organization of somatosensory cortex in primates. In: SCHMITT FO, WORDEN FG, ADELMAN G, DENNIS SG (eds) The organization of the cerebral cortex. MIT Press, Cambridge, pp 237-267

LAMB GD (1983) Tactile discrimination of textured surfaces: psychophysical performance measurements in humans. J Physiol 338: 551-565

LEDERMAN SJ (1974) Tactile roughness of grooved surfaces: the touching process and effects of macro- and microsurface structure. Percept. Psychophys 16: 385-395

LEMON RN, PORTER R (1976) Afferent input to movement-related neurons in conscious monkeys. Proc R Soc Lond [Biol] 194: 313-339

LEMON RN, HANBY JA, PORTER R (1976) Relationship between the activity of precentral neurones during active and passive movements in conscious monkeys. Proc R Soc Lond [Biol] 194: 341-373

McKENNA TM, WHITSEL BL, DREYER DA (1982) Anterior parietal cortical topographic organization in macaque monkey: a reevaluation. J Neurophysiol 48: 289-317

MORLEY JW, GOODWIN AW, DARIAN-SMITH I (1983) Tactile discrimination of gratings. Exp Brain Res 49: 291-299

MOUNTCASTLE VB (1975) The world around us: neural command functions for selective attention. Neurosci Res Program Bull [Suppl] 14: 1-47

POWELL TPS, MOUNTCASTLE VB (1959) The cytoarchitecture of the postcentral gyrus of the monkey *Macaca mulatta*. Bull Johns Hopkins Hosp 105: 108-131

RANDOLPH M, SEMMES J (1974) Behavioral consequences of selective subtotal ablations in the postcentral gyrus of *Macaca mulatta*. Brain Res 70: 55-70

SEMMES J, PORTER L, RANDOLPH MC (1974) Further studies of anterior postcentral lesions in monkeys. Cortex 10: 55-68

SOSO MJ, FETZ FE (1980) Responses of identified cells in postcentral cortex of awake monkeys during comparable active and passive joint movements. J Neurophysiol 43: 1090-1110

STRICK PL, PRESTON JB (1979) Multiple representation in the motor cortex: a new concept of input-output organization for the forearm representation. In: ASANUMA H, WILSON VJ (eds) Integration in the nervous system. Igaku Shoin, Tokyo, pp 205-221

TANJI J, EVARTS EV (1981) Anticipatory activity of motor cortex neurons in relation to direction of an intended movement. J Neurophysiol 39: 1062-1068

WENRICH M, WISE ST (1982) The premotor cortex of the monkey. J Neurosci 2: 1329-1345

WOOD JONES F (1926) The arboreal man. Arnold, London

Functional Surface Integration, Submodality Convergence, and Tactile Feature Detection in Area 2 of the Monkey Somatosensory Cortex

Y. Iwamura, M. Tanaka, M. Sakamoto, and O. Hikosaka

Department of Physiology, Toho University, School of Medicine, 5-21-16, Omori-Nishi, Ota-Ku, Tokyo, 143, Japan

Introduction

The first somatosensory cortex (SI) of the primate is cytoarchitecton-ically differentiated into three subdivisions: areas 3, 1, and 2 (Powell and Mountcastle 1959). While the anterior part of SI, area 3, is typically granular and the primary sensory receiving cortex, the posterior part of SI, area 2, has fewer granular and more pyramidal cells. Area 2 receives less thalamic input than area 3 but does re- ceive corticocortical fibers from area 3 (Vogt and Pandya 1977; Jones et al. 1978; Künzle 1978). Thus, there is a hierarchical order in the flow of information within SI, and area 2 is more an associative than a primary receiving area. Recent studies of single neuronal ac- tivity in unanesthetized monkeys have shown that the receptive fields of SI neurons become progressively larger and more complex in areas 1 and 2 than in area 3 (Hyvärinen and Poranen 1978; Iwamura and Tanaka 1978; Iwamura et al. 1978, 1981, 1983a, b; McKenna et al. 1982). As a result, the representation of the hand and fingers in area 2 was not discretely somatotopic but contained overlap (Iwamura et al. 1980). A number of specific neurons in areas 1 and 2 were found to respond to moving stimuli with directional selectivity or to respond to the presence of edges (Schwarz and Fredrickson 1971; Whitsel et al. 1972; Hyvärinen 1976; Iwamura and Tanaka 1978; Iwamura et al. 1978, 1981, 1983b; Costanzo and Gardner 1980; Gardner and Costanzo 1980). Thus, area 2 is more likely the site of information processing for higher-order perception than that of simple sensory reception.

In the present communication we give an overview of receptive field properties of area 2 neurons, recorded in conscious monkeys. We also

Experimental Brain Research, Suppl. 10
© Springer-Verlag Berlin · Heidelberg 1985

describe several types of tactile feature-detection neurons found in area 2 during systematic mapping. These feature-detection neurons provide strong neurophysiological support for the proposition derived from clinical and animal cortical ablation studies that SI is involved in aspects of perception, such as stereognosis (Corkin et al. 1970; Randolf and Semmes 1974; Roland 1976).

Methods

The present results are based on experiments done in five unanesthetized, conscious, and cooperative Japanese monkeys (*Macaca fuscata*). Only well-isolated, initially negative spikes were studied. Details of experimental methods are described elsewhere (Iwamura et al. 1983a).

Results

Submodality Types of Area 2 Neurons

Of 1068 neurons isolated in area 2 of six hemispheres, 37% were skin neurons, 22% were joint manipulation or other types of deep neurons, and 5% were activated by both skin and deep stimulation. In the rest (36%) receptive fields could not be identified by conventional somatic stimuli. The present results include a substantial number of these neurons, which were identified by other modes of stimulation.

Receptive Fields of Area 2 Skin Neurons

The receptive fields of area 2 neurons were studied both in the crown of the postcentral gyrus and in the anterior bank of the intraparietal sulcus. The receptive fields of skin neurons recorded in the middle portion of the crown were on fingers covering either the distal or the whole ventral or the dorsal surface. A crude topographical projection of fingers was recognized lateromedially, but this pattern was not as clear as that in area 3 because some area 2 neurons had receptive fields of the multifinger type. In the more caudal part of the crown, neurons with receptive fields of the multifinger type increased. The distal finger segments were the dominant receptive field position in this area.

In the anterior bank of the intraparietal sulcus, receptive fields of skin neurons became larger as the representations of the palmar and finger skin merged. The largest one covered the whole volar skin. Receptive fields were also found covering multiple fingertips or the ventral or dorsal multiple finger skin. Among these large receptive fields smaller ones were found. Neurons with similar receptive fields were recorded in a cluster with a range of 0.5 - 1.0 mm along the electrode track followed by a different cluster of neurons with an entirely different configuration of the receptive field. The receptive field study thus indicated that there is no somatotopic representation of individual fingers in area 2, as we reported previously (Iwamura et al. 1980).

Functional Surfaces

We noticed that large skin receptive fields, as described above, could be categorized into several types, each of which was commonly observed within the population of recorded neurons. We identified eight categories, each with distinctive receptive fields, as illustrated in the inset of Fig. 1. We labeled them as *functional surfaces* for the reasons described below. A large number of neurons with different functional surfaces were collected from eight hemispheres and plotted on one graph (Fig. 1). The solid line indicates the extent of the distribution of those surfaces labeled a, b, c, and j: all these included both fingers and palmar or hand dorsum skin. In this group, surface a was found more frequently among cells in the lateral part of the somatosensory cortex than in the medial part. Surface b was found only in the medial part. Surface c was found over a wider range, and surface j was found mainly in the mediocaudal part of the postcentral gyrus. Surfaces e and e, on the distal finger segments and nails respectively, were found among neurons in a relatively narrow and medially shifted core region of the cortex (encircled by dotted line). Surfaces d and i, on the ventral and dorsal finger skin respectively, defined the receptive fields of neurons scattered in mediate zone (encircled by a broken line). Figure 1 thus demonstrates that there is a topographic organization, that is, receptive field surfaces represent-

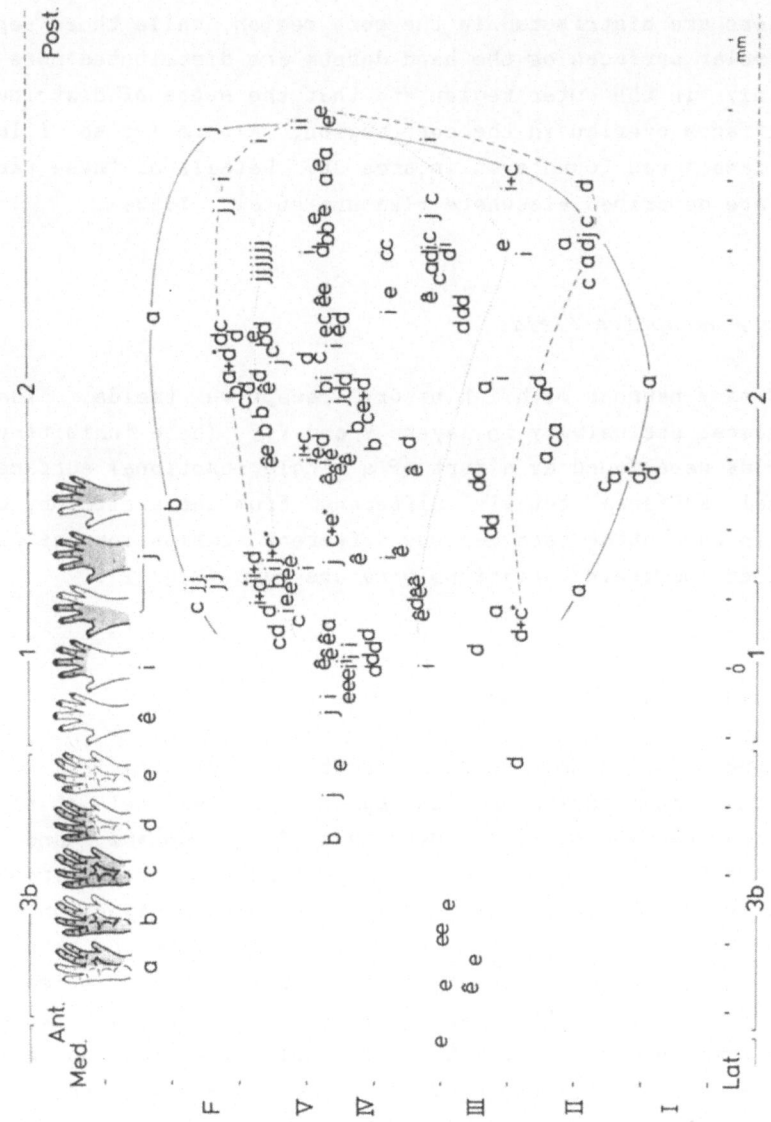

47

Fig. 1. Eight functional surfaces *(inset)* and their distribution on the unfolded SI map. The unfolding procedure is described elsewhere (Iwamura et al. 1983a). Data from seven hemispheres are superimposed. Surface ê represents receptive fields restricted to nails. In surfaces *a, b, c,* and *j,* the third finger is dispensable for unequivocal identification of these surfaces. Similarly, in surfaces *d, e, ê,* and *i,* one of the second or fifth fingers is dispensable. In the map three curves show the extent of the distribution of the eight functional surfaces, which were classified into three groups. *Solid line* indicates surfaces a, b, c, and j; *broken line* indicates surfaces d and i; and *dotted line* indicates e and ê. Letters with *asterisks* imply inhibitory receptive fields which were discarded when drawing the lines

ing fingers are distributed in the core region, while those representing the volar surfaces or the hand dorsum are distributed more widely but mainly in the outer region and that the areas of distribution of these surfaces overlap in the core region. Figure 1 also illustrates that surface e was found even in area 3b. Details of these particular neurons are described elsewhere (Iwamura et al. 1983a).

Inhibitory Receptive Fields

We found many neurons with inhibitory receptive fields. They were found almost exclusively in layers V and VI. These inhibitory receptive fields were found as a part of a single functional surface or as functional surfaces totally different from the excitatory counterparts. In the latter case various different combinations of functional surfaces and their variations were observed (Fig. 2).

Directionally Selective Neurons in Area 2

The presence of skin neurons with directional selectivity to moving stimuli has been described in monkey SI (Whitsel et al. 1972; Schwarz and Fredrickson 1971; Hyvärinen 1976; Costanzo and Gardner 1980). We found that about 20% of area 2 skin neurons responded preferentially or only to stimuli moving in a certain direction. The optimum method of stimulation was to rub or scrape the skin surface. This type of neuron was found throughout area 2 either in isolation or in clusters. In the latter case, the preferred direction varied among the clustered neurons. Directionally selective neurons were also found in area 1 (9% of skin neurons) but only rarely in area 3.

Joint Manipulation Neurons

Neurons responding to manipulation of joints or muscles were also mostly of the multifinger type and occurred in area 2. Neurons responding to manipulation of the metacarpophalangeal joints were scattered more widely than those responding only to distal phalangeal joint manipulation, which were distributed in the core region. Those

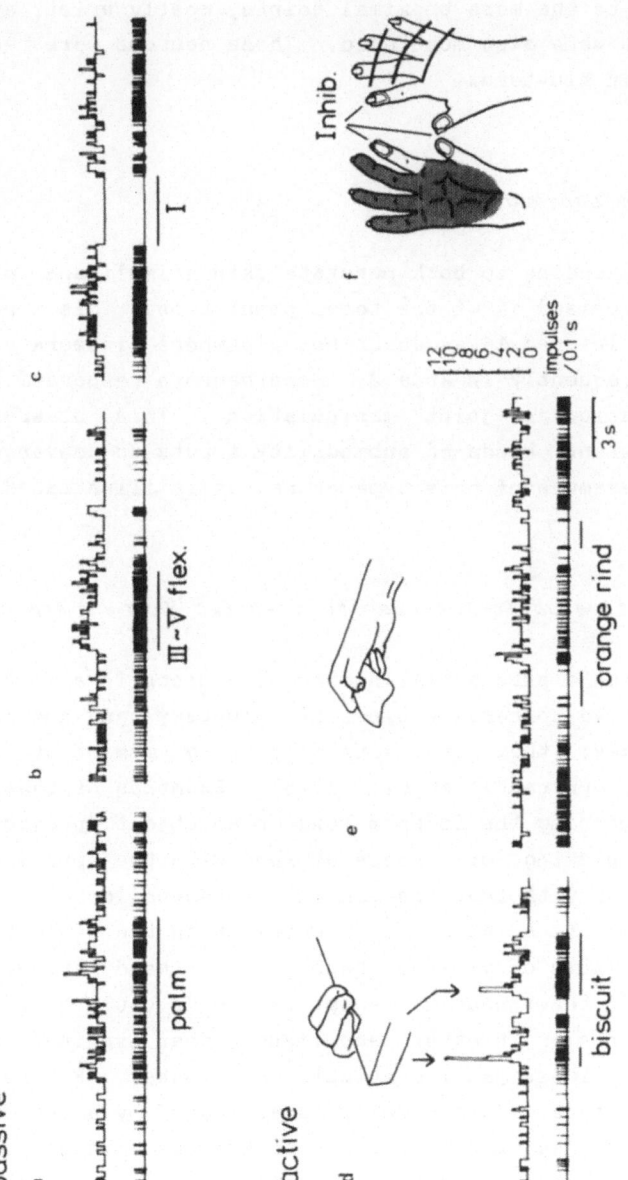

Fig. 2. A neuron with an inhibitory receptive field and possible sub-modality convergence. Responses to punctate skin stimuli **(a)**, flexion of phalangeal joints **(b)**, and a contact to the tip of the thumb **(c)**. Neuronal responses evoked by active finger actions **(d,e)**. f Receptive field determined by passive stimuli as described in a-c

responding to the more proximal joints, mostly wrist, and some shoulder or elbow were also scattered. These neurons were recorded in isolation or in clusters.

Submodality Convergence

Neurons responding to both punctate skin stimuli and manipulation of
joints comprised 5% of the total population. This type of neuron was
also found in area 1, as described elsewhere (Iwamura et al. 1983b),
but more frequently in area 2. Some neurons responded to hair or nail
bed stimulation and joint manipulation. This observation suggests
that different kinds of submodality inputs do converge on these neurons. An example of this type of neuron is illustrated in Fig. 2.

Neurons Activated in Relation to Specific Hand or Arm Movements

We found that a substantial number of neurons in area 2 were specifically related to certain types of voluntary hand and finger movements
of the monkey: they fired maximally at the moment of arm or finger
movement (Mountcastle et al. 1975). Examples of these movements included stretching the forearm towards an object (projection or reach
neurons), picking up a piece of food with the opposing thumb and the
2nd finger or with four fingers in a prehensile position, exploring
the inside of a hole or a bottle with the fingertips, holding or
grasping objects of various shapes, etc. (hand manipulation neurons).
In some of these neurons receptive fields could be identified on the
skin or joints or in other deep tissue, when examined passively, but
responses to these passive stimuli were weaker. Figure 2 shows a neuron of this type. This neuron was activated by a punctate stimulus on
palmar skin (Fig. 2a) or flexion of multiple finger joints (Fig. 2b),
and its background activity was inhibited by a punctate stimulus on
the first or second fingertip (Fig. 2c). Thus, the convergence of
skin and deep submodalities was suggested. This neuron fired more vigorously when the monkey grasped a biscuit in a prehensile position
(Fig. 2d) and was inhibited when the monkey picked up a similar object
(an orange rind) by opposing two fingertips (Fig. 2e). Thus, this
neuron could be concerned with the discrimination of one hand posture
from the other.

Edge-Detection Neurons

We found a total of 16 neurons which responded optimally when a sharp edge or a narrow object contacted a part of the receptive field in a particular orientation. Among these neurons the degree of selectivity to the edge varied considerably. Generally speaking, the larger the receptive field, the stronger the selectivity of the neuronal response to the edge, and the weaker the response to stimuli other than the contact of an edge. These edge-detection neurons are described below according to their receptive field positions, each of which could be referred to one of the functional surfaces described above.

Edge-Detection Neurons Related to the Thumb and Thenar Pad

A cluster of edge-detection neurons was found along a single penetration in a relatively lateral part of the area 2 finger region. The receptive fields were on the thenar pad or also included the thumb (surface a). A tonic response was evoked when an edge was applied to the thenar pad. The preferred orientation was mostly longitudinal to the hand axis but was transverse to it in one neuron. The contact of a broad surface, such as a ball or a thick cylinder, evoked little or no response in these neurons.

Edge-Detection Neurons Related to Volar Surfaces

Edge-detection neurons were found in a more medial region of area 2 and had receptive fields either on the ulnar half (surface b) or on the entire volar skin (surface c), or on the skin of one or more fingers (surface d). We never encountered edge-detection neurons whose receptive fields were on the combined fingertips (surface e) or on the dorsal surfaces (surfaces i or j). The preferred orientation of the edge was either transverse or diagonal to the hand axis (Iwamura and Tanaka 1978). No response or an inhibitory response was obtained when a large surface contacted the receptive field.

52

Neurons Activated by the Contact of a Large Surface

Two neurons were activated best by the contact of broader surfaces. They did not respond or responded poorly to edges or punctate stimuli. Receptive fields of these neurons were on the volar skin. They were found in isolation or together with the edge-detection neurons.

Neurons Activated by Active Grasping of Thin or Thick Objects

In two neurons whose receptive fields covered the entire volar skin, passive contact of a transverse edge was excitatory, while that of a large surface was inhibitory to their background activity. In these neurons, active grasping of a thin object by the monkey evoked larger and longer-sustained responses than corresponding passive stimuli (Fig. 3). In another two neurons, active grasping of either thin or thick objects was the only way to activate these neurons. They did not fire when the monkey grasped or contacted objects without these features. The passive contact of objects was not effective at all. We have previously reported a pair of neurons which were activated only when the monkey actively grasped a round or a square object respectively (Iwamura and Tanaka 1978). These examples indicate that there are neurons in area 2 which detect the presence of three-dimensional features or allow discrimination of the shape or size of objects by active grasping.

Neurons Responding Differently to the Contact of Different Materials

We found, in area 2, three neurons which were preferentially activated when the skin was rubbed by rough materials, such as felt, but not by smooth ones, such as an acrylic plate. One of them was directionally selective. In contrast, another four neurons were selectively activated by stationary contact of soft materials, such as a bottle brush, a piece of animal fur, or a pile of soft toilet paper (Fig. 4). These neurons did not respond to the contact of a hard surface, such as an acrylic plate, a metal plate, or a wood block. Conversely, there were four neurons which responded preferentially to the contact of hard surfaces. Receptive fields of these neurons were on either the ventral (surfaces c, e) or dorsal (surface i) surfaces of the hand

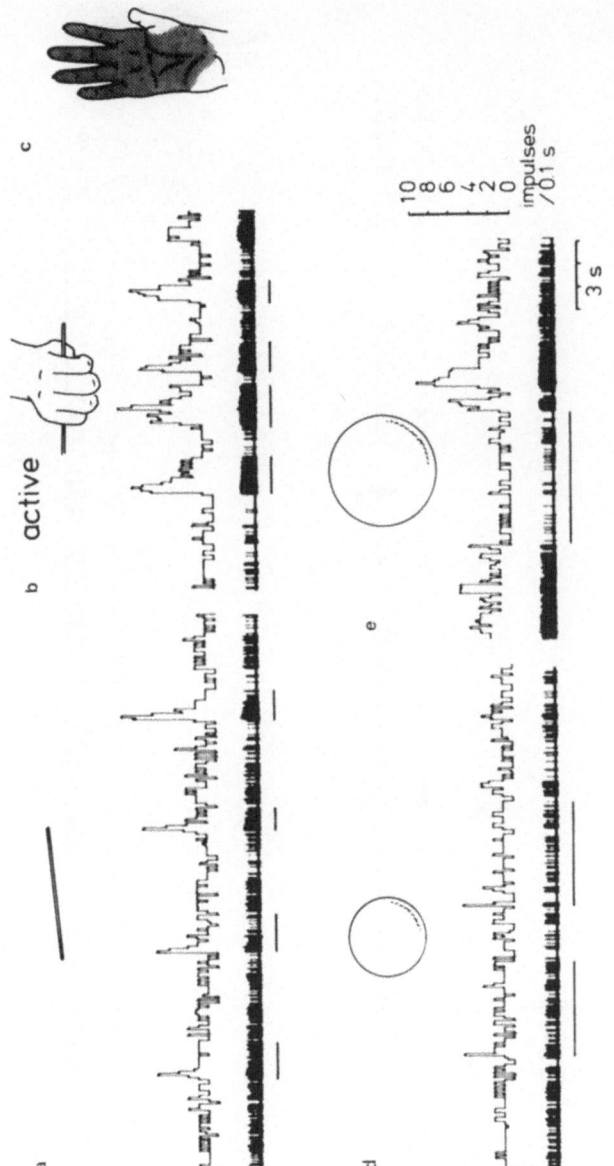

Fig. 3. An edge-detection neuron responding to contact with the palmar skin **(c)** of a thin rod transversely **(a)** and a smaller (3.8 cm) **(d)** or a larger ball (5.1 cm) **(e)**. More sustained responses were obtained when the monkey grasped the rod actively **(b)**. Note also that the background activity was inhibited when a larger ball contacted the palmar skin **(e)**

54

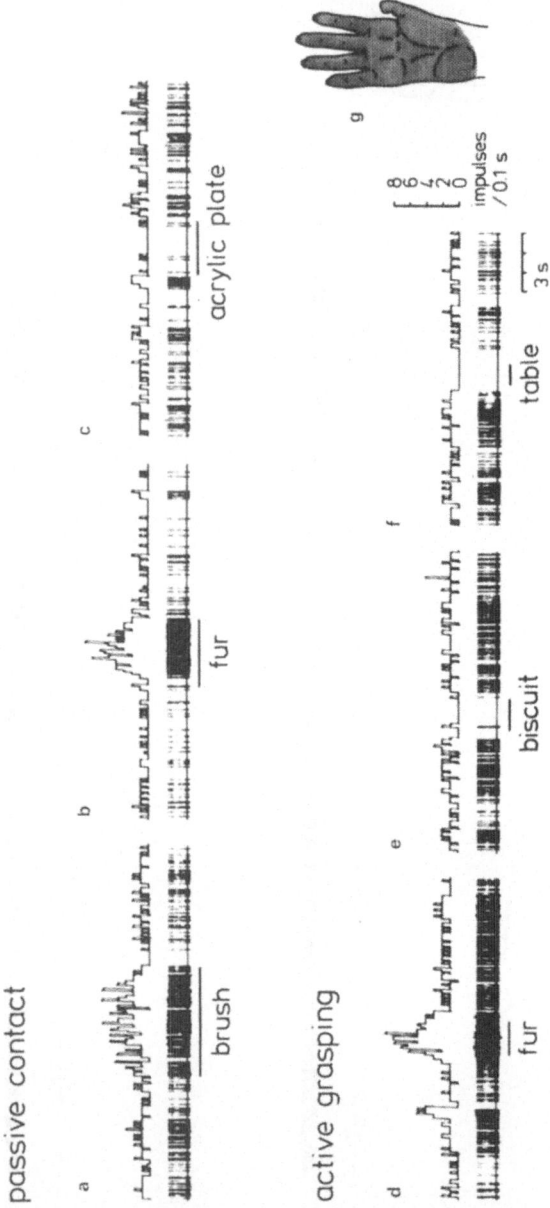

Fig. 4. An area 2 neuron responding preferentially to soft materials placed on the volar surface **(a–c).** Its background activity was inhibited by passive contact **(c)** or active grasping **(e,f)** of hard objects. **g** The receptive field of this neuron determined by punctate stimuli

and fingers. They were activated by passive contact of these materials, but in some of them, active grasping augmented the differential responses to different materials (Fig. 4), while others were simply activated or inhibited at the moment of active hand manipulation, regardless of material differences. These observations suggest that kinesthetic information is converging on these neurons in addition to information from the skin.

The cortical recording sites of these feature-detection neurons were determined in relation to the distribution of several functional surfaces. Edge- or large-surface-detection neurons were scattered in a wider region than those of material discrimination, which were found in the core region. In the core region, information from various functional surfaces, finger joints, and wrist joints overlapped.

Discussion and Conclusion

In area 2 we found nearly 40% of neurons had cutaneous receptive fields. However, the strict somatotopic representation of fingers was not seen (Iwamura et al. 1980). The majority of skin neurons had multifinger receptive fields, and they could be categorized into eight functional surfaces. As Fig. 1 indicates, these functional surfaces were distributed in a loose topography, although there was a complex mixture in the middle region. We think that this is a new mode of topographic organization in area 2, instead of digitotopy, and that this organization reflects the unique mode of information processing in this area for tactual perception. These surfaces may be the results of systematic integration through corticocortical connections within SI (Iwamura et al. 1983b, 1984).

We have presented evidence that these functional surfaces represent the skin areas which come in contact with objects most effectively in various manipulative actions. Receptive fields of neurons for feature detection were large and also coincided with one of these functional surfaces. Edge detection was correlated with several types of surfaces, one restricted to the thumb and thenar pad (the major part of surface a), the other involving four fingers and palm skin (surfaces b, c and d). In contrast, receptive fields of neurons that respond differently to different materials had receptive fields defined by surfaces e, c, or i. In active exploratory maneuvers, the hand and fingers form impressions of objects, which may be obtained by the simultaneous contact of multiple surfaces (surfaces d and i, for example) or by changing the contact surfaces one by one successively. In these circumstances the information collected by the multiple surfaces in various combinations will be transmitted to the brain simultaneously or successively. Information concerning one object but coming from the multiple surfaces of the hand and fingers could be processed efficiently if these surfaces were represented closely to each other on the cortical surface. Since there are various ways for the hand to explore the tactual world, functional surfaces would be in action in various combinations, and thus the representation of each surface must be multiplied.

The information from finger joints and also that from proximal joints (mostly the wrist joint) overlapped with that from the finger skin. Thus, different modes of information, those from the skin and those from the deep receptors located at the same or different body sites, are mixed. Some of these signals may reach area 2 after being transformed to provide more complex information, such as that concerning integrated movements. We found that active hand manipulation influenced the responses of some feature-detection neurons. This is consistent with the earlier statement that the tactual discrimination of objects is complete only when it is actively performed (Gibson 1962). We postulate that there exist different levels of complexity in feature-detection processes found in this region of the cortex, from the simplest edge detection to the discrimination of shape or size, or from the discrimination of different materials to the hard-soft discrimination. The higher-level feature detection may require more information to be combined and integrated. For example, while simple discrimination of different materials is performed on the basis of the surface stimulus, the hard-soft discrimination may require information from the deep receptors as well as from the skin.

References

CORKIN S, MILNER B, RASMUSSEN T (1970) Somatosensory threshold. Arch Neurol 23: 41-58

COSTANZO RM, GARDNER EP (1980) A quantitative analysis of responses of direction-sensitive neurons in somatosensory cortex of awake monkeys. J Neurophysiol 43: 1319-1341

GARDNER EP, COSTANZO RM (1980) Neuronal mechanisms underlying direction sensitivity of somatosensory cortical neurons in awake monkeys. J Neurophysiol 43: 1342-1354

GIBSON JJ (1962) Observations on active touch. Psychol Rev 69: 477-491

HYVÄRINEN J (1976) Cellular mechanisms in the parietal cortex in alert monkey. In: ZOTTERMAN Y (ed) Sensory functions of the skin in primates with special reference to man. Pergamon, Oxford, pp 241-259

HYVÄRINEN J, PORANEN A (1978) Receptive field integration and submodality convergence in the hand area of the postcentral gyrus of the alert monkey. J Physiol 283: 539-556

IWAMURA Y, TANAKA M (1978) Postcentral neurons in hand region of area 2: their possible role in the form discrimination of tactile objects. Brain Res 150: 662-666

IWAMURA Y, TANAKA M, HIKOSAKA O (1978) Functional organization of neurons in area 2 of monkey somatosensory cortex (SI). Abst Soc Neurosci 4: 554

IWAMURA Y, TANAKA M, HIKOSAKA O (1980) Overlapping representation of fingers in area 2 of the somatosensory cortex of the conscious monkey. Brain Res 197: 516-520

IWAMURA Y, TANAKA M, HIKOSAKA O (1981) Cortical neuronal mechanisms of tactile reception studied in the conscious monkey. In: KATSUKI Y, NORGREN R, SATO M (eds) Brain mechanisms of sensation. Wiley, New York, pp 61-70 (III International symposium, division of brain science, The Taniguchi Foundation)

IWAMURA Y, TANAKA M, SAKAMOTO M, HIKOSAKA O (1983a) Functional subdivisions representing different finger regions in area 3 of the first somatosensory cortex of the conscious monkey. Exp Brain Res 51: 315-326

IWAMURA Y, TANAKA M, SAKAMOTO M, HIKOSAKA O (1983b) Converging patterns of finger representation and complex response properties of neurons in area 1 of the first somatosensory cortex of the conscious monkey. Exp Brain Res 51: 327-337

IWAMURA Y, TANAKA M, SAKAMOTO M, HIKOSAKA O (1984) Comparison of the hand and finger representation in areas 3, 1 and 2 of the monkey somatosensory cortex. (to be published)

JONES EG, COULTER JD, HENDRY SHC (1978) Intracortical connectivity of architectonic fields in the somatic sensory, motor and parietal cortex of monkeys. J Comp Neurol 181: 291-348

KÜNZLE H (1978) Cortico-cortical efferents of primary motor and somatosensory regions of the cerebral cortex in *Macaca fascicularis.* Neuroscience 3: 25-39

McKENNA TM, WHITSEL BL, DREYER DA (1982) Anterior parietal cortical topographic organization in macaque monkey: a reevaluation. J Neurophysiol 48: 289-317

MOUNTCASTLE VB, LYNCH JC, GEORGOPOULOS A, SAKATA H, ACUNA C (1975) Posterior parietal association cortex of the monkey: command functions for operations within extrapersonal space. J Neurophysiol 38: 871-908

POWELL TPS, MOUNTCASTLE VB (1959) The cytoarchitecture of the postcentral gyrus of the monkey *Macaca mulatta*. Bull Johns Hopkins Hosp 105: 108-131

RANDOLF M, SEMMES J (1974) Behavioral consequences of selective subtotal ablations in the postcentral gyrus of *Macaca mulatta*. Brain Res 70: 55-70

ROLAND PE (1976) Astereognosis. Arch Neurol 33: 543-550

SCHWARZ DWF, FREDRICKSON JM (1971) Tactile direction sensitivity of area 2 oral neurons in the rhesus monkey cortex. Brain Res 27: 397-401

VOGT BA, PANDYA DN (1977) Cortico-cortical connections of somatic sensory cortex (areas 3, 1 and 2) in the rhesus monkey. J Comp Neurol 177: 179-192

WHITSEL BL, ROPPOLO JR, WERNER G (1972) Cortical information processing of stimulus motion on primate skin. J Neurophysiol 35: 691-717

Sensorimotor Cortex Responses to Vibrotactile Stimuli During Initiation and Execution of Hand Movement

R. J. Nelson

Department of Anatomy, University of Tennessee, Center for the Health Sciences,
875 Monroe Avenue, Memphis, TN 38163, USA

The responses of neurons in the somatosensory cortex (SI) of primates are well documented for stimuli presented when the hand is used to sense a stimulus but not when the sensing hand is also used to make a movement in response to the stimulus (see, for example, Mountcastle et al. 1969). There are, however, reasons to believe that changes in sensory responsiveness occur in the context of active movement. Several investigators have demonstrated alterations of sensory responses in the dorsal column nuclei and medial lemniscus (Ghez and Lenzi 1971; Ghez and Pisa 1972; Coulter 1974), the ventrobasal thalamus (Coquery 1978), and in primary somatosensory cortex (Chapin and Woodward 1982; Coquery 1978) prior to active movement of the limb. It has been suggested that the sensorimotor cortex is involved in these alterations because of the premovement neuronal activity that occurs in sensorimotor cortex in relation to the initiation of movement (see Chapin and Woodward 1982) and because it projects upon the dorsal column nuclei (Jones and Wise 1977; Kuypers and Lawrence 1967).

The present study examined responses of SI neurons during a goal-oriented motor task that involved the active maintenance of a position, the detection of a vibrotactile cue, and the execution of a movement involving the stimulated hand. The purpose of the study was to compare the responses of neurons during vibrotactile stimulation prior to movement with responses to identical stimuli that were not followed by an active movement of the sensing hand.

Experimental Brain Research, Suppl. 10
© Springer-Verlag Berlin · Heidelberg 1985

Experimental Procedures

Two adult rhesus monkeys (*Macaca mulatta*) were trained to perform movements in response to vibrotactile stimuli delivered to the glabrous surface of the hand. In this task, the monkey's hand rested on a smooth aluminum handle attached to the axle of a DC torque motor. The axle was placed directly beneath the wrist joint to allow flexion and extension of the hand about the wrist. A constant torque was applied to the axle in the upward direction requiring a flexion force of 0.06 Nm at the tip of the handle to maintain a centered position. The angular position of the handle was monitored and amplified and the resultant displayed to the animal by a single row of 31 light-emitting diodes (LED).

Each trial began when the animal positioned the handle so that the central LED of the display was illuminated. This position corresponded to horizontal positioning of the hand so that there was neither flexion nor extension about the wrist. The animal was required to maintain this hold position for a period (0.5, 1.0, or 1.5 s; computer randomized) without deviating more than 0.5° in either direction. Upon successful completion of this required hold period, the handle was vibrated by a low amplitude sine wave signal (27, 57, or 127 Hz), which resulted in a displacement of the torque motor axle of less than 0.02°.

This vibrotactile stimulation was the trigger for flexion or extension of the hand about the wrist. The appropriate direction for movement was indicated by a small LED displayed to the left of the main array. When illuminated, it indicated that the monkey was to extend; otherwise the monkey was to flex. The vibration remained on until the monkey had moved 5° from the hold position. If the monkey made the requested movement, fruit juice reward was given; if not, the trial went unrewarded. In either case, the vibration was turned off after each 5° movement, and the next trial could begin once the monkey returned the handle position to the central hold zone. After nine consecutive, rewarded trials, the movement request was changed to the other direction.

Activity in the pre- and postcentral cortex was recorded by conventional means. A stainless steel recording chamber was implanted over the forelimb region of sensorimotor cortex contralateral to the trained extremity. Platinum-iridium microelectrodes were advanced in transdural penetrations. Neuronal discharge was discriminated using a time-amplitude window discriminator. Discriminator pulses and the analog position signal were digitized by an on-line data collection routine run by a PDP-11 microcomputer. This routine also controlled the behavioral paradigm and marked the occurrence of significant events.

After recording the activity of each cortical unit, peripheral receptive fields were determined for each cell that remained well isolated. The cutaneous surfaces of the forelimb were examined with hand-held probes and air puffs; hairs were moved and joints manipulated in an attempt to find sources of input capable of modulating the neuronal discharge. During this procedure, the monkeys sat quietly and tolerated the passive stimulation without resisting. If joint mainpulation caused a change in discharge and if a cutaneous receptive field could not be determined, the cells were classified as being sensitive to noncutaneous input.

Prior to the sacrifice of the animals, electrolytic lesions were made by passing current (10 μA for 10 s) through a recording electrode. The animals were deeply anesthetized and perfused intracardially with 10% formol-saline. The cortex was sectioned at 50 μm intervals in the sagittal plane. Histological verification of electrode penetrations indicated that records were predominantly from areas 3a, 3b, 1, and 2; some recordings were from area 4 and these were used in this study for comparison with recordings from cells in somatosensory cortex.

Observations

The activity of 619 task-related units in areas 3a, 3b, 1, and 2 was recorded. Of these, 188 were chosen for detailed analysis. Criteria for selection were a short latency response (<30 ms) related to the onset of the vibrotactile stimulus and changes in activity correlated with some aspect of the movement. Those that met these criteria were divided into two categories. Cells in the first category (VIB-) exhibited a profound decrease in firing rate just after stimulus onset. Cells in the second category (VIB+) showed a marked increase in activity in relationship to the stimulus. Included in this second category are 39 neurons which were entrained to the stimulus frequency. Ten cells were included in both categories because they exhibited one type of response at one stimulus frequency and the other at a different frequency. The number of each type is shown in Fig. 1.

Units with Decreased Activity After Stimulus Onset (VIB-)

A profound decrease in firing rate was observed in 87 SI neurons within 30 ms of stimulus onset. Each such unit was tested for peripheral input by passively moving joints, palpating muscle bellies, and mapping cutaneous receptive fields with a hand-held probe. The majority (47/87, 54%) had well-defined responses to the manipulation of a single joint. A smaller number (21/87, 24%) had small, punctate cutaneous receptive fields most often located on the glabrous distal phalanges of the digits. For 12/87 (14%) of the VIB- units, a complete set of behavioral trials was recorded, but the neuron was lost during attempts to define the source of peripheral input. The response profiles of these 12 neurons for active movement were nearly identical to those of the 47 VIB- units that responded to joint manipulation, and so the 12 were included with the 47 to form a group of 59 for the purpose of statistical analysis. These 59 showed activity changes approximately 20 ms prior to movement onset, in addition to the decrease in discharge rate observed at stimulus onset. In contrast, VIB- neurons with cutaneous receptive fields often changed firing rates at stimulus onset and just after movement onset. Seven VIB- neurons (8% of the total) were responsive to peripheral stimulation, but no peripheral site could be unequivocally defined. These neurons, therefore, were classified simply as having no clear receptive field.

Two examples of the discharge patterns of VIB- neurons are illustrated in Fig. 2 for the response of a neuron in area 1 and in Fig.3 for the response of an area 3a neuron. In both instances, the neurons were excited by passive extension of a single joint and showed substantial discharge during the active maintenance of the hold position. Each showed a profound decrease in firing rate just after the onset of the vibrotactile stimulus. The average latency of this decrease was 22.5 ms for the total sample of 87 units, with no significant differences between those neurons receiving noncutaneous or cutaneous input. The neuron whose response is shown in Fig. 3 discharged prior to active wrist flexion and passive wrist extension. Since the direction of active movement was opposite to that which excited the cell during passive movement, it was reasoned that this neuron probably received

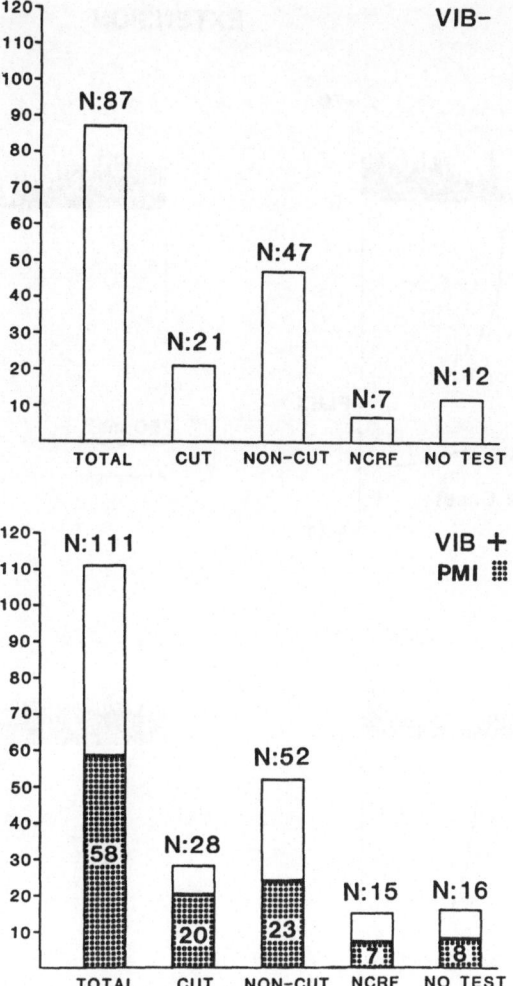

Fig. 1. Totals of each type of neuron with short latency responses to the vibrotactile stimulus. *Bars* signify the number of cells with that type of receptive field *(RF)*; *CUT*, cutaneous RF, *NON-CUT*, noncutaneous RF; *NCRF*, no clear RF; *NO TEST*, not tested for RF. VIB- cells decrease their firing rate in response to the stimulus, while VIB+ cells increase their activity. *Stippled bars* indicate VIB+ neurons showing premovement inhibition *(PMI)*

FLEXION　　　　　　　　　**EXTENSION**

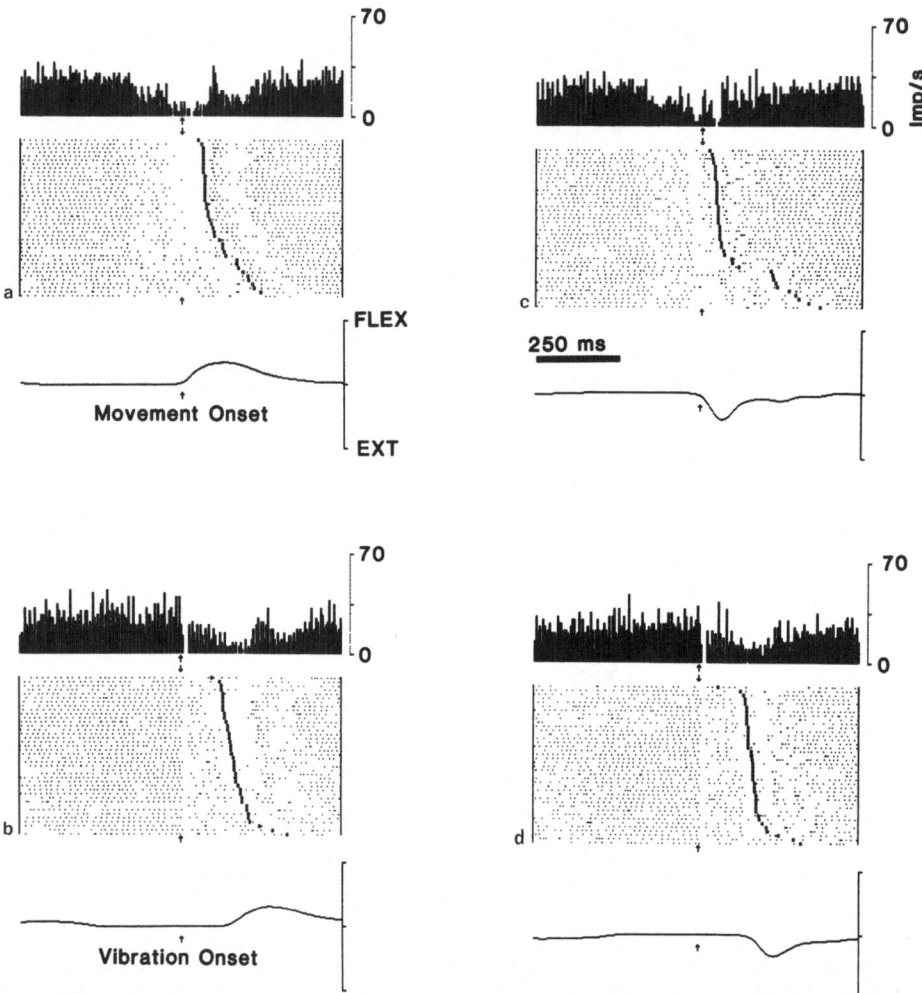

Fig. 2. Activity of a VIB- neuron in area 1. Rasters, histograms, and average position trace centered on movement onset in *upper panels* and vibration onset in *lower panels* for several trials requiring flexion *(left)* or extension *(right)* of the wrist. *Bottom panels:* trials have been ordered according to reaction time. *Top panels:* ordered according to time from movement onset to 5° deviation of position trace. Each is indicated by the *dark mark* in the raster. This cell was excited by passive extension of the fourth digit at the metacarpophalangeal joint

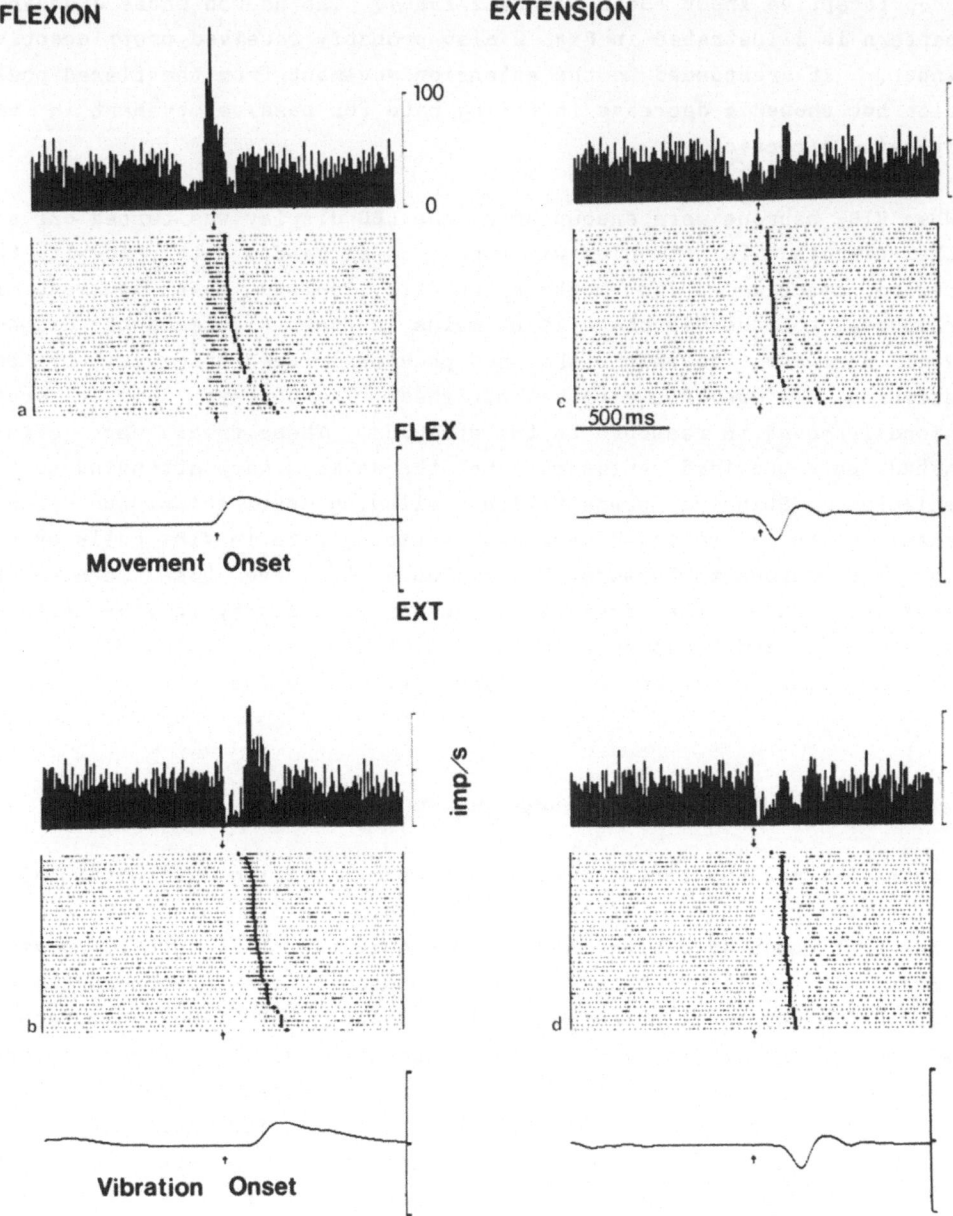

Fig. 3. Pattern of discharge for a VIB- neuron in area 3a excited by passive wrist extension. Format as in Fig. 2

proprioceptive input (Soso and Fetz 1980). The neuron whose discharge
pattern is illustrated in Fig. 2 also probably received proprioceptive
input.   It responded to the extension movement from the flexed position but showed a decrease in firing rate for passive movement in the
opposite direction.

When VIB- neurons were encountered, the LED display was turned off and
the vibrotactile stimulus was turned on at random intervals with the
reward disabled.   The monkeys quickly learned that movement in
response to the vibrotactile stimulus in these circumstances did not
yield a reward. Movements that had previously been triggered by the
stimulus, in general, were extinguished. The monkeys, however, occasionally moved in response to the stimulus. These trials were eliminated but provided evidence that the animals were attending to the
stimulus.   Stimulus presentation following extinction no longer
resulted in a prolonged decrease in firing rate in VIB- cells except
in a few neurons that were later found to have been located in
area 3b.   Thus, the dramatic reduction of activity in VIB- cells in
areas 3a, 1, and 2 depended on the set of the monkey to move the sensing hand upon detection of the vibrotactile stimulus.

Neurons with Increased Discharge Following Stimulus Onset (VIB+)

Of the 111 neurons that showed an increase in firing rate within 30 ms
of stimulus onset, 53/111 (48%) maintained a discharge rate greater
than that observed during the hold phase of the task until movement
onset. An example of this kind of response profile is seen in the records of a neuron illustrated in Fig. 4a and c. This unit was recorded deep in the penetration and was adjacent to cells receiving cutaneous input; it was later found to have been located in area 3b. While
no cutaneous receptive field could be determined for this cell, it was
exquisitely sensitive to passive stimulation of the surface of the
hand.   Its response profile during active movement was characteristic
of other 3b neurons that received cutaneous or deep input, in that
these neurons, as a whole, did not significantly decrease firing rate
prior to movement onset.

In contrast to area 3b neurons, most neurons in areas 1 and 2 that had
short-latency excitatory responses to the vibrotactile stimulus also
showed a significant decrease in firing rate 60-80 ms prior to move-

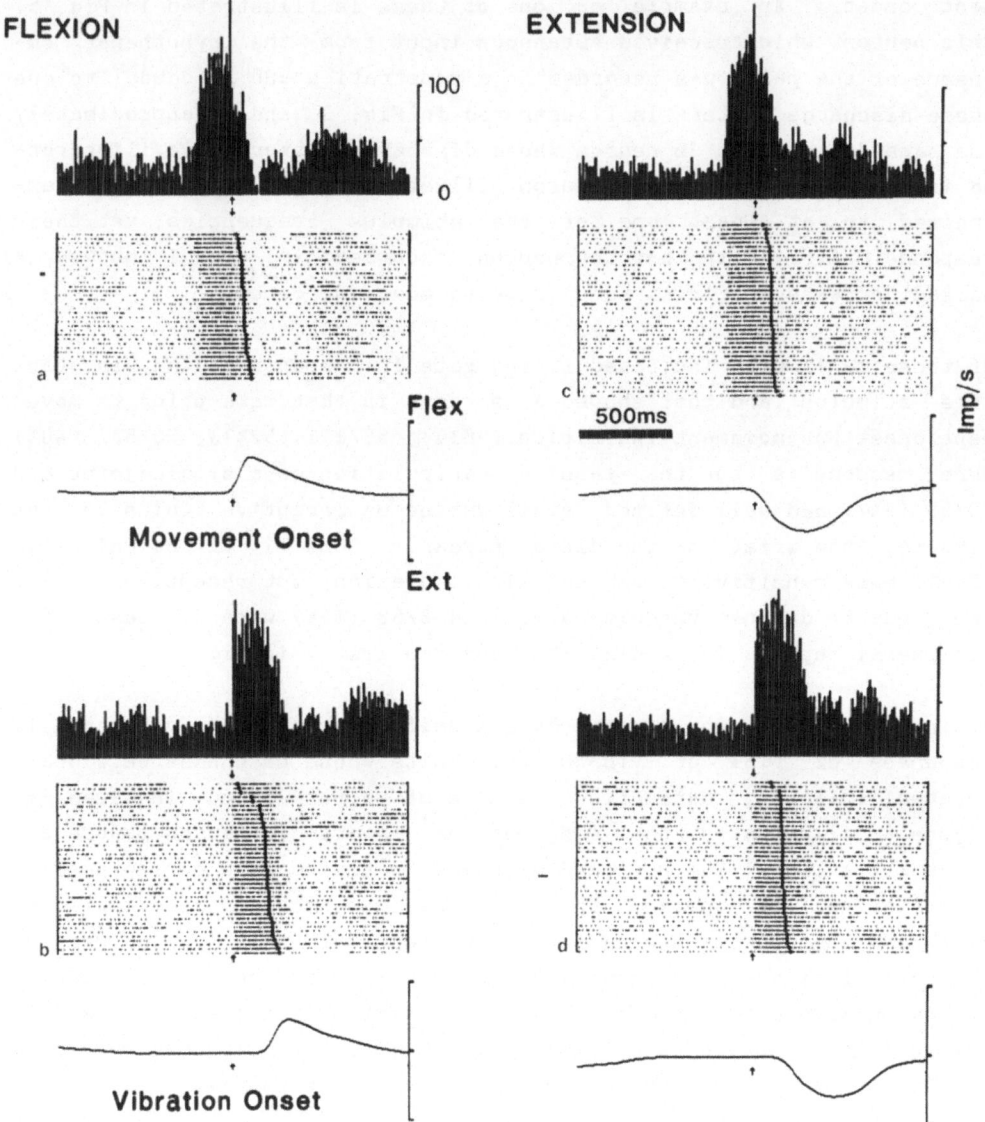

Fig. 4. Activity of VIB+ neuron in area 3b which was sensitive to passive stimulation of the hand, but for which no clear-cut RF could be defined. Note increased dicharge continues after movement onset *(lower panel, mark)*. Format as in Fig. 2

ment onset. An example of one of these is illustrated in Fig. 5. This neuron, which received cutaneous input from the hypothenar eminence of the palm, was recorded in a penetration 500 μm caudal to one whose discharge pattern is illustrated in Fig. 3, and at approximately the same depth. The 3b neuron whose discharge patterns are illustrated in Fig. 4 and the area 1 neuron illustrated in Fig. 5 were entrained to at least one of the stimulus frequencies, yet their response differed, in that the neuron recorded in area 1 showed a marked decrease in firing rate prior to movement onset.

Of those cells with increased firing rate in response to the vibrotactile stimulus and that showed a decrease in that rate prior to movement onset [premovement inhibition (PMI): 58/111 (52%)], 23/58 (40%) were responsive to the passive manipulation of a single joint and 20/58 (34%) had well-defined, small cutaneous receptive fields on the fingers, the wrist, or the distal forearm. Some 12% of the PMI cells (7/58) were sensitive to peripheral stimulation, but receptive fields could not be defined unequivocally, and 8/58 (13%) were not tested for peripheral input. These distributions are listed in Fig. 1.

Extinction of movement did not have a uniform effect on vibrotactile responses of VIB+ or VIB- units. Units whose responses were unalterated following extinction were most often found in area 3b. Figure 6 illustrates the response of an area 3a neuron before and after extinction. The profound decrease in firing rate during passive presentation, as opposed to the increased discharge preceding movement, was one common observation. Another was that there was no change in firing rate in some VIB- neurons in response to the stimulus following movement extinction. Neurons located in areas 3a, 1, and 2, unlike those in area 3b, often showed strikingly different responses to the stimulus, depending on the behavioral contingency.

Recordings from Motor Cortex

Twenty-four units were recoded in area 4 during the performance of this task, the purpose being to compare the timing of VIB responses in somatosensory cortex with the timing of activity in area 4. Figure 7 illustrates the firing pattern of an area 4 neuron that was sensitive to passive flexion of wrist. Intracortical microstimulation at the recording site resulted in movement of the hand. This neuron fired

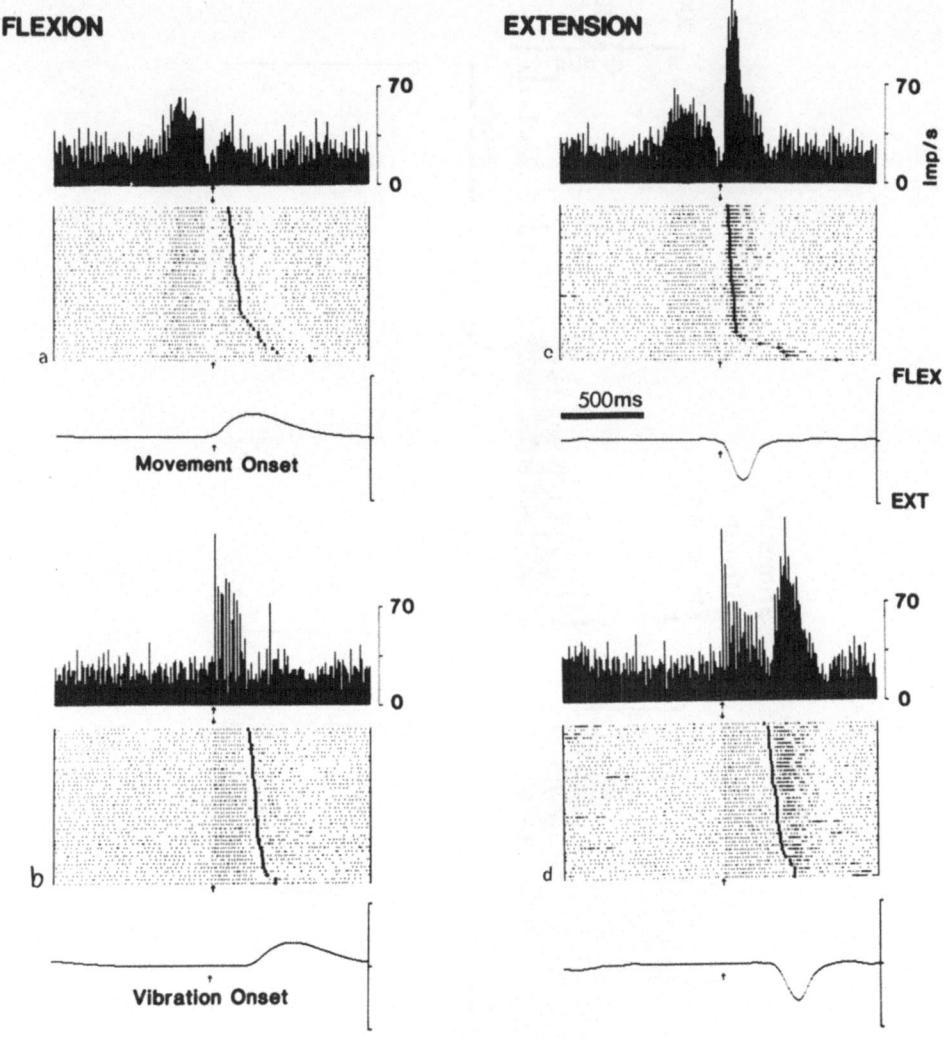

Fig. 5. Discharge pattern of an area 1 VIB+ neuron with small punctate RF on the hypothenar eminence. The discharge of this neuron was entrained to the stimulus frequency (57 Hz). Note the profound decrease in firing rate approximately 60 ms prior to movement onset (premovement inhibition) in both flexion and extension trials. Format as in Fig. 2

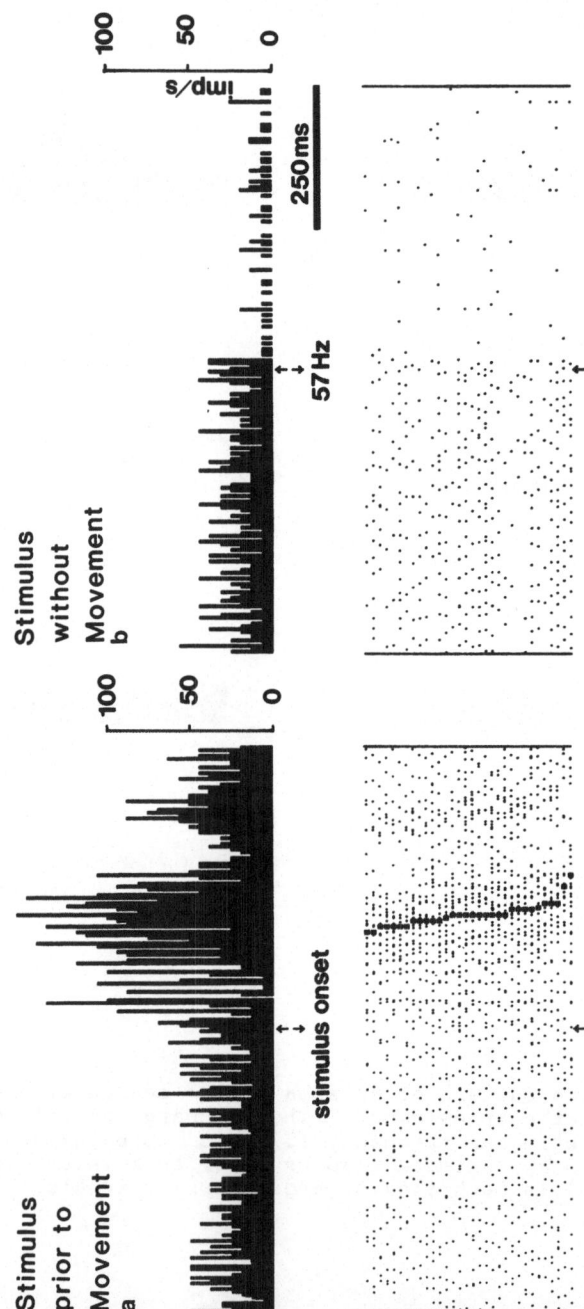

Fig. 6. *Left:* rasters and histograms of the activity of an area 3a
neuron during 32 trials requiring wrist extension. *Right:* presentation of the same stimulus without requiring movement. Note that in
the *left panel* there was an increase in discharge just after stimulus
onset (this neuron was weakly entrained to the stimulus frequency).
When movement was not required, stimulus presentation caused a profound decrease in the firing rate

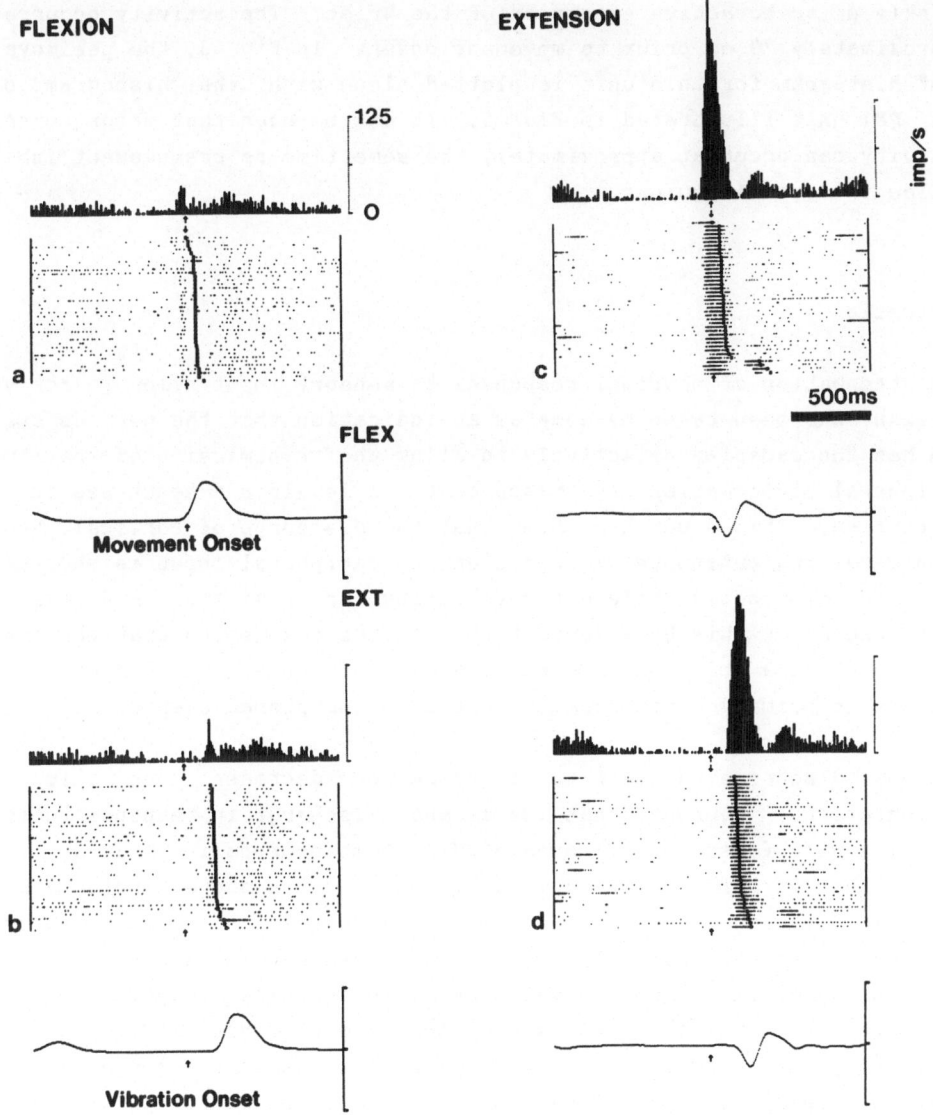

Fig. 7. Activity of an area 4 neuron during 40 trials of flexion and extension. This neuron was responsive to passive wrist flexion. Format as in Fig. 2

briskly prior to active extension of the wrist. The activity occurred
approximately 70 ms prior to movement onset. In Fig. 8, the perimove-
ment histogram for this unit is plotted along with the histogram of
the PMI unit illustrated in Fig. 5. It can be seen that motor cortex
activity can occur at approximately the same time as premovement inhibition in some SI neurons.

Discussion

The attenuation of neuronal responses to sensory input prior to active
movement has been taken by some as an indication that the nervous system has the capacity selectively to allow the transmission of salient
peripheral information to reach cortical levels and hence result in
perception. This study has shown that the discharge of cortical neurons receiving cutaneous and noncutaneous peripheral input is modulated in two distinctly different ways during the initiation and execution phases of this behavioral task. It has also shown that the context in which a stimulus is presented may influence the response of
neurons in primary somatosensory cortex to peripheral events.

Neuronal discharge was seen to increase or decrease upon stimulus
presentation. Neurons that decreased discharge in response to the
stimulus were found in each area of SI. Some neurons in areas 3a, 1,
and 2 that were initially excited by the stimulus showed profound decreases in firing rate before movement onset. In contrast, neurons
located in area 3b commonly maintained increased discharge rates from
stimulus onset until the removal of the vibrotactile stimulus. That
is, area 3b neurons rarely showed premovement inhibition. However,
neither type of modulation is well correlated with the peripheral receptive fields of the neurons. Neurons receiving either cutaneous or
noncutaneous input were found to have VIB+ and VIB- type responses.

An important observation is that neurons in areas 3a, 1, and 2 that
had excitatory peripheral receptive fields often exhibited decreases
in discharge rate shortly following the onset of vibrotactile stimulation only in the context of impending movement. When the stimulus was
presented after the movement was extinguished, these neurons often
showed no change in discharge rate in response to the stimulus. Some
neurons with excitatory responses to the stimulus triggering wrist
movement appeared to be inhibited by the same stimulus when presented

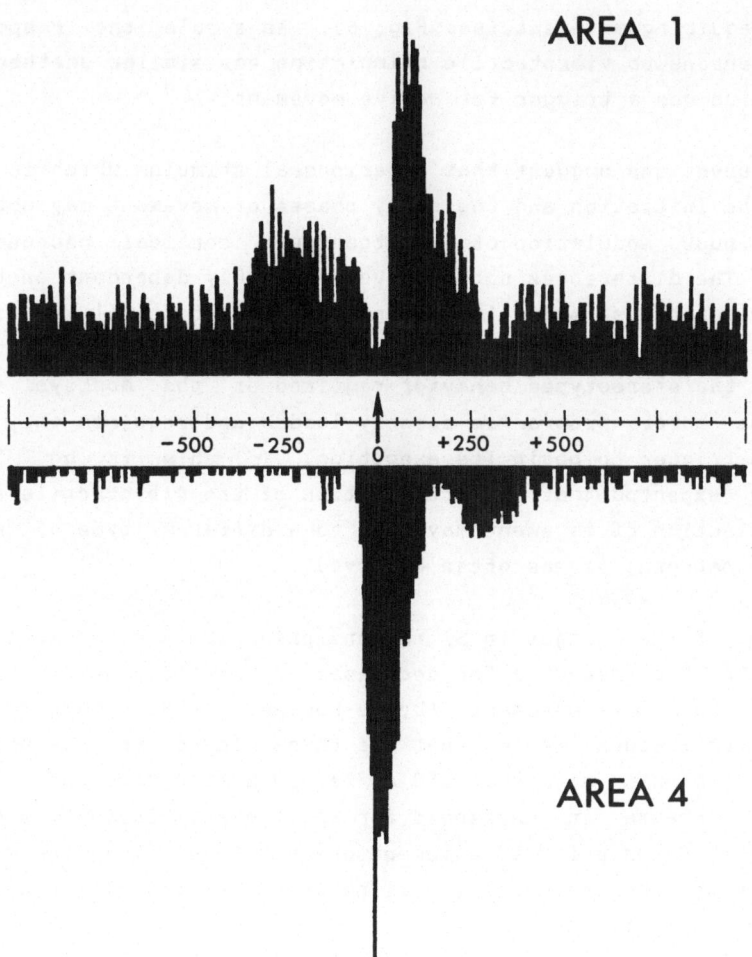

Fig. 8. Histograms of the activity of the neuron illustrated in Fig. 7 compared with the activity of the premovement inhibition (PMI) neuron illustrated in Fig. 5. Histograms centered on movement onset. Note the temporal correlation between increased activity in the area 4 neuron and the decrease in firing rate in the area 1 PMI cell

without requiring movement (see Fig. 6). As a rule, the response of area 3b neurons to vibrotactile stimulation was similar whether or not the stimulus was a trigger for active movement.

These observations suggest that a peripheral stimulus which is present during the initiation and the early phases of movement may not result in a continuous modulation of somatosensory cortical neuronal discharge. The differences noted above are highly dependent on the context in which the stimulus is presented and the cortical area in which the neurons are located. It is reasonable to assume that in the execution of the stereotyped behavior required of the monkeys in this task, the motor program is already loaded and requires only the appropriate trigger to begin its execution. If this is so, then it might be expected that the presentation of the vibrotactile stimulus after extinction of movement may lead to a different type of neuronal discharge pattern, as was often observed.

The timing of the changes in SI neurons prior to active movement is similar to that reported for decreases in tactile sensitivity in humans prior to active movement (Dyhre-Poulsen 1978; Coquery 1978). These investigators found that the threshold for tactile perception increased beginning more than 150 ms before active movement. A substantial increase in threshold for tactile stimulation was observed beginning at 60 ms prior to movement onset. These times are well correlated with the decrease in firing rate seen for VIB- and VIB+ PMI cells, respectively.

The exact nature of these modulatory effects and the location at which these influences occur are unknown. The close temporal correlation between motor cortical activity during this task and the modulation of somatosensory cortical neurons suggests that the motor cortex may play a role in afferent input modulation via the mechanisms of corollary discharge (see Evarts 1971) exerted perhaps at the level of the dorsal column nuclei (Jones and Wise 1977; Kuypers and Lawrence 1967) or by corticocortical connections with the somatosensory cortex (Jones et al. 1978; Vogt and Pandya 1978; Jones and Porter 1980). It is tempting to speculate that premovement inhibition may be the result of direct corticocortical connections from area 4 to areas 3a, 1, and 2, but not to area 3b. The apparent sparing of 3b from these influences suggests this. Alternatively, if the decrease observed is caused by influences mediated at the level of the dorsal column nuclei, then it might be necessary to assume that corticofugal connections are segregated to those cells whose processes are destined to relay in the thalamus and project exclusively to areas 3a, 1, and 2, and not to area 3b.

Changes in neuronal discharge patterns dependent upon the context in which the stimulus is presented have been observed previously in primate sensorimotor cortex (Evarts and Tanji 1976; Hyvarinen et al. 1980; Poranen and Hyvarinen 1982; Seal et al. 1983; see also Darian-Smith 1982 for discussion of topic). These studies found context-dependent changes in sensory responses that occurred in neurons located in areas other than those that comprise SI, or that changes in SI neurons were related to the monkey's attention to the stimulus. What is striking is that these influences are also observed in primary somatosensory cortex and that there appears to be a significant difference between area 3b and areas 3a, 1, and 2 in the response of neurons to a peripheral stimulus and its presentation context.

Acknowledgement. The author wishes to thank Drs. E.V. Evarts, J.P. Donoghue, V.A. Jennings, and S.P. Wise for their comments on initial versions of this report.

References

CHAPIN JK, WOODWARD DJ (1982) Somatic sensory transmission to the cortex during movement: gating of single cell responses to touch. Exp Neurol 78: 654-669

COQUERY J-M (1978) Role of active movement in control of afferent input from skin in cat and man. In: GORDON G (ed) Active touch: the mechanisms of object manipulation: a multidisiplinary approach. Pergamon, Oxford, pp 161-169

COULTER JD (1974) Sensory transmission through lemniscal pathway during voluntary movement in the cat. J Neurophysiol 37: 831-845

DARIAN-SMITH I (1982) Touch in primates. Annu Rev Psychol 33: 155-194

DYHRE-POULSON P (1978) Perception of tactile stimuli before ballistic and during tracking movements. In: GORDON G (ed) Active touch: the mechanisms of object manipulation: a multidisiplinary approach. Pergamon, Oxford, pp 171-176

EVARTS EV (1971) Feedback and corollary discharge: a merging of the concepts. Neurosci Res Program Bull 9 [1]: 86-112

EVARTS EV, TANJI J (1976) Reflex and intended responses in motor cortex pyramidal tract neurons of monkeys. J Neurophysiol 39: 1069-1080

GHEZ C, LENZI GL (1971) Modulation of sensory transmission in cat lemniscal system during voluntary movement. Pflugers Arch 323: 273-285

GHEZ C, PISA M (1972) Inhibition of afferent transmission in cuneate nucleus during voluntary movement in the cat. Brain Res 40: 145-151

HYVARINEN J, PORANEN A, JOKINEN Y (1980) Influence of attentive behavior on neuronal responses to vibration in primary somatosensory cortex in the monkey. J Neurophysiol 43: 870-882

JONES EG, PORTER R (1980) What is area 3a? Brain Res Rev 2: 1-43

JONES EG, WISE SP (1977) Size, laminar and columnar distribution of efferent cells in the sensory-motor cortex of monkeys. J Comp Neurol 175: 391-438

JONES EG, COULTER JD, HENDRY SHC (1978) Intracortical connectivity of architectonic fields in the somatic sensory, motor and parietal cortex of monkeys. J Comp Neurol 188: 113-136

KUYPERS GJM, LAWRENCE DG (1967) Cortical projections to the red nucleus and the brain stem in the rhesus monkey. Brain Res 4: 151-188

MOUNTCASTLE VB, TALBOT WH, SAKATA H, HYVARINEN J (1969) Cortical neuronal mechanisms in flutter-vibration studied in unanesthetized monkeys. Neuronal periodicity and frequency discrimination. J Neurophysiol 32: 452-484

PORANEN A, HYVARINEN J (1982) Effects of attention on multiunit responses in the somatosensory region of the monkey's brain. EEG EMG 53: 525-537

SEAL J, GROSS C, DOUDET D, BIOULAC B (1983) Instruction-related change of neuronal activity in area 5 during simple forearm movement in the monkey. Neurosci Lett 36: 145-150

SOSO MJ, FETZ EE (1980) Responses of identified cells in postcentral cortex of awake monkeys during comparable active and passive joint movements. J Neurophysiol 43: 1090-1110

VOGT BA, PANDYA DN (1978) Cortico-cortical connections of somatic sensory cortex (areas 3, 1 and 2) in the rhesus monkey. J Comp Neurol 177: 179-192

Tactual Roughness Perception in Human:
A Psychophysical Assessment of the Role of Vibration

S.J.Lederman

Department of Psychology, Queen's University, Kingston, Ontario K7L 3N6, Canada

Let us begin with a little phenomenology. When you wish to know some-thing about surface texture, you typically move your fingers back and forth, laterally, across the surface in question. As you do so, you may notice the tiny vibratory impulses that are being generated within the skin. What role, if any, do such vibrations play in the human perception of roughness? I shall focus primarily on roughness percep-tion, since it is one of the salient tactual aspects of texture and has received the most attention to date.

David Katz (1925; for English summaries, see Krueger 1970, 1982) ar-gued strongly for the importance of these vibrations in the perception of surface texture, or "modifications of the surface," as he described it. In fact, Krueger (1970) stated that "for Katz... the vibration sense determines the modifications of the surface" (p. 339). According to Katz, without vibration, there can be no perception of texture. It certainly appears that when a person lightly rests his or her fingers on a surface without lateral motion (and thus, without vi-bration), texture is only fleetingly experienced, if at all. Unfortunately, it is not entirely clear what Katz intended by the term "vibration." If, as seems possible, he was referring specifically to the dynamic or "ac" aspects of the signal (e.g., rate of skin dis-placement, temporal frequency, etc.), his interpretation of the stimulus determinants of roughness may be described as "temporal." As additional support for his position, Katz demonstrated that a subject could discriminate some textures almost as well when he or she exam-ined surfaces by means of a pencil held in the hand as when the finger made direct contact. Furthermore, performance was greatly reduced when the vibrations were attenuated by wrapping the pencil in cloth.

Experimental Brain Research, Suppl. 10
© Springer-Verlag Berlin · Heidelberg 1985

These observations suggest that the skin need not be deformed by the surface contours in order for roughness to be perceived. It is unclear, however, whether the situations used by Katz are representative of roughness sensations or discrimination under normal conditions of touch. Nonetheless, a temporal coding position is certainly, on the face of it, a reasonable one to propose. R.H. Gibson (Taylor et al. 1973), for example, has argued that vibratory frequency is the critical determinant of roughness perception. Moreover, Darian-Smith and his colleagues (e.g., Darian-Smith and Oke 1980; Darian-Smith et al. 1980) have demonstrated that the stimulus temporal frequency is indeed neurally coded in the response of individual mechanoreceptor units.

Taylor and Lederman (1975) have argued differently. The perception of roughness may disappear when a static finger is used, just as visual patterns fade when the micromovements of the retinal image are prevented (Pritchard 1961). Thus, vibration is just a by-product of the lateral motions required for roughness perception. They believe that the dynamic aspects of the vibratory signal are relatively unimportant. Their interpretation focuses rather on the contribution of various static or "dc" characteristics of skin deformation (e.g., depth of penetration, volume of skin deformed, skin stress, skin strain, etc.). Some of the experimental results that will be reviewed here form the original data base for a quasi-static skin deformation model of roughness perception developed by Taylor and Lederman (1975). The model has sometimes been described as "spatial" (in contrast with a temporal position). While spatial factors may ultimately play a role in this model, strictly speaking, the current position does not permit the recovery of the invariant spatial patterns from the code; hence, the term "intensity-dependent" is probably more appropriate.

With various associates, Lederman has assessed the role of vibration in the tactual perception of roughness using three different psychophysical methods. Together, the results argue very strongly against Katz' position that vibration is a necessary condition for the perception of roughness.

General Features of the Psychophysical Experiments

In all of the experiments that will be reported in this paper, the stimulus surfaces consisted of linear metal gratings. A balance-like apparatus was used to control finger force. The stimuli were placed in a tray positioned at one end of a balance arm with counterweights at the other. A magnitude estimation procedure was used in which subjects were instructed to assign any positive, nonzero number to the magnitude of the roughness perceived.

Method 1: Spatial Period Altered, Hand Speed Constant

In the first series of experiments (Lederman and Taylor 1972; Lederman, to be published), temporal factors (e.g., fundamental pulse frequency) were evaluated by altering the spatial geometry of the gratings while maintaining hand speed constant. In the earlier work, a set of gratings was used in which the spatial period (1 groove + 1 ridge) was increased by increasing groove width (ridge width was constant). The results indicated that perceived roughness markedly increased with increasing fundamental spatial period (see Fig. 1, groove-varying set). For purposes of later comparisons with other data also shown here, perceived roughness (log magnitude estimates) is plotted as a function of groove width for two controlled force conditions. Note that increasing finger force also increased the magnitude of the roughness felt (as also shown in Fig. 2). However, since spatial period was completely confounded with groove width, a second set of plates was presented - fundamental spatial period was varied exactly as before, but the previous groove and ridge width values were now interchanged. Thus groove width was constant while ridge width increased (along with fundamental spatial period). If fundamental spatial period were a major determinant of roughness perception, the data from the second set should have been very similar, if not identical, to those from the first set. However, perceived roughness tended to decrease slightly as a function of increasing fundamental spatial period (and ridge width). Therefore, to the extent that the fundamental pulse frequency in this study is determined by fundamental spatial period, the data argue against a temporal coding model of roughness.

Moreover, this conclusion is further strengthened by additional data, which were recently rediscovered and reported by Lederman (to be published). They are based on the results obtained from two other sets of gratings also presented in this earlier work. In one set, the duty cycle was varied to maintain a constant spatial period throughout. In the other set, the groove: ridge width ratio remained constant at 1:1

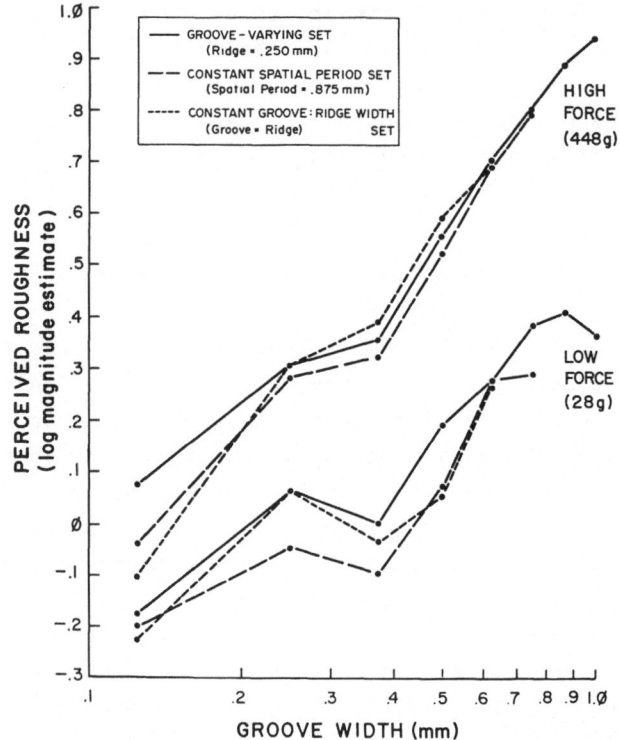

Fig. 1. Perceived roughness (log magnitude estimates) as a function of groove width and fingertip force. For both of the 28- and 448-g-force conditions, the data for the groove-varying, constant spatial period, and the constant groove: ridge sets are plotted separately. (Lederman, to be published)

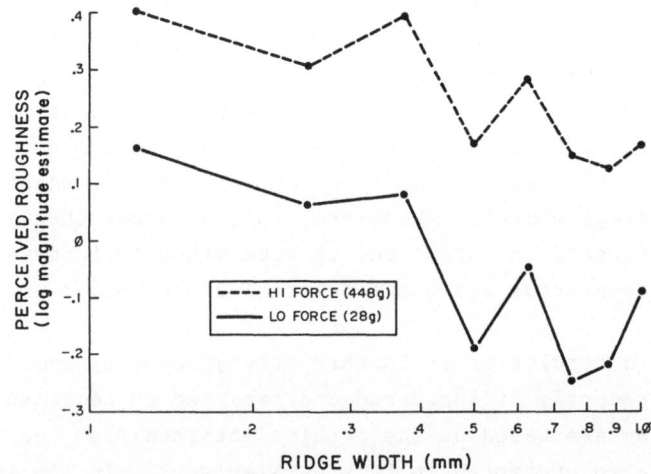

Fig. 2. Perceived roughness (log magnitude estimates) as a function of rigde width and applied force (ridge-varying set). (Lederman, to be published)

across plates (i.e., groove and ridge width were equal). The results of the three sets of gratings in which groove width varied are presented together in Fig. 1. Particularly significant to the main argument is the fact that within each force condition, the three curves are indistinguishable. When groove width varies and fundamental spatial period remains constant, perceived roughness varies. Conversely, when groove width remains constant and fundamental spatial period varies, perceived roughness remains unchanged. Together, the results strongly argue against the importance of both fundamental spatial period and fundamental temporal pulse frequency as determinants of human roughness perception.

Method 2: Hand Speed Altered, Spatial Period Constant

A second way of altering temporal factors, such as the temporal pulse frequency, is to vary the speed of motion between skin and surface while holding the spatial period of the gratings constant. Lederman (1974) originally demonstrated that when subjects moved their hands actively across a set of groove-varying gratings, speed had relatively little effect on perceived roughness. Thus, for a 25-fold increase in hand speed, felt roughness decreased only very slightly; moreover, the effect was negligible when compared with that of groove width.

It is of course possible that because subjects knew how fast they were moving their hands, they simply took speed into account when using temporal frequency. Accordingly, two additional experiments were conducted (Lederman, to be published) to assess the validity of this "constancy" explanation. In the first experiment, subjects either moved their hands actively (active touch) at different speeds (12-fold range) over stationary gratings, or the latter were moved at corresponding speeds across the subjects' stationary fingers (passive touch). Regardless of whether touch was active or passive, the results were remarkably similar to those in the original study. The results are shown in Fig. 3. Perceived roughness is plotted as a function of groove width for the three speed conditions. Figure 3a shows the results of the active touch condition, and Fig. 3b represents the passive touch condition. While the effect of speed is somewhat better highlighted in the passive touch condition, as a speed constancy interpretation would have predicted, the proportion of variability accounted for by the speed factor (both main and interaction

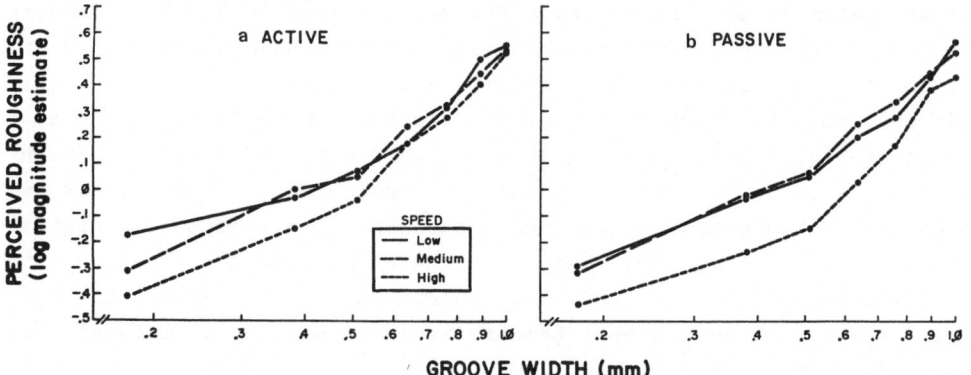

Fig. 3a,b. Perceived roughness (log magnitude estimates) as a function of groove width. Data are shown separately for each rate condition under **(a)** active and **(b)** passive modes of touch. (Lederman, to be published)

effects) was less than 1% (compare 40% for the main effect of groove width). Moreover, virtually identical results were obtained in yet another experiment in which subjects used *only* passive touch. In this study, the subjects would have had still less information about speed. In conclusion, the data obtained with the second method also argue against the importance of dynamic aspects of vibration in the perception of roughness.

Method 3: Selective Vibrotactile Adaptation

In this third method, a selective vibrotactile adaptation technique was used to evaluate the role of vibration in the perception of roughness. If, as Katz argued, the sense of vibration does in fact underlie the perception of roughness, then a procedure that affects the perception of vibrotactile stimulation should also affect the perception of surface roughness.

In 1968, Verrillo proposed a duplex model of mechanoreception for vibrotactile detection based on psychophysical evidence. He suggested that sensitivity to high-frequency stimulation (with maximum response at around 250-300 Hz) is mediated by the pacinian system (PC), while sensitivity to lower frequencies (particularly below about 40 Hz) is mediated by a separate pathway with associated end-organs and referred to simply as the nonpacinian system (NP). (More recently, Capraro et al. (1979) have suggested the participation of at least two nonpacinian receptor populations.) The psychophysical work has been reasonably well complemented by physiological data on glabrous skin of monkey obtained by Talbot et al. (1968). Their data show that the low and high branches of the human psychophysical threshold curves for vibration (as a function of frequency) correspond well to the frequency tuning curves of the quickly adapting units (QA), presumed to end in Meissner's corpuscles, and pacinian afferents, respectively. More recently, masking and adaptation studies by Gescheider et al. (1979), Labs et al. (1978), and Verrillo and Gescheider (1977) have added to the psychophysical evidence by demonstrating a fair degree of functional independence between the NP and the PC systems.

Accordingly, in the current set of experiments (Lederman et al. 1982), a selective adaptation technique was used in an attempt to alter the perceived magnitude of suprathreshold vibrotactile stimuli.

According to the logic that underlies the current research, it was necessary to demonstrate a strong effect in the experiment before one could properly evaluate the role of vibration in the perception of roughness. In the control condition, subjects made estimates of the magnitude of sensations of vibrotactile test stimuli varying in intensity (5 to 40 dB SL), presented at both 20 and 250 Hz. The finger simply rested lightly on the static contactor for 10 min. The same sets of judgements were obtained following selective adaptation to both 20- and 250-Hz adapting stimuli presented at 40 and 45 (or 50) dB SL, respectively. The results are shown in Fig. 4. The data are plotted separately for the individual subjects. Perceived magnitude of vibration (log magnitude estimates) is plotted as a function of vibrotactile intensity for the control, low-, and high-frequency adaptation conditions (Fig. 4a-c). The data are shown for the low-frequency test condition. In each display, the low-frequency adaptation curve lies considerably below that of the control and high-frequency adaptation curves; the high-frequency adaptation curves lie either on top of or a little below the static control. In Fig. 4d-f, corresponding data are shown for the high-frequency test stimuli. Here, the high-frequency adaptation curve lies considerably below both the control and low-frequency adaptation curves; while there also appears to be some adaptation in the latter curves, it is considerably less than that obtained in the high-frequency adaptation conditions. Thus, strong selective vibrotactile adaptation effects were obtained with supraliminal stimulation. However, following identical conditions of selective adaptation, the magnitude estimates of the *roughness* of linear gratings were virtually unaffected. The results are shown in Fig. 5. Perceived roughness (log magnitude estimates) is plotted as a function of log groove width for the three adaptation conditions. Once again, the data for individual subjects are shown in separate panels. To repeat, while the sense of vibration was strongly affected by the selective vibrotactile adaptation procedure, the perception of roughness was not. Counter to Katz' claim, therefore, vibration *per se* is not a primary determinant of roughness.

This conclusion is now strongly supported by the results of three converging operations: the relative contribution of temporal factors has been evaluated by varying the spatial geometry of the stimulus surface, by varying the speed of relative skin surface motion, and by using a selective vibrotactile adaptation procedure. Clearly, the temporal aspects of vibration do not underlie the perception of roughness.

20 HZ TEST

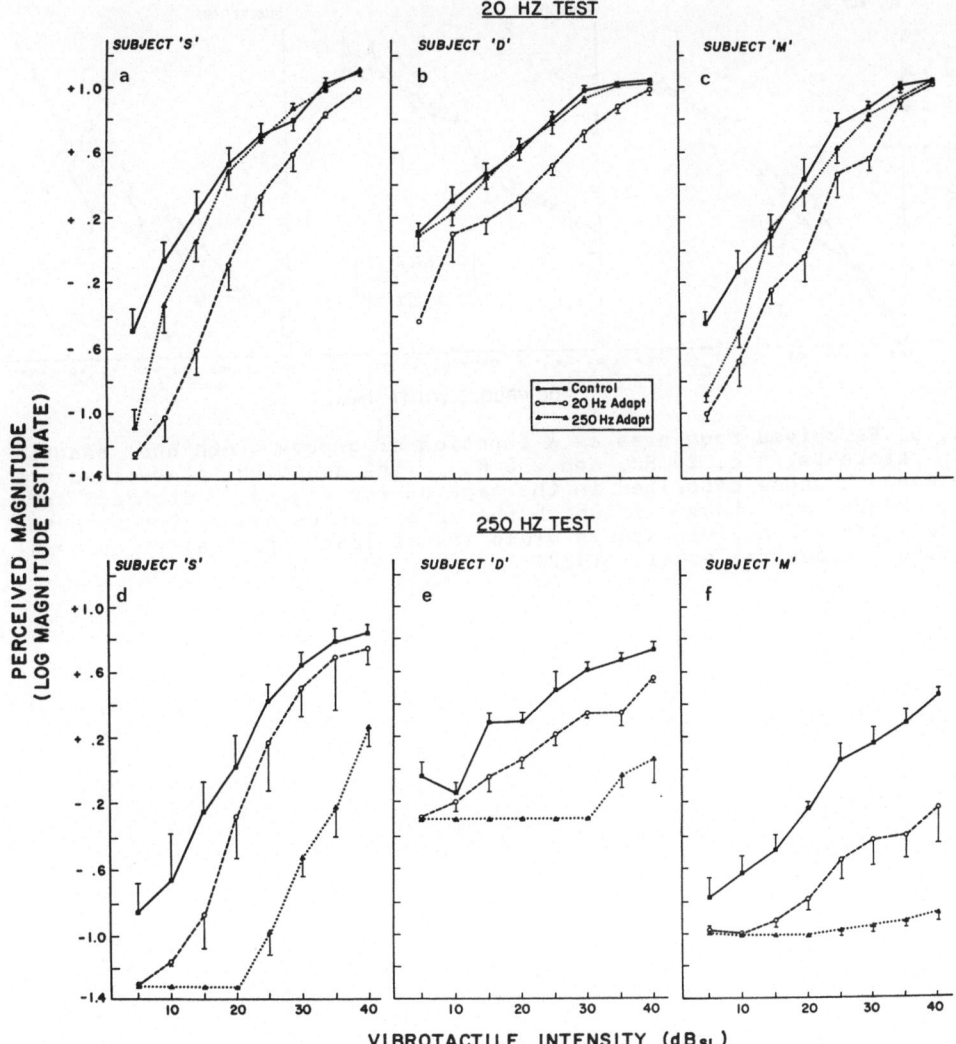

Fig. 4a-f. Perceived magnitude of vibration and associated standard errors as a function of vibrotactile intensity and adaptation state (static, 20 Hz, and 250 Hz). The adapting intensity of the 20-Hz signal was 40 dB SL for all subjects; the adapting intensity of the 250-Hz signal was 45, 50, and 50 dB SL for subjects S, M, and D, respectively. (Lederman et al. 1982)

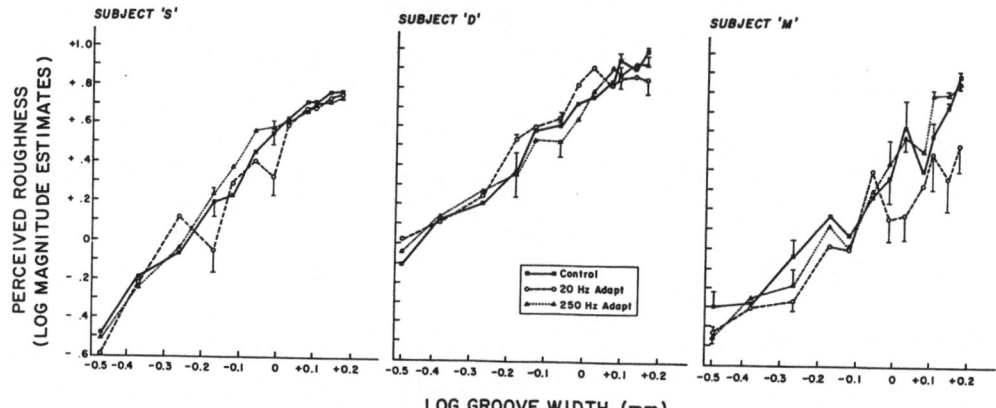

Fig. 5. Perceived roughness as a function of groove width and adaptation state (static, 20 Hz, and 250 Hz). The adapting intensities are the same as those described in the caption for Fig. 4. Standard errors shown are those in which the error ranges associated with the three means for a given groove width are at least partially nonoverlapping. (Lederman et al. 1982)

Taylor and Lederman's model of roughness (1975), however, focuses on
intensity-related aspects of skin deformation. The model proposes
three deformation parameters that best predict known psychophysical
data concerning the roughness of linear gratings (e.g., Lederman
1974). One is the depth to which the finger penetrates the grooves
(mainly determined by the spacing among the elements that form the invariant spatial pattern and by the applied force). The two other aspects of skin deformation that the model suggests are important (and
possibly better) predictors of perceived roughness are two measures of
the cross-sectional area of skin penetrating the tiny grooves summed
across the entire fingertip. Actually, volume, rather than area, is
the more appropriate measure. None of these factors depends upon the
dynamic aspects of vibration, e.g., temporal pulse frequency and rate
of skin displacement.

Implications for Sensory Coding of Human Roughness Perception

The perception of surface texture is clearly a multidimensional
phenomenon, with components such as roughness, spatial density, pattern, jaggedness, oiliness, and slipperiness all relevant. What can
be said at this time concerning the neural mechanisms that specifically underlie human roughness perception? As little is yet known about
this particular dimension, the following discussion must be considered
speculative and is mainly confined to the periphery.

Knibestol and Vallbo (1970) have identified four functionally distinct
types of mechanoreceptors in the glabrous skin of the human hand. Two
rapidly adapting types are the PC units and RA (or QA) units (presumed
to end in Meissner's corpuscles, and likely a major determinant of the
NP response). Two other types of slowly adapting (SA) receptors have
also been identified and may contribute to the responses of Verrillo's
NP system. SAI units are believed to end in Merkel's cell neurite
complexes, situated on the deep end of the intermediate ridges that
project into the dermis. SAII units are presumed to end in Ruffini's
cylinders. The two classes of slowly adapting units have not been reliably differentiated in monkey and are sometimes jointly referred to
as SA units (Talbot et al. 1968).

How might these mechanoreceptor populations contribute to the neural
code(s) that underlie the stimulus dimensions shown to be important in
human roughness perception?

Temporal Stimulus Dimensions

Darian-Smith and his colleagues (Darian-Smith and Oke 1980; Darian-Smith et al. 1980) have demonstrated that individual neurons temporally modulate their firing according to the stimulus temporal frequencies generated by gratings moved across monkey fingertip, the three mechanoreceptor populations modulating their activity best within different parts of the stimulus frequency range. Actually, the temporally modulated activity across spatially distributed populations would be required to reflect the invariant spatial structure. As the psychophysical evidence argues against the importance of temporal aspects of vibration in the human perception of roughness, it is concluded that although at least one form of neural encoding is available, it is not used.

Spatial Stimulus Dimensions

The psychophysical results clearly suggest that groove width and to a considerably lesser extent, ridge width, are the most important spatial parameters of the stimulus gratings for human roughness perception. Neither spatial period nor groove-to-ridge width ratio has been shown to be a crucial determinant. What are clearly needed now are recordings of neural activity in mechanoreceptors and cortical units (SA, QA, and PC) in response to independent manipulation of the grooves and ridges.

Phillips and Johnson (1981) have examined neural responses of SA, QA, and PC units to gratings described in terms of their spatial period. They found no evidence of spatial modulation in PC units. SA and QA units showed good spatial modulations, but the latter did so only for gratings with spatial periods of more than about 3 mm. Johnson and Davidson (1981) also found good spatial modulation in SA units, but only for spatial periods above 2 mm. In other words, below that value, there does not appear to be a spatial code available in any of the populations discussed above. These investigators speculated that the discrimination of texture may involve a nonspatial coding mechanism, i.e., the relative activity in the three main afferent classes (perhaps four in human). This notion has considerable appeal, resembling the color-coding system in vision. Nevertheless, the results of Lederman et al.'s selective adaptation experiments question the validity of such a coding mechanism. Note that all of the gratings used in the psychophysical work reported in this paper are less than 2 mm. Hence, even if a spatial code is available, it is considerably limited by the spatial period.

Intensity-Based Stimulus Dimensions

The Taylor and Lederman (1975) model of roughness perception considers intensity-based factors of the stimulus at the proximal level. It determines the combined effects of groove and ridge width, as well as finger force, on a number of different parameters of static skin deformation. It best predicts perceived roughness as a function of a cross-sectional area of skin deviating from the mean resting position summed across the entire area of skin contact. As mentioned before, because it is not possible to record the invariant properties of the distal surface, the term "intensity-based" is the more appropriate description of the roughness model at this time. Which mechanoreceptor populations might encode the intensive aspects of the proximal stimulus described above? The selective adaptation studies (Lederman et al. 1982) suggest that PCs play no role in the human perception of roughness. Moreover, there is nothing known as yet about the response profile of RA and SA units to changes in the skin deformation parameters described above.

On the basis of the following evidence, it is interesting to speculate that SA units might be important. Lederman (1978) has suggested a role for SAI units on the basis of their potential for being stimulated by shear and normal forces applied to the freely moving intermediate ridges, which, in turn, might function as microlever systems. However, the possibility that SAII units may code shear forces applied to the skin (Johansson 1978) suggests the further possibility that they may play a role in coding the effects of shear on the perception of roughness (Lederman 1978). Vierck (1979) has suggested that SAI and SAII units may contribute to the coding of texture in the cat; he demonstrated a strong preferential sensitivity for punctate stimulation and for edges presented to a portion of the receptive fields of these units, as compared with a smooth disk presented to the entire field. The edge effect was considerably more pronounced in SA than in RA units. A similar result was recently reported by Johansson et al. (1982), who recorded activity in RA and SAI units of human subjects. These last two experiments may be directly relevant to the interpretation of the selective adaptation studies reported earlier. As the contactor in the Lederman et al. work (1982) was a large smooth disk, SA units might have been considerably less affected than either RA or PC units. And because SA units respond well to edges and points, it is possible that they coded roughness in the selective adaptation experiments. If this interpretation proves true, then to the extent that SA units are a part of the NP system, the latter becomes implicated in tactual roughness perception, because the NP system is not dependent on vibrotactile frequency (Verrillo 1968).

In summary, the psychophysical experiments reported in this paper suggest that human roughness perception is not determined by the sense of vibration[1], specifically the dynamic aspects. While such evidence questions the validity of Katz' interpretation, it underlies and supports the quasi-static model of roughness perception proposed by Taylor and Lederman (1975). The temporal, spatial, and intensity-based stimulus features implicated by the psychophysical experiments reported in this paper have been outlined, and the possible involvement of various mechanoreceptor populations in the glabrous skin considered.

Acknowledgements. The author would like to thank Dr. David Foster for his comments on an earlier draft of this paper. The writing of this paper was supported by a Natural Science Engineering Research Council of Canada grant.

---

[1] It has been shown (Darian-Smith and Oke 1980) that information about the velocity of object movement is available in the cutaneous activity profiles of spatially distributed mechanoreceptor populations. The active touch condition might still be at some advantage because of additional motor and proprioceptive information pertaining to velocity. Nevertheless, to the extent that speed constancy is possible with both modes of touch, perhaps method 2 is not a sufficiently powerful test of the role of vibration in the perception of roughness. The data have been included, nonetheless, because they further confirm that there is no meaningful difference between active and passive touch in the perception of roughness (Lederman 1981, Lamb 1983)

References

CAPRARO A, VERRILLO RT, ZWISLOCKI J (1979) Psychophysical evidence for a triplex system of cutaneous mechanoreception. Sens Processes 3: 334-352

DARIAN-SMITH I, OKE L (1980) Peripheral neural representation of the spatial frequency of a grating moving at different velocities across the monkey's finger pad. J Physiol 309: 117-133

DARIAN-SMITH I, DAVIDSON D, JOHNSON K (1980) Peripheral neural representation of the two spatial dimensions of a textures surface moving across the monkey's finger pad. J Physiol 309: 135-146

GESCHEIDER GA, FRISINA RD, VERRILLO RT (1979) Selective adaptation of vibrotactile thresholds. Sens Processes 3: 37-48

JOHANSSON RS (1978) Tactile sensibility in the human hand: receptive field characteristics of mechanoreceptive units in the glabrous skin area. J Physiol 281: 101-123

JOHANSSON RS, LANDSTROM U, LUNDSTROM R (1982) Sensitivity to edges of mechanoreceptive afferent units innervating the glabrous skin of the human hand. Brain Res 244(1): 27-32

JOHNSON K, DAVIDSON I (1981) A somesthetic coding mechanism for texture like color coding in vision. Soc Neurosci Abstr 7: 273

KATZ D (1925) Der Aufbau der Tastwelt. Zeitschrift für Psychologie. Barth, Leipzig

KNIBESTOL M, VALLBO A (1970) Single unit analysis of mechanoreceptor activity from the human glabrous skin. Acta Physiol Scand 80: 178-195

KRUEGER LE (1970) David Katz's *Der Aufbau der Tastwelt* (The World of Touch): a synopsis. Percept Psychophys 7: 337-341

KRUEGER L (1982) Tactual perception in historical perspective: David Katz's world of touch. In: SCHIFF W, FOULKE E (eds) Tactual perception: a sourcebook. Cambridge University Press, New York

LABS SM, GESCHEIDER GA, FAY RR, LYONS CH (1978) Psychophysical tuning curves of vibrotaction. Sens Processes 2: 231-247

LAMB G (1983) Tactile discrimination of textured surfaces: peripheral neural coding in the monkey. J Physiol 338: 567-587

LEDERMAN SJ (1974) Tactile roughness of grooved surfaces: the touching process and effects of macro- and microsurface structure. Percept Psychophys 16: 385-395

LEDERMAN SJ (1978) Heightening tactile impressions of surface texture. In: GORDON G (ed) Active touch: the mechanism of recognition of objects by manipulation. A multidisciplinary approach. Pergamon, Oxford

LEDERMAN SJ (1981) The perception of roughness by active and passive touch. Bull Psychonomic Soc 18: 253-255

LEDERMAN SJ (to be published) Tactual roughness perception: spatial and temporal determinants. Can J Psychol

LEDERMAN SJ, TAYLOR MM (1972) Fingertip force, surface geometry, and the perception of roughness by active touch. Percept Psychophys 12: 401-408

LEDERMAN SJ, LOOMIS JM, WILLIAMS DA (1982) The role of vibration in the tactual perception of roughness. Percept Psychophys 32: 109-116

PHILLIPS J, JOHNSON K (1981) Tactile spatial resolution. II. Neural representation of bars, edges and gratings in monkey primary af-ferents. J Neurophysiol 46: 1204-1225

PRITCHARD RM (1961) Stabilized images on the retina. Sci Am 204: 72-78

TALBOT WH, DARIAN-SMITH I, KORNHUBER H, MOUNTCASTLE V (1968) The sense of flutter-vibration: comparison of the human capacity with response patterns of mechanoreceptive afferent from the monkey hand. J Neuro-physiol 31: 301-334

TAYLOR MM, LEDERMAN SJ (1975) Tactile roughness of grooved surfaces: a model and the effect of friction. Percept Psychophys 17(1): 23-36

TAYLOR MM, LEDERMAN SJ, GIBSON RH (1973) Tactual perception of tex-ture. In: CARTERETTE EC, FRIEDMAN MP (ed) Biology of perceptual sys-tems. Academic, New York, pp 251-272 (Handbook of perception, vol 3)

VERRILLO RT (1968) A duplex mechanism of mechanoreception. In: KENSHALO D (ed) The skin senses. Thomas, Springfield

VERRILLO RT, GESCHEIDER GA (1977) Effect of prior stimulation on the vibrotactile thresholds. Sens Processes 1: 292-300

VIERCK CJ Jr (1979) Comparisons of punctate, edge and surface stimula-tion of peripheral, slowly adapting, cutaneous, afferent units of cats. Brain Res 175: 155-159

Somatosensory Detection in Man

P. E. Roland

Department of Clinical Neurophysiology, Karolinska Hospital, Box 60500, 10401 Stockholm, Sweden

Detection of the presence of a signal requires one bit of information. Three neurons in series are sufficient to transfer the information to the primary somatosensory area (SI). In addition, three or four neurons in series would be sufficient to transfer the information to a detector outside SI. Hensel and Boman (1960) showed that a single impulse in a peripheral nerve fiber could be detected in normal young subjects. This does not imply that a single impulse in the brain can be detected. A single afferent impulse might be multiplied in the caudate nucleus or the ventrolateral thalamus (Poggio and Mountcastle 1963). A single afferent impulse might also be transmitted to many cortical afferents. Thus, a single impulse could be amplified either by multiplication along a single line of transmission or by parallel transmission of single impulses along several lines of transmission. If the noise in the somatosensory system is modest, the somatosensory detector can distinguish among multiples caused by one, two, or three peripheral impulses, and somatosensory detection would be quantal. If the noise is stronger, detection would consist of distinguishing – on statistical grounds – among a large number of impulses caused by the afferent impulse plus noise and a large number of impulses caused by noise alone (Green and Swets 1966).

In this report the question asked is whether somatosensory detection is quantal. Also, how might single impulses in a few afferent fibers be amplified? Finally, to which areas does this one bit of information spread, and to how great a territory within the cortical areas is this one bit of information transmitted?

In contrast with the signal detection experiment, somatosensory identification of complex objects has in principle no upper limit for the amount of information that can be used to achieve a precise picture of the palpated object. Unlike signal detection, identification of complex objects requires integration of sequentially sampled information. The efficiency of this identification depends on the efficiency of the sampling of somatosensory information. In this report the following problems are analyzed: (a) how is somatosensory information about complex objects sampled, (b) how might the somatosensory shape detector function, and (c) which are the cortical areas participating in the sampling, and which are the areas partcipating in the reconstruction of the image of the object palpated? The conclusion from these analyses is that whether a single impulse or the neural discharge elicited by a complex object is reconstructed by the somatosensory system and the central cortical detectors, vast populations of neurons in SI and the prefrontal cortex are activated.

Somatosensory Detection of Threshold Signals

If somatosensory detection is quantal, the somatosensory detector must be able to distinguish inputs originating from one, two, or three impulses in a peripheral nerve. This means that the relation between probability of detection and stimulus amplitude, the psychometric curve, would be a staircase function and not a monotonically increasing function.

The psychometric curve was measured in 80 normal subjects between 20 and 67 years of age. Two ring electrodes, soaked in concentrated sodium chloride, were placed on the middle phalanx of the index finger. The finger was held stretched out in the air. The stimulus was a 1 ms square pulse from a constant-current stimulator. The pulse amplitude and pulse shape were measured and controlled continuously. The psychophysical method was a two-alternative forced-choice method (2AFC) in which the signal was presented in either the first (SN presentation) or the second (NS presentation) observation interval. The amplitude of stimulation was randomized. There were 20 trials with random NS and SN presentations for each amplitude. The probability of detection was $Pd = 1/2 \ (C_{SN}/A_{SN} + C_{NS}/A_{NS})$, in which C was the number of correct responses and A the number of presentations.

The psychometric curves for two normal subjects are shown in Fig. 1. Figure 1a shows a one-quantum function, that is, the probability of

detection jumps from chance level to 0.95 with an increase in stimulus
amplitude of less than 0.05 mA. One-quantum functions were seen in 7%
of all normals or 15% of all normals from 20 to 40 years of age. The
rest of the subjects had two (23%), three (36%), or four (10%) quanta
functions, that is, staircase curves with statistically significant
jumps in Pd to increases in stimulus amplitude of less than 0.05 mA.
The multiquanta functions of 18% were best fitted by a straight line.

These statistically significant jumps in probability of detection
strongly indicate that the central detector in the somatosensory system responds quantally to increments in the afferent input. This is
not proof of quantal detection, but the variance of the increment in
amplitude at the detector in the central end of the somatosensory system was so small that in practice the distinction of the quantal increments was noise free. An amplitude of 0.6 mA would correspond to
zero to two active fibers in the median nerve with one impulse each
(Buchthal and Rosenfalck 1966). Thresholds below 1 mA occurred in 70%
of all subjects less than 40 years old. The probability that the
points in Fig. 1 could belong to the upper half of a normal distribution integral, as predicted by the theory of signal detection (Green
and Swets 1966), was less than 0.001.

These results support the findings of Johansson and Vallbo (1979a)
that somatosensory detection is noiseless. In contrast with Johansson
and Vallbo, who proposed a normal probability integral centered around
a threshold value, the psychometric functions in the majority of the
subjects were best fitted by a staircase curve. The discrepancy might
be due to a minor variance in electrical stimulation of afferent
fibers compared with stimulation of mechanoreceptors.

Amplification of Weak Somatosensory Signals: Somatotopical Tuning

One way to measure neuronal activity is to record the electrical
changes associated with neuronal function. Another way to measure
neuronal activity is to record the metabolic changes associated with
neuronal function. Since the regional cerebral blood flow (rCBF) is
linearly related to the regional cerebral oxygen consumption, rCBF can
also be used to measure neuronal activity indirectly. The method for
measuring the rCBF of the cerebral cortex has been described in detail
previously (Roland and Larsen 1976; Sveinsdottir et al. 1977).

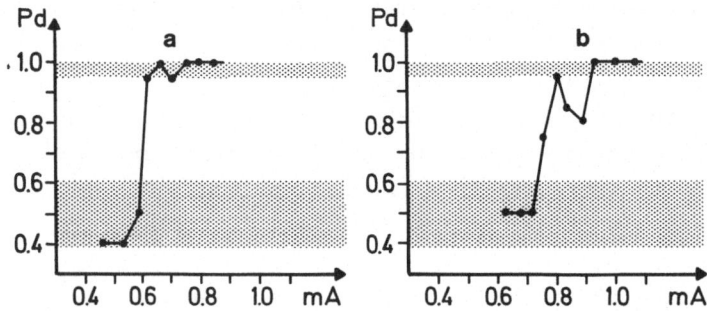

Fig. 1a,b. Quantal detection by two normal subjects. *Abscissa*, stimulus amplitude in mA. *Ordinate*, probability of detection. For each point the standard deviation equals $\sqrt{1/20\ Pd\ (1-Pd)}$. *Shaded areas* show two standard deviations for the point of no detection (Pd=0.5) and for Pd=0.975. Each point represents 20 two-alternative forced-choice trials. **a** One-quantum function. Pd jumps from chance level to 1.00 with an increment in stimulus amplitude < 0.05 mA. This occurred in 7% of normals. **b** Three-quanta function. Three statistically significant jumps to increments < 0.05 mA. This occurred in 36% of normals

In a recent study the rCBF was measured during rest (Roland and Larsen 1976) and also when the subjects focused their attention on the tip of the index finger (Roland 1981). The subjects expected a threshold touch on the tip of the index finger but, in fact, were not stimulated. The only difference between the rest and attention states of the subject was that during the latter he was required to focus attention on the tip of the index finger. The external conditions during the two rCBF measurements were in every respect identical. Selectively attending to the tip of the index finger gave rise to a 25% rCBF increase in the contralateral SI finger area (Fig. 2). In addition the posterior superior prefrontal cortex and the cortex of the midfrontal gyrus were activated contralaterally. Ipsilaterally, the posterior superior prefrontal cortex was activated, but otherwise the activity was confined to heterotopical areas.

This increase of rCBF in the SI finger area represented a considerable local increase in the metabolism of the cerebral cortex. This was probably not due to postsynaptic inhibition, since GABA agonist-induced postsynaptic inhibition powerfully decreases the rCBF in the cortex (Roland and Friberg 1983). Strong median nerve stimulation at an intensity at which more than 80% of the myelinated fibers are recruited causes only a 12.7% rCBF increase in the sensory hand area, and stimulation of the digital nerve at an intensity at which

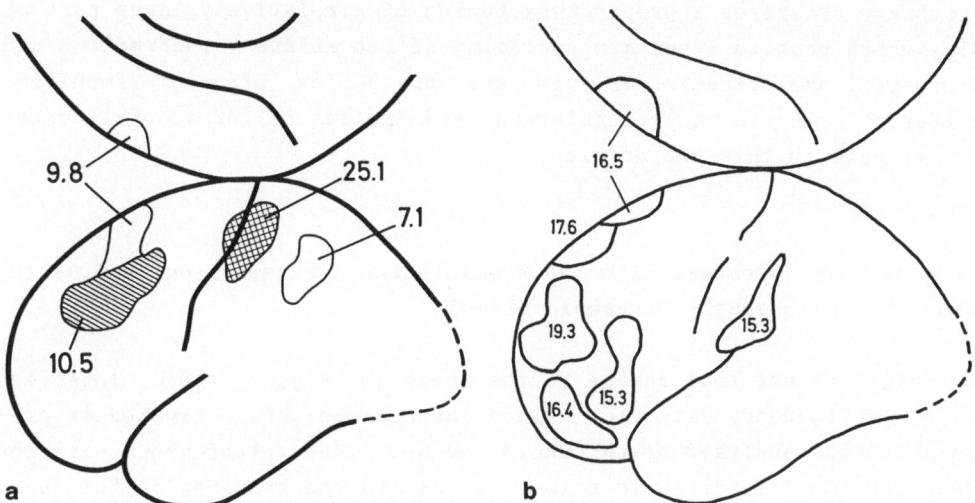

Fig. 2a,b. Mean increase of rCBF in percent when subjects focused their attention on the tip of the index finger which, however, was not stimulated. The size and location of each focal increase was the geometrical mean of the individual foci. **a** Mean of eight subjects focusing their attention upon the *contralateral* tip of the index finger, preparing themselves to receive a threshold touch with a von Frey hair. For the *crosshatched area*, the SI index finger area, the increase of rCBF was significant at the 0.0005 level (Student's *t* test, Dunnet's correction). *Hatched area* showed significant increase at the 0.005 level; other areas showed rCBF increase significant at the 0.05 level. (Roland 1981) **b** Mean of three subjects preparing themselves to receive a threshold touch with a von Frey hair on the *ipsilateral* tip of the index finger (Roland, unpublished results). The specific tuning of the SI finger area was strictly contralateral. The posterior superior prefrontal cortex was activated bilaterally; otherwise, the areas activated in the ipsilateral prefrontal cortex were heterotopical to the areas activated on the contralateral side. The increase in the supramarginal region might in part originate from the parietal operculum (Roland 1983)

80% of the fibers are recruited causes only a 4.8% rCBF increase in the sensory finger area (Foit et al. 1980). Even if only a fraction of the 25% increase in rCBF in the SI finger area were due to impulses arriving via the lemniscal route, because of noise in the system such a number of action potentials, if they arrived at the detector, would be incompatible with the psychometric functions observed in 82% of normal subjects. The tentative explanation was that the increase in rCBF in the SI finger area was mainly due to the metabolic work connected with excitatory postsynaptic potentials (EPSPs). The mechanism was called somatotopical tuning. The tuning was assumed to be controlled by the superior or midfrontal cortex, most likely the superior prefrontal cortex (Roland 1981, 1982). Thus, prior to the stimulation the brain organizes a preparatory tuning of a relatively large part of SI, which permits a certain spreading of the afferent information and a powerful amplification of even one or a few afferent impulses. Different areas in the contralateral and ipsilateral prefrontal cortex participate in this preparation.

Lesions that Interfere with the Demodulation, Transmission, and Detection of Somatosensory Threshold Stimuli

One might expect that damage to the areas in Fig. 2 would interfere with somatosensory detection, since these areas were activated in preparation for analysis of threshold touches. The interference with the demodulation of threshold stimuli in SI and the transfer of the quantal information between SI and other cortical areas was studied in 92 patients with circumscribed lesions of the cerebral hemispheres. The 92 patients were selected from 800 patients whose lesions were studied and anatomically measured by the author while they underwent a craniotomy. The selection criteria were:

1. A unilateral disease process was totally removed macroscopically from the brain.
2. The vascular damage during the operation was restricted to capillary bleeding and clipping or coagulation only of arterioles supplying diseased tissue.
3. After removal of the diseased tissue an inspection of the exposed brain showed no ischemia or venous stasis.
4. The patient had no history or clinical signs of arteriosclerosis or previous damage to the nervous system.

Fig. 3a-h. Mean detection threshold in patients with circumscribed lesions of the hemispheres. The brain was divided into subsectors according to a proportional stereotaxic system (Talairach et al. 1967). The mean detection threshold, a_t, in mA for patients with lesions invading more than 50% of a subsector was calculated. For each subsector a_t was compared with a_t of a group of normals with identical distribution of age and sex (t test for multiple comparisons, Dunnet's correction). The subsector for which the difference in means was statistically significant, $p < 0.05$, was marked in the figure with a_t in mA. **a** *FO* shows the positions of the semidiagramatic frontal sections. *NL*, no lesion of this subsector. **b** Lesions of the left middle frontal gyrus *(F3)*, **(c)** the superior frontal gyrus *(F4)*, **(f)** the SI finger area *(F7)*, **(e,f)** the thalamocortical radiation *(F6, F7)*, and **(f, g, h)** the hippocampus *(F7, F8, F9)* produced significant increases in thresholds. In addition, lesions of the frontal centrum semiovale produced threshold increases (see Fig. 5)

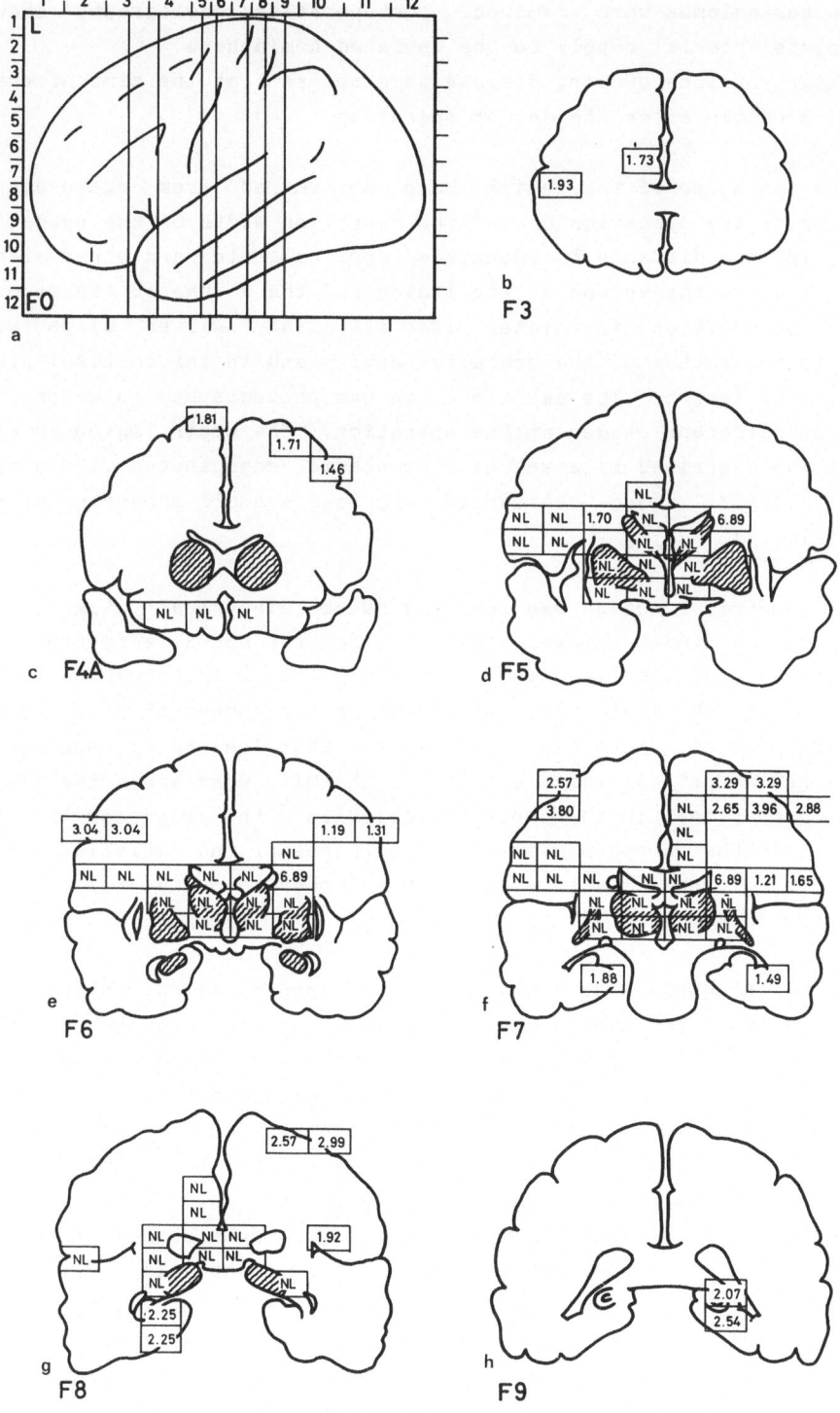

5. When hemangiomas were removed, postoperative angiography showed complete arterial supply to the operated hemisphere.

6. No signs of intercurrent disease were apparent at the time of testing, 3 months after the day of operation.

The size and shape of the lesion were determined from measurements made during the operation of (a) the depth and width of the operative cavity, (b) the distance to identified cortical sulci and other landmarks, and (c) the volume of the lesion and the volume of the removed tissue. In addition, for later identification, silver clips were fixed to the bottom of the operative cavity and in the cortical periphery of the lesion. The exposed brain was photographed in color and drawn at different stages of the operation. The brain lesion in each patient was described by a set of stereotaxic coordinates (Talairach et al. 1967). Each coordinate comprised a small subsector of the brain (Fig. 3).

The psychometric function was measured using electrical stimuli applied to the index finger, as described for normal subjects. The best-fitting straight line was determined by the least squares method and a_t, the threshold, was determined as the number of milliamperes for which Pd=0.75. From Fig. 3 it is seen that lesions of subsectors in the thalamocortical radiation to SI, the SI finger area, the middle frontal gyrus, the frontal centrum semiovale, the superior frontal gyrus, and the hippocampus caused increases of the detection thresholds. Usually the causative lesion was larger than the few subsectors shown in Fig. 3.

Lesions of the whole middle third of the postcentral gyrus elicited the highest thresholds of 6 mA. The area of maximal overlap of larger lesions of SI associated with increased thresholds corresponded to the horizontal sectors 2, 3, and 4 in Fig. 3. This area also corresponded to the area that was tuned when the attention was focused on the tip of the index finger. The mediolateral extension of this SI index finger area corresponded to the free area in Fig. 4a. The lesions of the patients with normal thresholds extended considerably into this area and left only a small free area on the right side (Fig. 4b). This small free area might be the center of the SI index finger area. Lesions of this center caused only a slightly increased threshold of 2-4 mA. Larger lesions caused higher thresholds. This indicated that the information from the few peripheral afferents excited at the

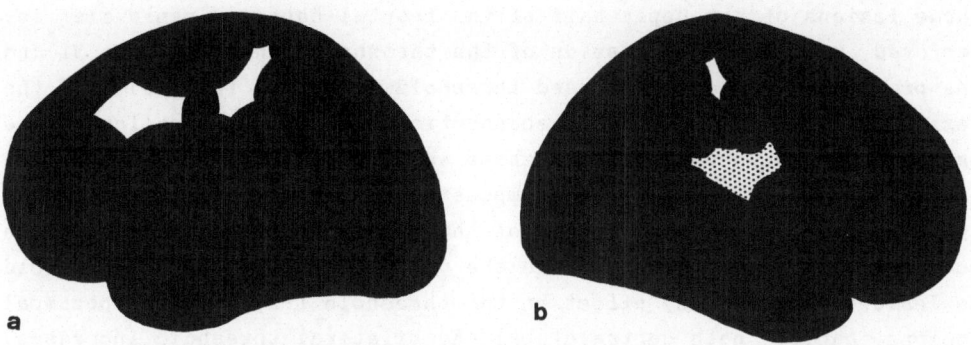

Fig. 4a,b. Lesions of patients with normal somatosensory signal detection. The area covered by these 71 lesions is shown in *black*. **a** There are 28 lesions of the left hemisphere and **b** 43 lesions of the right hemisphere. *Dots*, no lesions within this area. The left middle frontal gyrus and SI finger area were spared. Patients with lesions invading these areas all had an increased detection threshold, a_t. Due to the lack of symmetry in the distribution of lesions of SI, the area spared on the right was bigger than the area spared on the left. Minor lesions of the middle frontal gyrus, the superior frontal gyrus, or the peripheral parts of SI finger area did not affect somatosensory detection. There were no minor lesions of the left middle frontal gyrus.

threshold was spread out over the whole SI index finger area. It could not be excluded, however, that within the range of normal thresholds both skin and joint afferents were stimulated and that the skin afferents might project anteriorly in SI and the joint afferents posteriorly and that this combined projection might have contributed to the large anterior-posterior extension of the SI index finger area. As was the case for the somatotopical tuning, the threshold increase after SI lesions was strictly contralateral.

Outside SI, minor lesions of one of the participating cortical areas did not elicit any increased thresholds. There were no minor lesions of the left middle frontal gyrus (Fig. 4). Lesions of this gyrus increased the thresholds, but the psychometric curves were still quantal. If a lesion of the middle frontal gyrus (right or left) was combined with a lesion of the inferior frontal gyrus and the lateral orbitofrontal cortex, then the increase in thresholds was bilateral and the subjects showed difficulties in maintaining their attention. Destruction of the superior frontal gyrus in front of the supplementary motor area increased the thresholds. The psychometric curves were flat and noisy, but there were no general changes in attentive behavior.

Large lesions of the upper half of the frontal centrum semiovale interfered with the transmission of the threshold signal between SI and the prefrontal cortex and caused threshold increases (Fig. 5). If the lesion was combined with a disconnection of the corpus callosum, the psychometric curve was the same shape as in normal subjects, but the threshold was raised. It was impossible on this basis to locate the transmission path for the tuning of the SI finger area. The caudate nucleus was not the target, since the whole head of this nucleus could be lesioned without any effect on the threshold (Fig. 5). Hippocampal lesions caused both contralateral and bilateral threshold increases. The psychometric curves looked like those seen after orbitofrontal lesions. Rather strangely, somatosensory discrimination of complex objects using the exact same 2AFC paradigm was normal in these patients (Roland 1983). Hippocampal lesions thus affected only detection of near-threshold stimuli.

From comparison with the tuning experiment, it was clear that with the exception of the posterior parietal areas, interference with somatosensory signal detection appeared after lesions to all cortical areas that were found to prepare for the reception of threshold stimuli. This means that the brain prepared the areas that participated in the information processing prior to the reception of the afferent signals. It was also clear that transfer of one bit of information engaged a large area in SI, many fibers in the frontal lobe, white matter, and large prefrontal cortical areas.

Somatosensory Detection of Complex Objects: A Model

It is impossible to measure input-output relations of the human somatosensory system without making assumptions about the nature of somatosensory detection. Since somatosensory detection was quantal and the noise in the somatosensory system negligible, these findings were included in the model. The model also included the fact that both simple signals as well as complex objects have to be detected on the basis of discrete events: the stream of afferent action potentials. This model is more elaborately described in a forthcoming paper (Roland and Mortensen, in preparation).

We assumed that the somatosensory detector was an optimal linear detector. If the detector is optimal and the sampling of information

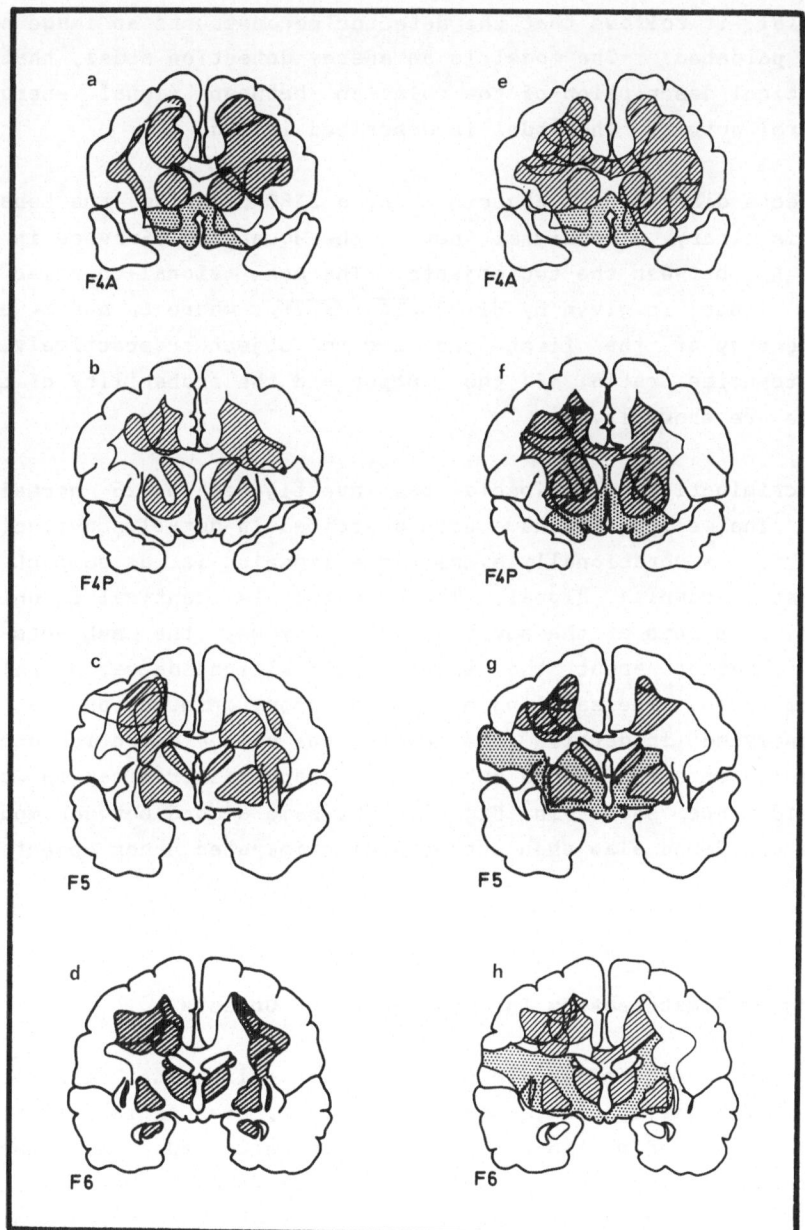

Fig. 5a–h. Lesions of the frontal lobe white matter. **a–d** Patients with lesions associated with normal somatosensory detection. **e–h** Patients with lesions associated with increased somatosensory detection thresholds. In comparison, it is seen that the lesions impairing detection were larger and comprised more of the upper half of the frontal centrum semiovale and in some cases extended into the anterior limb of the internal capsule **(g)**. *Dots*, no lesions within this area

sequential, it follows that the detector reconstructs an image of the object palpated. The model is an energy detection model, that is, a mathematical description of the relation between signal energy and behavioral output. The model is described in Fig. 6.

If subjects discriminate objects using a 2AFC paradigm, the behavioral output is binary. The signal then is the squared difference in signal energy, Δ^2, between the two objects. The mean signal-to-noise ratio of the input is given by $SI/NI = \sqrt{\Sigma \Delta^2}/\sqrt{S_1 + S_2}$, where S_1 and S_2 are the signal energy of the first and second object respectively. The signal-to-noise ratio of the output and the probability of correct response are shown in Fig. 7.

The discrimination of ellipsoids was investigated in 25 normal subjects. The ellipsoids have been described in detail previously (Roland 1975). A rotationally symmetric ellipsoid is a compact (non-redundant), complex signal. The curvature is identical in only four symmetrical points of the surface. The only way the subjects could get information about the shape of the ellipsoids was by palpating their surfaces. The sampling of somatosensory information was therefore analyzed in detail. The predictions of the detection model and the results obtained by the subjects in shape discrimination of the ellipsoids are shown in Fig. 7. The same accord between model and results was found also when subjects discriminated other objects (Roland 1983).

Sampling of Somatosensory Information About Objects

Twelve naive subjects discriminated sets of ellipsoids, sets of rectangular parallelepipeds, and sets of spheres (Roland 1975). The manipulations used by the subjects were unrestricted. All movements were recorded on videotape and analyzed in detail. There were three different basic modes of somatosensory information sampling about objects: static encompassings, rolls, and dynamic digital palpation. With the "static encompassings," the object was encompassed by the glabrous skin of the palm and fingers in a steady grasp. With the "rolls," the object was held between the fingers and then rolled 180° – 360° over the adjacent skin surface. With the "dynamic digital palpation," the individual fingertips were independently slid across the object surface. The sampling was sequential. Between the modes,

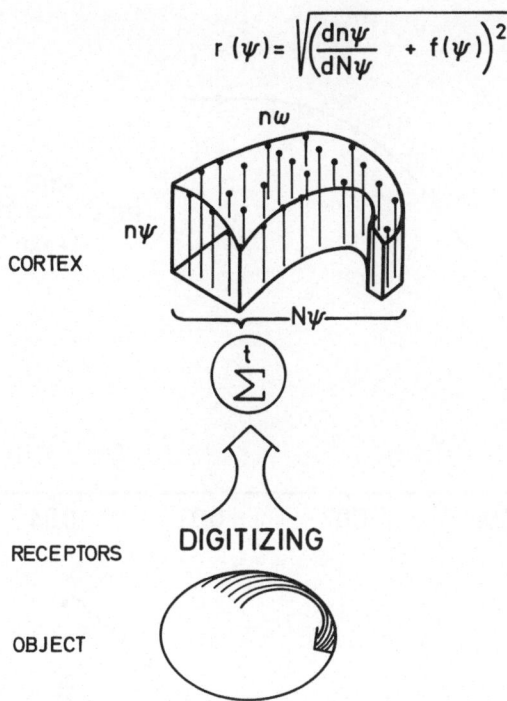

$$r\,(\psi)= \sqrt{\left(\frac{dn\psi}{dN\psi}\ + f(\psi)\right)^2}$$

Fig. 6. Energy detection model (Roland and Mortensen, in preparation). The subject palpates the object (ellipsoid) by moving a finger along the path shown. The mechanoreceptors of the skin digitize this stimulus energy into a stream of afferent action potentials. The slowly adapting mechanoreceptors (SA) are assumed to give rise to a mean frequency of $k_1 a^{b1}$, in which a is the identation amplitude and k and b are constants (Knibestöl 1975). Rapidly adapting mechanoreceptors (RA) are assumed to give rise to a mean frequency of $k_2 v^{b2}$, in which v is the velocity (Knibestöl 1973). The impulses from RA and SA are assumed to be summated by separate populations of cortical neurons, $N\psi$. RA is also assumed to provide information about the arc length n_ω when the number of impulses are summated by directionally sensitive neurons in SI. The detector is assumed to possess information about the relative position of the receptive fields. The number of impulses summated by each cortical neuron is n_ψ. n_ψ is assumed to have a gamma distribution with a variance that was equal to or less than the value of a poisson distribution. The detector then reconstructs the palpated surface by laying an envelope over the spatial field $N\psi$ in such a way that the only information used is the spatial derivative. $r\psi$ is the spatial envelope and $f(\psi)$ includes the noise term. In this way the shape detector is insensitive to the indentation amplitudes

106

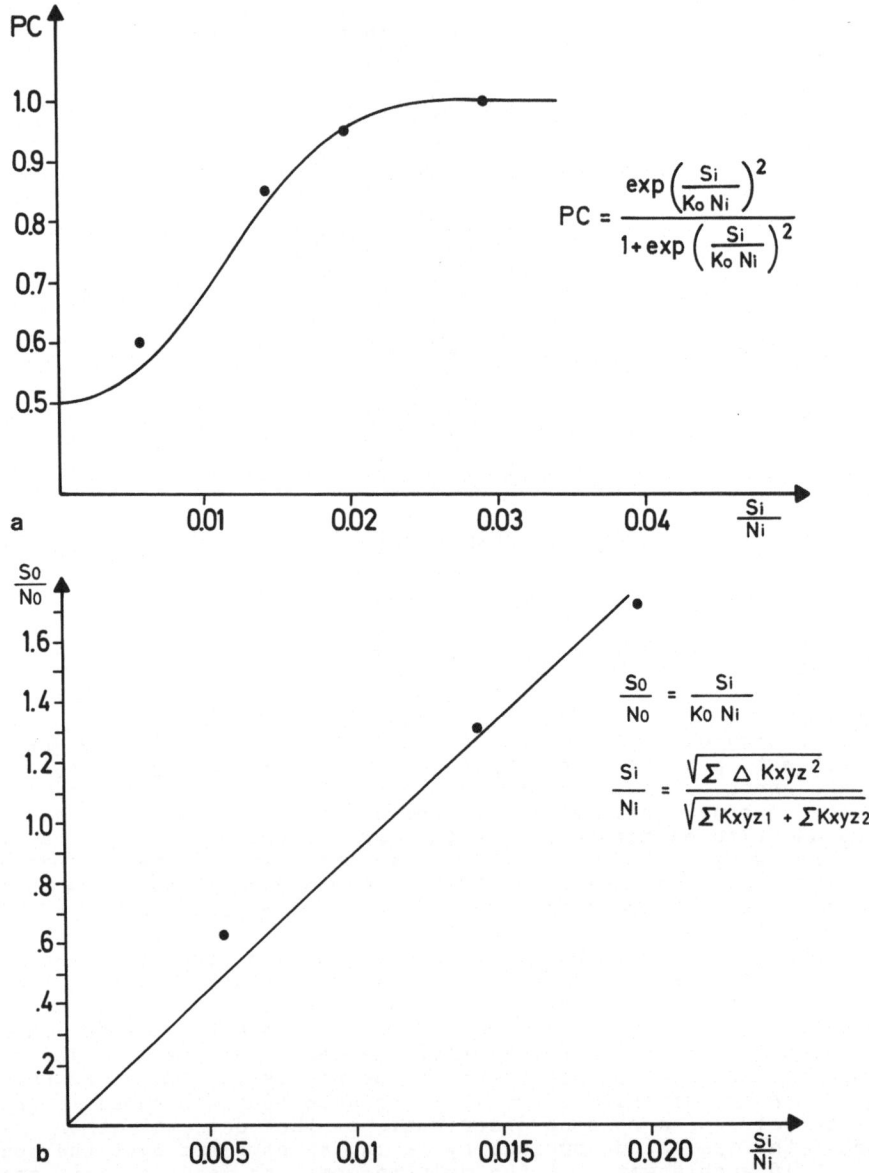

Fig. 7a,b. Discrimination of ellipsoids by normal subjects. a The
curve shows the predicted relation between probability of correct
response *(PC)* in a two-alternative forced-choice discrimination and
the input mean signal-to-noise ratio (S_I/N_I). The points were mean PC
values from 27 normal subjects discriminating between a sphere and ellipsoids with increasing eccentricity (Roland 1975). b Relation
between the input and output of the biological discrimination was
linear with a slope K_0^{-1}. The points were mean signal-to-noise outputs from the 27 subjects discriminating the ellipsoids in a. $\Delta kixyz$
was the difference in curvature of corresponding points of the two objects. *Kxyz 1*, the curvature of object$_1$ in surface point *xyz*.
Despite noise, the So/No > S_I/N_I due to amplification of the afferent
signals by the somatosensory system

the object was rotated and an unexplored surface exposed. The sampling lasted from 2000 ms to several thousand milliseconds per object.

When the subjects discriminated ellipsoids they used, with each explored object, on average 0.85 encompassings, 6.75 rolls, and 12.70 dynamic digital palpations with one or two fingers. Static encompassing of compact objects is a suboptimal sampling procedure because the spatial resolution is limited. There are two main reasons for this: the contact area is stationary in relation to the object, and the object is in contact with the palm and bases of the fingers, which have a lower density of mechanoreceptors than the fingertips (Johansson and Vallbo 1979b). Furthermore, the conformation of the skin surface to the object is uneven and incomplete. Rolls are optimal. The instantaneous burst rate of the previously unstimulated skin RA mechanoreceptors was utilized (Knibestöl 1973), and the object was rolled over the most sensitive parts, i.e., the fingertips (Johansson and Vallbo 1979b).

Dynamic digital palpation was also an optimal sampling strategy. The contact between the fingertips and the object was optimal. A steady, moving contact activated both RA and SA mechanoreceptors. The fingertips moved individually along independent paths across the surface. The path a single fingertip followed was also optimal in several respects. First, it almost always included the pole of the ellipsoids where the largest difference in curvature between the two objects occurred. Second, the path was a movement in three planes, which meant that the maximal number of surface points with different curvatures were touched. Third, the lenght was usually one quadrant (Fig. 6), which was also optimal, since the objects were symmetrical along two planes. Fourth, the axis of the fingertip was rotated along the path (Fig. 6). The last movement might improve the spatial resolution in a special way. Orthogonal to the palpation path the spatial resolution was determined by the distance between the mechanoreceptors. Along the palpation path the spatial resolution was determined by the quantification error. That is, the length of palpation corresponding to a difference of one impulse from an afferent fiber connected to a receptor in a skin surface in contact with the object. Since the quantification error presumably is less than the error due to the spatial density of the mechanoreceptors, the spatial resolution might improve from such a turn of the fingertip.

The surface velocity also had to be within a certain range. The quantification error would increase if velocity was either too high or too
low. The optimal velocity is unknown, but our subjects had a mean
surface velocity of 110 mm sec^{-1} with a 95% range of
65 - 200 mm sec^{-1}. In conclusion, naive subjects, when unrestricted
in their manipulation of objects, showed optimal somatosensory sampling. The sampling was sequential. The optimal way of utilizing the
sampled information would be to use the total amount of sampled information and compare that with the total amount of sampled information
about the other object. Also the detector had to know what information about object 1 had to be compared with what information about object 2. Consequently, the detector must reconstruct both objects before the comparison if it is an optimal detector.

The optimal sampling of somatosensory information requires a detailed
collaboration between the cortical areas that organize voluntary movements and the cortical areas that iteratively reconstruct the object
from the sampled information.

**Cortical Areas that Participate in the Control of Sampling of Somato-
sensory Information About Objects, and Areas that Participate in the
Internal Reconstruction of Palpated Objects**

The rCBF was measured in two subjects while they discriminated ellipsoids of different curvature by a 2AFC procedure (Fig. 8). They
rolled the ellipsoids and digitally palpated the ellipsoids dynamically. The rCBF increased in the contralateral SI hand area, the midfrontal cortex, and the superior prefrontal cortex. These were the
areas that were participating in the preparatory organization and the
tuning of SI. In addition, there were increases in the rCBF in the
cortex lining the postcentral sulcus and in the supplementary sensory
area. Damage to all these areas might interfere with the reconstruction of complex objects (Roland 1983).

The control of manipulation contingent on the sensory information gave
rise to increases in the premotor cortex and the supplementary motor
area. Finally, the verbal answers gave rise to rCBF increases in
Broca's area and the motor mouth area.

These results confirmed the findings of Roland and Larsen (1976). We
found an 11% - 20% increase in the rCBF in the contralateral midfron-

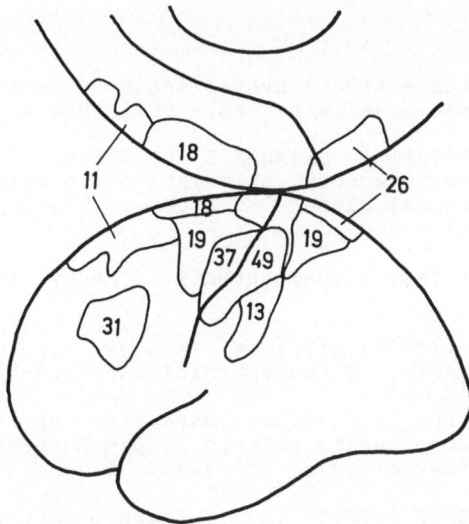

Fig. 8. Mean extent and mean increase in percent of rCBF for two subjects discriminating ellipsoids of different curvature by the
two-alternative forced-choice procedure. The rCBF increases in the
motor mouth area and Broca's area due to the verbal answering are not
shown. With the two-dimensional single-photon technique used in this
investigation, it was not possible to monitor the SII and the retroin-
sular cortex, which might be of some importance for the reconstruction
of objects perceived through the somatosensory system (Roland 1983)

tal and superior prefrontal cortex when the subjects were discriminating the shape of rectangular parallelepipeds. The increase in the SI
hand area was 60%. If the subjects manipulated the objects, there
were increases of 30% - 40% in the motor hand area and 11% - 20% in
the other motor areas. We did not take any recordings of the brain
from vertex and therefore we missed the increase in the supplementary
sensory area.

Acknowledgements. These studies were supported by the Foundation for
Experimental Neurological Research and the Danish Medical Research
Council.

References

BUCHTHAL F, ROSENFALCK A (1966) Evoked action potentials and conduction velocity in human sensory nerves. Brain Res 3: 1-122

FOIT A, LARSEN B, HATTORI S, SKINHØJ E, LASSEN NA (1980) Cortical activation during somatosensory stimulation in voluntary movement in man: a regional cerebral blood flow study. Electroencephalogr Clin Neurophysiol 50: 426-436

GREEN DM, SWETS JA (1966) Signal detection theory and psychophysics. Wiley, New York

HENSEL H, BOMAN KKA (1960) Afferent impulses in cutaneous sensory nerves in human subjects. J Neurophysiol 23: 564-578

JOHANSSON RS, VALLBO ÅB (1979a) Detection of tactile stimuli. Thresholds of afferent units related to psychophysical thresholds in the human hand. J Physiol 297: 405-422

JOHANSSON RS, VALLBO ÅB (1979b) Tactile sensibility in the human hand: relative and absolute densities of four types of mechanoreceptive units in the glabrous skin area. J Physiol 286: 283-300

KNIBESTRÖL M (1973) Stimulus-response functions of rapidly adapting mechanoreceptors in the human glabrous skin area. J Physiol 232: 427-452

KNIBESTRÖL M (1975) Stimulus-response functions of slowly adapting mechanoreceptors in the human glabrous skin area. J Physiol 245: 63-80

POGGIO GF, MOUNTCASTLE VB (1963) The functional properties of ventrobasal thalamic neurons studied in unanesthetized monkeys. J Neurophysiol 26: 775-806

ROLAND PE (1975) Some principles and new methods of tactile stimulation. Behav Res Methods Instrum 7: 333-338

ROLAND PE (1981) Soamtotopical tuning of postcentral gyrus during focal attention in man. A regional cerebral blood flow study. J Neurophysiol 46: 744-754

ROLAND PE (1982) Cortical regulation of selective attention in man. J Neurophysiol 48: 1059-1078

ROLAND PE (1983) Cortical areas in man participating in somatosensory discrimination of microgeometric surface deviations and macrogeometric object differences. In: v EULER C, FRANZEN O, LINDBLOM U, OTTOSON D (eds) Somatosensory mechanisms. MacMillan, London

ROLAND PE, FRIBERG L (1983) Are cortical rCBF increases during brain work in man due to synaptic excitation or inhibition? J Cerebr Blood Metabolism [Suppl] 3: S244-S245

ROLAND PE, LARSEN B (1976) Focal increase in cerebral blood flow during stereognostic testing in man. Arch Neurol 33: 551-558

SVEINSDOTTIR E, LARSEN B, ROMMER P, LASSEN NA (1977) A multidetector scintillation camera with 254 channels. J Nuclic Med 18: 168-174

TALAIRACH J, SZIKLA G, TOURNOUX P, PROSSALENTIS A, BORDAS-FERRER M, COVELLO L, IACOB M, MEMPEL E (1967) Atlas d'anatomic stéréotaxique du télencéphale. Masson, Paris

Coordinated Control Programs for Movements of the Hand

M. A. Arbib, T. Iberall, and D. Lyons

Center for Systems Neuroscience and Laboratory for Perceptual Robotics, Department of Computer and Information Science, University of Massachusetts, Amherst, MA 01003, USA

1 Introduction

We use schema theory to analyze perceptual structures and distributed motor control. In the present paper, we study the control of an articulated manipulator (primate hand or robot gripper). Actions performed by the manipulator will provide new sensory input, which can be used to modify further actions, thus constituting the action/perception cycle (Neisser 1976). Reaching and grasping movements have been used widely in both the psychological (Howarth and Beggs 1981; Jeannerod and Biguer 1982; Paillard 1982) and neurophysiological (Smith and Bourbonnais 1981; Humphrey 1979; Muir and Lemon 1983) literatures. There also exists detailed literature on the shaping of the human hand for grasping (Napier 1956, 1962).

Following Arbib (1981), we distinguish perceptual and motor schemas. A *perceptual schema* is an internal model of a section of the environment with which the organism has preconstructed plans for interaction. The contents of the perceptual schema are dictated by the interaction plan, which is called a *motor schema*. An object is, in basic cases, only perceived in terms of the task in which it is being used.

Every motor schema has an embedded perceptual schema, with which it can parameterize movements. Parameters to a motor schema represent properties about facets of the environment: size, location, orientation, relative motion, etc. The motor schemas provide a unit of motor control within an overall *coordinated control program*. A motor schema is a control system, continually monitoring feedback from the system it controls to determine the appropriate pattern of action to achieve its goals. The embedded perceptual schema provides an "identification procedure," which estimates parameters relevant to the controlled system.

The *activation level* of a perceptual schema represents the credibility of the hypothesis that the task represented by the schema is indeed afforded by the environment. The activation level of a motor schema is an indication of how useful this schema is in dealing with the present environment as perceived. A motor schema may have its activation level affected by other motor schemas as well as by dynamic perception. Cooperating schemas will increase each other's activity; competing schemas will attempt to decrease each other's activity. The overall behavior of the system is the combined behavior of all of its component schemas.

A coordinated control program is the structure that interweaves activation of motor schemas to control action. We first give a broad-brush view of such a program for a human reaching to grasp an object (Fig. 1). We have a ballistic movement towards the target concurrently with which the fingers are adjusted to the size of the object and the hand is rotated to the correct orientation. When the hand is near the object, a final feedback adjustment is made in the position of the hand. Tactile feedback then shapes the hand to the object.

The spoken instructions to the subject drive the planning process that creates the appropriate plan of action, which we here hypothesize to take the form shown in the lower half of Fig. 1. The perceptual schemas hypothesized in the upper half of the figure need not be regarded as a separate part of the program; rather they are invoked to pass the proper parameter values to the motor schemas per se. Analysis of visual input locates the target object within the subject's reaching space. Information about the location is fed to the control surface of the hand-reaching control system (i.e., it is not the job of the system to choose the target). On activation, the hand-reaching system directs a ballistic movement towards the target and activates a tuning mechanism to utilize visual and tactile feedback (this is referred to as "discrete-activation feedforward," cf Sect. 4). Prior to the actual reaching, however, analysis of visual input also extracts the size and orientation of the target object and feeds this information to the grasping schema. Next, finger and wrist adjustments are coactivated. This is followed by an inactivity lasting until contact,

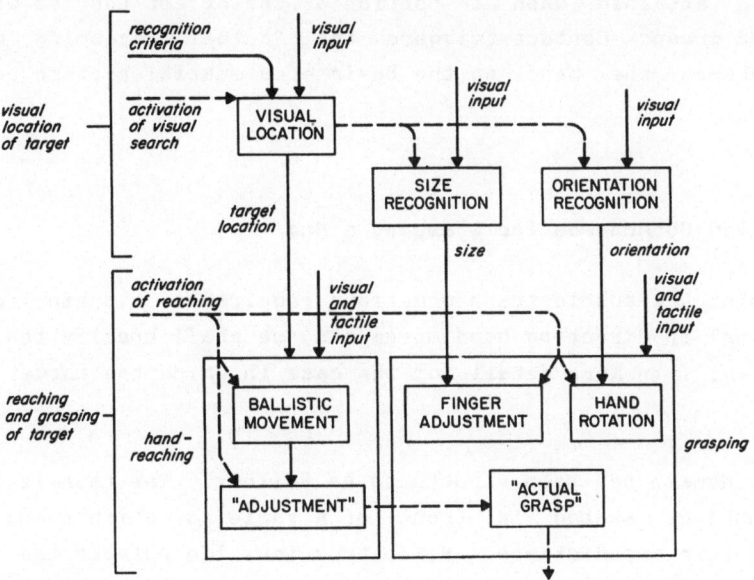

Fig. 1. Hypothetical coordinated control program for human's visually directed reaching to grasp an object. *Dashed lines* carry activation signals; *solid lines* transfer data. (Arbib 1981)

which is attained when a portion of the object touches within the
preshaped grasp. Contact triggers the "actual" grasping movement,
which shapes the hand on the basis of a spatial pattern of tactile
feedback.

2   Detailed Subschemas for Grasping a Mug

To pinpoint the subtle transformations required of a brain (or robot
controller) in directing hand movements, we shall specify the subschemas of Fig. 1 in more detail for the case in which the target is a cup
or mug.

The task domain has been structured as follows. The task is performed
by a subject seated in front of a table, on which a mug has been
placed within arm distance. No obstructions lie between the mug and
the subject's hand. Figure 2 shows a sequence of photographs docu-
menting the progress of the task being performed. From a resting position in Fig. 2a, the hand preshapes while the arm reaches
(Fig. 2b,c). In Figure 2d, the grasp begins, and by Fig. 2e, the mug
has been actually grasped.

The preshaping of the hand involves object parameters, such as a visually determined estimation of the handle's size and orientation. In
this task, the five fingers of the hand have three major functions:
to provide a downward force from above the handle, to provide an upward force from within the handle, and, if necessary, a third force to
stabilize the handle from below. We hypothesize that each of these
functions can be represented as the task of a *virtual finger*. The
fingers within a virtual finger move in conjunction with and have the
same characteristics as real fingers. This both limits the degrees of
freedom to those needed for a given task and provides an organizing
principle for task representation at higher levels in the brain. It
is then a subtask at a lower level to perform the actual mapping to
real fingers, making task implementation somewhat "tool free." We note
studies of other tasks performed quite well in spite of imposed variant mappings, e.g., standard handwriting with pen, chalk, or a pen on
a pole.

As evidence for the concept of virtual finger, consider the behaviors
pictured in Fig. 3. In all of them, the thumb is mapped into VF1, the

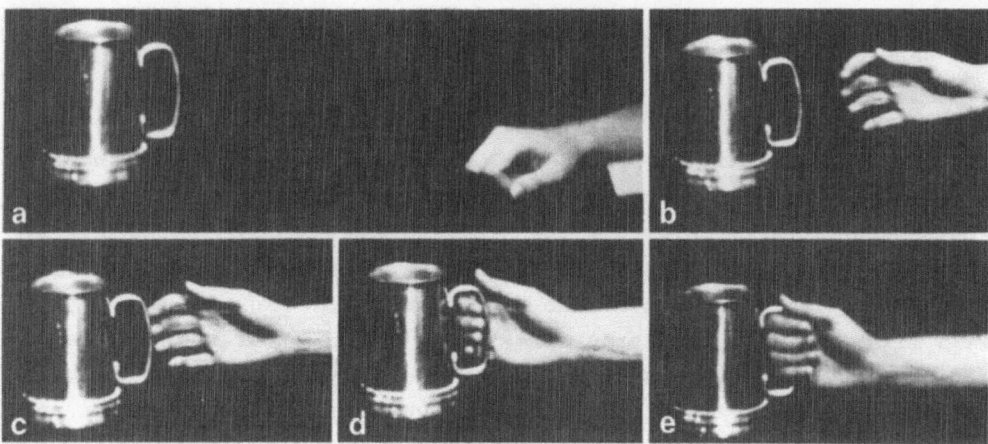

Fig. 2. Grasping a mug involves object parameters (handle size, orientation. etc.), a target (within handle), and body parameters (wrist reference point, hand orientation, hand size, etc)

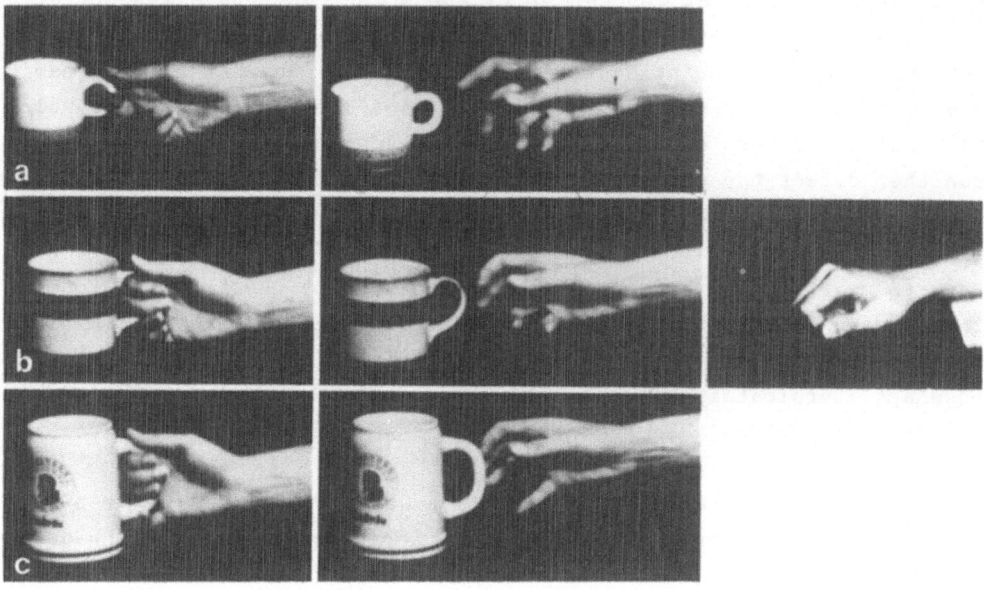

Fig. 3. Various combinations of real fingers can be mapped into virtual fingers for different size objects

first virtual finger function, which is to provide a force from above the handle.

For a teacup with a very small handle (Fig. 3a), only one finger will fit inside the handle. During the preshaping, only the forefinger is mapped into VF2, the second virtual finger function, that of providing an upward force from within the handle. The rest of the fingers become VF3, virtual finger three, to provide support for cup stabilization. For a coffee mug of the kind pictured in Fig. 3b, two fingers will fit within the handle to form VF2, with the other two fingers becoming VF3. For the mug of Fig. 3c, VF2 will comprise three fingers, while Fig. 2 demonstrates the case in which all four fingers are mapped into VF2, with an empty mapping to VF3.

The arm movement needs a visually determined target parameter, A, which in this task we assume to be inside the hole defined by the handle. However, the body referent for this movement, which we posit could be located at the center of the wrist, must be aimed not at A but at a point displaced from A by the vector linking the wrist to the intended contact point on the hand. The hand, no matter its dynamic shape or what it is carrying, will thus have a size parameter to which the motor schema controlling the arm move must adjust. This schema can then direct the wrist reference point to a target which is offset from the target by the dynamic hand size.

The arm move and the hand preshape are concurrent. Both work from a feedforward model supplied by the *grasp schema*. The move terminates when the hand is at the required working offset from the mug. The preshape terminates when a satisfactory shape (a model-determined set of finger positions) has been assumed, or when the move terminates.

Schemas can be generated from the task description in a top-down fashion. The highest level schema under consideration is the grasp schema.

Grasp Mug Schema

The embedded perceptual schemas for grasp will provide three visual cues: a target near the inside of the handle, the size of the handle, and the orientation of the handle. These can be used as parameters to the lower-level motor schemas necessary for carrying out the task.

Grasp activates in parallel a "reach" schema to move the hand to a position near the mug, ready for the grasp to begin, and a "preshape"
schema. Preshape must first activate a schema to map the task's virtual fingers into real fingers. If any of these schemas fail, then
the grasping process fails, perhaps calling for replanning at a higher
level. One reason for failure might be contact with another object
before the mug is near. We describe these three subschemas before
describing the program for the final stage of the movement. These are
as follows.

Virtual Finger Mapping Schema

The virtual finger mapping schema maps virtual fingers onto real
fingers, depending on the visually determined size of the handle, as
described in our discussion of Fig. 3. All other schemas will deal
with virtual fingers. If this schema remains active as long as the
hand is preshaping, the fingers can be remapped dynamically (i.e., on
finding that the initial estimate of the handle size was wrong).

Preshape Hand Schema

This schema defines the hand in terms of three virtual fingers, one of
each of the forces needed in the grasping of the handle. The preshape
hand schema positions the three virtual fingers in anticipation of the
following roles:
• VF1 provides an opposing force on the top of the handle against VF2
• VF2 grasps the handle and provides stabilization
• VF3 provides stabilization against the bottom of the handle, if
  needed
This schema is provided with the visually defined target, handle
orientation, and handle size as parameters, and it will call upon
low-level subschemas to move the virtual fingers into some standard
preshaped pattern relative to the handle. It will also provide other
schemas with the current preshaped size of the hand; in this task it
will provide the reach schema with the vector from the tip of VF2 to
the wrist reference point. It will terminate with success when the

shape is achieved or when the tip of VF2 comes into contact with the handle. Actual contact is not really necessary, and so it also terminates when the tip gets to a near enough location. Preshape fails if there is contact anywhere else, letting higher level schemas know. Once preshape terminates, other schemas will call upon the same low-level subschemas to move the virtual fingers. This provides a smooth transition from contact-free preshaping to contact-oriented grasping movements.

Reach Schema

This motor schema will handle the details of moving the wrist reference point toward the mug. The grasp schema passes the visually defined target for the move to this schema, as well as the handle orientation, and an expected contact point on the body. The preshape schema passes the size of the hand (from the tip of VF2 to the wrist reference point) to it. The reach schema will then call the move wrist schema to move the wrist reference point to the target, and it also calls the orient hand schema to point the hand in the right direction. It will terminate successfully when either contact is made on the body referent (in this case, it is also the tip of VF2) or else when it is near the target. The current location of the body referent redefines the target location. If contact is made anywhere else, this schema will terminate with a failure.

We now briefly expand the description of the two subschemas embedded within the reach schema. The *orient hand schema* is responsible for aligning the hand with the axis of the handle of the mug. It rotates the wrist until virtual finger 2 is in a position to slide through the mug handle. It must call elementary motor-servo routines to control the actual points in the hand. The *move wrist schema* will inspect the position of the desired object in body-centered coordinates. The coordinates are mapped onto arm extension coordinates, and the wrist is moved to a location near the target, but offset by the hand size. If any tactile contact is made in the duration of this schema, it will return indicating failure. It is up to the higher-level reach schema to decide if this return is acceptable. This schema must call motor-servo routines to actually move the arm joints.

Once the hand is preshaped, and in a position to grasp, the grasping process begins. This consists of the parallel actions of hooking the second virtual finger around the handle, bringing the first virtual finger down onto the top of the handle, and bringing the third virtual finger upwards to a position of support. The visually defined target provided by the embedded perceptual schema becomes tactilely defined after the successful completion of the reach schema. Subschemas activated in the grasping process can use the redefined target.

Hook Virtual Finger Schema

This will incrementally close all joints of a virtual finger until all have achieved tactile contact. Once tactile contact is achieved for a joint, that joint stops moving. When all joints are stopped, the schema terminates. The schema curls the finger towards the handle even prior to tactile stimuli. This schema is used to hook VF2 around the handle.

This schema may activate the reach schema if it has determined that the wrist reference point must be moved backward in order to achieve contact. It will do so using the inside of VF2 as the expected contact point, as well as using the body referent in aiming.

Oppose Virtual Finger

VF1 is brought to a tactilely defined target. In this case, VF1's target is where VF2 is touching below the top of the handle. The movement is made as a "curl" of the virtual finger until it makes contact on its tip; contact anywhere else is a failure. It successfully terminates on contact with the handle, before it actually reaches its target.

Extend Virtual Finger

If there is a VF3, it is extended towards the underside of the handle. When contact is made with the handle or when the virtual finger cannot move anymore, successful termination is made.

3 Schema Language and Simulation

The informal description of schemas given here is being extended to provide a formal schema-based language (Iberall and Lyons 1983) to be used to control robot hands as well as to describe primate hand control. Further, to test the ideas postulated in this paper, a computer simulation has been constructed. This program simulates parallel schema interactions. The output of the program is a graphic model (Fig. 4) of a human or robot hand (Lyons and Iberall 1983). The model simulates the kinematic structure of the hand, but not its dynamics at present. Joint angles are controlled directly from active schemas without reference to a muscular structure. Included in the simulation is a world model which maintains a data base of the hand and any defined objects. The world model then feeds back simulated sensory information corresponding to visual, tactile, and proprioceptive stimuli for a real hand movement. The simulation can be used to provide statistical data for comparison of schema-generated movements with recorded human behavior.

Our work in providing visual and tactile systems for robots is reviewed by Arbib et al. (1984) while Overton (1983) presents algorithms for static and dynamic processing of the output of tactile array sensors, and analyzes schemas that integrate visual and tactile cues in the control of a robot in an assembly task.

4 Neural and Behavioral Correlates of Schemas

The description of coordinated control programs given above is essentially phenomenological, based on informed observations of human performance. However, it has been guided by the constraint that the schema language be precise enough to be used for robot programming and to help refine the language used in specifying the search for neural mechanisms of hand control. To the latter end, we now discuss probable neural and behavioral correlates of the schema concepts presented in this paper. For a discussion of possible neural correlates for the processes whereby coordinated control programs are *planned*, see Arbib (1981), Brooks (1979), and Evarts (1981).

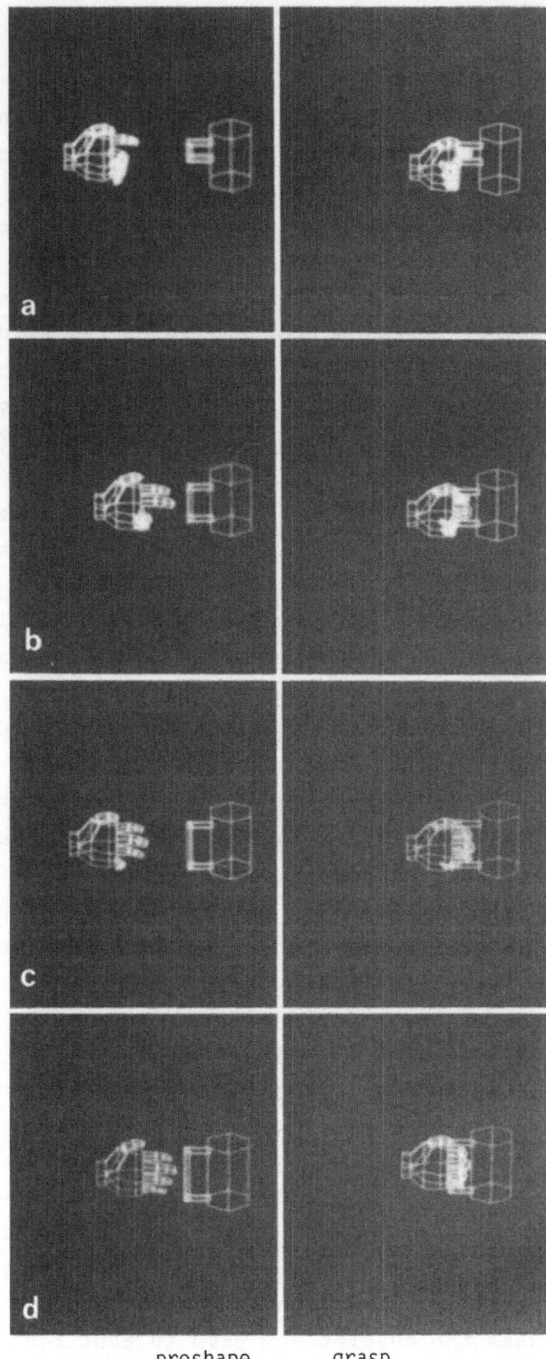

preshape   grasp

Fig. 4. Computer simulation of the handle-dependent virtual finger formation, preshaping, and grasping behaviors of Fig. 2 and 3

Concurrency and Localization of Schema Activation

Our coordinated control program of Fig. 1 was designed to formalize the observations of Jeannerod and Biguer (1982) on the concurrent initiation of reaching and hand preshaping. As Jeannerod (1981) notes, our program shows that the existence of separate perceptual schemas activated in parallel to extract different, task-related parameters (such as shape and location in space) of an object does not deny the existence of an overall program which can orchestrate the use of these channels. Moreover, there is neurophysiological evidence for dividing the act into a reaching component (proximal muscles) and a preshaping and manipulation component (distal muscles). Figure 5 is based on Brinkman and Kuypers (1972) and Haaxma and Kuypers (1974). They were able to show that finely coordinated, visually guided behavior involved the cooperative computation of two different systems. A pathway involving the brain stem controls the undifferentiated hand movements akin to the simple grasping schema of Fig. 1. A pathway from visual cortex to precentral gyrus and thence directly via the pyramidal tract to motor neurons controls the distal musculature and is responsible for the control of relatively independent finger movements. With interruption of either the corticocortical connections or of the pyramidal tract, the animal was unable to shape its hand in such a way as to dislodge a pellet from a groove whose orientation could be visually determined. Instead, the animal could reach for the pellet but without preshaping its hand, and would then move its hand back and fourth under tactile control until by chance the pellet was dislodged. At that time, the tactile feedback sufficed to allow the animal to grasp the pellet efficiently and bring it to his mouth.

Muir and Lemon (1983) add further support to the hypothesis that direct corticomotoneuronal connections confer the ability to perform discrete finger movements, but their work further suggests that cells of motor cortex may be seen as not related to specific muscle contractions so much as to the activation of muscles within the execution of a specific motor schema. They identified a subpopulation of pyramidal tract neurons (PTN) with direct influences on motor neurons of small hand muscles, finding that these PTNs are active in the "precision grasp schema" when the muscles are used to position the fingers independently, but not in the "power grasp schema" when a generalized co-contraction flexes all muscles together. The power grasp schema must thus be represented by neurons elsewhere. (For more on the power and precision grips, their anatomical basis, and how the decision

about which grasp to use depends on the task rather than the shape of
the tool, see Napier (1956, 1962)).

Distributed Motor Control

Pitts and McCulloch (1947) offered a model of the superior colliculus
in which each collicular neuron was connected to motoneurons whose
firing would cause a contraction of oculomotor muscles that would turn
the eye in such a way as to center gaze in the direction corresponding
retinotopically to the given point in the superior colliculus.
Crucially, the model predicts that the response to a complex visual
stimulus will be such as to drive the gaze to the center of gravity of
the visual pattern. This led Arbib (1972, p. 160) to postulate a to-
pographically structured "distributed motor controller" as a basic
component of motor systems.   The controller has an input surface,
stimulation at a point of which will be transfomed into the appropriate sequence of motoneuron commands for movement to a spatially corresponding target position. An array of inputs would then yield movement to the average of the encoded motor targets (see the corresponding discussion of "maps as control surfaces" in Arbib (1981)).   The
hypothesis, then, is that when we look for motor schemas as instanti-
ated in the brain, we should look for a brain region in which inputs
are coded retinotopically or somatotopically, and in which activity of
an input population yields movement to an average target. Such structures have been found by McIlwain (1982) and Georgopoulos et al.
(1983a, b), working independently of each other and apparently, of the
earlier theoretical studies.

McIlwain (1982) finds that microstimulation in the intermediate gray
layer of cat superior colliculus yields *widespread* synaptic activation
of the layer. Yet such stimuli evoke saccades whose metrics seem to
depend primarily on the location of the stimulating electrode. This
leads McIlwain to postulate, in what may be seen as an updating of the
Pitts-McCulloch model in the light of the findings of Robinson (1972,
1981) in the monkey, that the spatial densities of the cells projecting to vertical and horizontal generators of the saccadic system vary
systematically beneath the retinotopic collicular map.

In studies more closely related to hand control, Georgopoulos et al.
(1983a, b) recorded neurons in motor cortex of rhesus monkeys con-

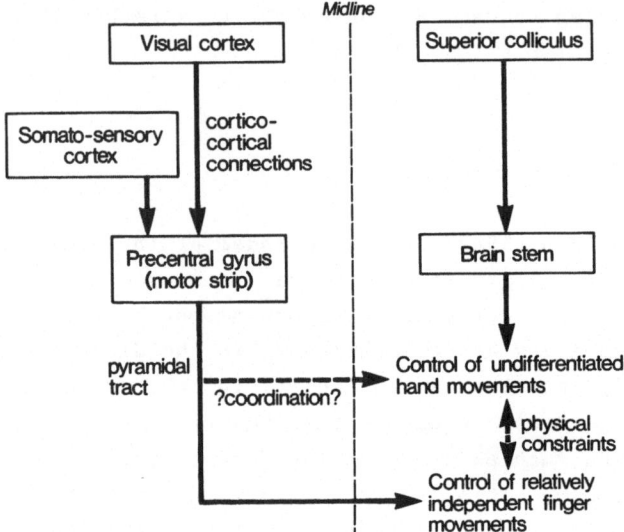

Fig. 5. Pyramidal pathway supports schemas for differentiated finger movements, while the extrapyramidal system provides the substrate for a rough grasp schema

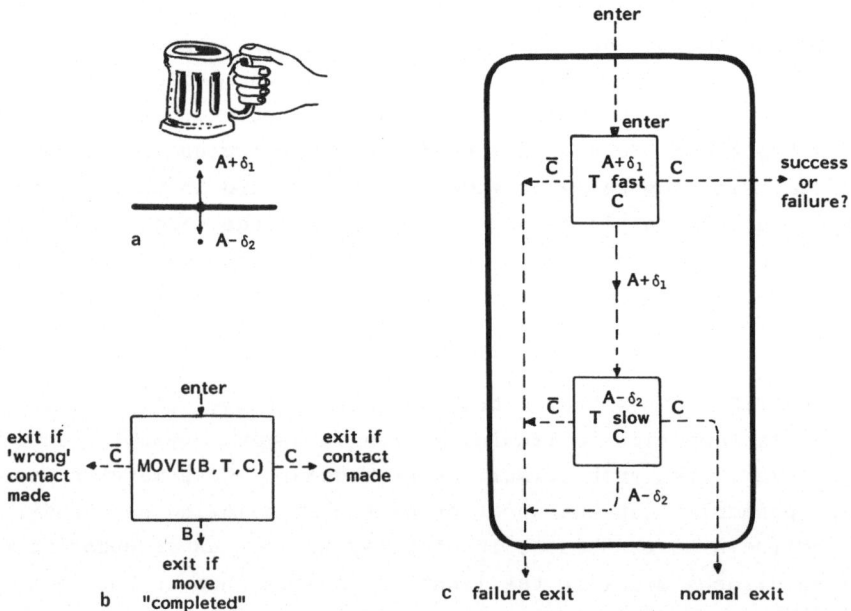

Fig. 6. **a** The placing of a mug on a table may be seen as composed of two movements: a fast movement to a target $A+\delta_1$ above the intended resting point A on the table (with the "safety undershoot" increasing with an estimate of decreasing accuracy of the movement), and a slow movement towards a point $A-\delta_2$ just below the table, designed to terminate under feedback control on contact. **b** The basic move schema. **c** The coordinated control program for the movement of **a**, involving two calls of the schema of **b**, each with different parameters

tralateral to the arm engaged in moving a manipulandum to capture a visual target. They found that 75% of the task-related cells discharged with higher frequencies with movements in a particular direction and at progressively lower frequencies with movements made in directions further and further away from the preferred one. Thus, a large population of cells was involved in each movement. In their variant of the Pitts-McCulloch hypothesis, the vector model, Georgopoulos et al. advanced the hypothesis that each cell should be viewed as casting a "vote" for movement in its preferred direction, with the weight of that vote given by the amount (positive or negative) by which its firing rate exceeded the average rate; they were then able to show that the corresponding vector sum of neurally coded directions did indeed closely match the direction of arm movement.

Note, however, that their results suggest that for this motor schema it is the *direction* of movement and not its end point that is the principal determinant of cell discharge. Since Polit and Bizzi (1979) have posited that the end point of the movement is the controlled spatial variable, this suggests that we must search for a re-coding process elsewhere than in motor cortex. Of course, the problem is ever more subtle, as can be seen by examining Fig. 3 of Georgopoulos et al. (1983b), which shows the time course of EMG activity of 13 muscles during movements in one direction. We note that there is not only a (frequently bimodal) time course for each EMG trace, but that most muscles have different resting levels of EMG before and after the movement. We may thus suggest that these traces exhibit the activity of two motor schemas, the "move" (which will involve both acceleration and deceleration) and the "hold," with the "hold" perhaps coextensive with the second schema of the two-phase analysis to which we now turn.

Two-Phase Analysis of a Motor Schema

In an informal study of the motor skills involved in drinking beer, Lyons (personal communication) observed that the placing of a beer mug could be separated into two phases: a fast movement to a target above the table, followed by a slower movement lowering the mug gently to make contact with the table top (Fig. 6a).

Moreover, the distance δ_1 of the first target above the table seemed to increase with decreasing sobriety – suggesting a deliberate undershoot to insure that the first, rapid movement would not end in a too-sharp contact with the table top. This leads us to posit that such a movement can be seen as involving two activations of a basic motor schema, but with different parameters. The basic schema MOVE(B,T,C) is shown in Fig. 6b: move to target B with timing parameters T and expected contact C. In the beer mug example, C is the subtle spatiotemporal tactile pattern to be experienced on the hand when the mug hits the table from above. (Note, again, the need for a complex visual-tactile transformation to anticipate C.) Just what constitutes the timing parameters T is a matter for experiment: candidates include movement duration, peak velocity, average velocity, peak force, etc. All that matters for the present discussion is that we can speak of "fast T" and "slow T." Note that MOVE(B,T,C) has three paths whereby activation can be transferred to other schemas: B (move completed, though probaply near B, rather than exactly at B - thus the need for feedback tuning); C (if contact is made with an expected tactile pattern); and C (if contact is made with the "wrong" tactile pattern).

Figure 6c shows the coordinated control program for the two-phase movement of Fig. 6a. Activation of the overall program activates MOVE($A+\delta_1$, T fast, C), which should normally terminate by reaching the position (near to) $A+\delta_1$. If contact C is made, there is a "failure exit" - some error-correction schema must be activated - while contact C may be treated either as failure (completion unexpected; check for spilled beer) or success (mug on table). On normal exit from the first schema, control is transferred to MOVE ($A-\delta_2$, T slow, C) with a normal exit occuring on contact C (termination under tactile feedback "in the right ballpark"). Both C (unexpected contact) and $A-\delta_2$ (the table is not where it was thought to be!) are failure exits from the overall schema. We may compare the first schema to a ballistic phase and the second schema to a feedback phase, refining the discrete-activation feedforward schema of Arbib (1981, Fig. 11); compare Navas and Stark (1968) for the idea that the low-velocity terminal part of the trajectory is a guided phase, while the earlier, high-velocity phase is ballistic. Jeannerod (1981, 1983) found that prehension movements (cf Fig. 1) involved a fast-velocity initial phase and a low-velocity final phase, with (but not in all subjects) the peak velocity highly correlated with amplitude and total duration independent of amplitude. Moreover, he posits that the reaching and grasping components are temporally coordinated - the fingers, having formed the pregrasp, begin to close in anticipation of contact at the

transition from the high-velocity to the low-velocity phase of reaching. However, where Jeannerod would suggest that the timing of the
two components is achieved by a centrally generated temporal pattern,
we would hypothesize that the activation signal (Fig. 6c) from the
first subschema to the second schema of the reach movement also serves
to activate a second subschema in the grasp - which is precisely the
hypothesis (though with a more primitive analysis of the subschemas)
embodied in the activation arrow from "adjustment" to "actual grasp"
in Fig. 1.

5 Conclusions

This paper has looked at a common hand task and, using the schema architecture postulated by Arbib (1981), has proposed a small set of in-
teracting motor schemas which will perform the task in a robust way.
In the process a schema language was outlined, which allows the description of hand tasks in a formal manner. Two crucial concepts of
our theory are that the schemas for hand movements are at the level of
virtual fingers, and that discrete-activation feedforward may involve
the consecutive activation of two instantiations of a given subschema,
but with different timing parameters and exit conditions. We have
also indicated a few behavioral and neural correlates which test our
schema concepts. Other papers will describe the formalism of our
schema language (Iberall and Lyons 1983), and its use in computer
simulation of hand movements (Lyons and Iberall 1983). Future work
will proceed in two mutually supportive, but somewhat divergent,
directions: the full articulation of a programming language for distributed control of robots with dynamic sensing, and the use of schemas to develop precise models of human and animal behavior subject to
behavioral and neurophysiological testing.

Acknowledgements. Preparation of this paper was supported in part by
grant NS14971-65 from the National Institutes of Health and grant
ECS-8108818 from the National Science Foundation. We thank Judy
Franklin for lending us a hand.

References

ARBIB MA (1972) The metaphorical brain: an introduction to cybernetics as artificial intelligents and brain theory. Wiley-Interscience, New York

ARBIB MA (1981) Perceptual structures and distributed motor control. In: BROOKS VB (ed) The nervous system. II. Motor control. American Physiological Society, Bethesda, pp 1448-1480 (Handbook of physiology)

ARBIB MA, OVERTON KJ, LAWTON DT (1984) Perceptual systems for robots. Interdis Sci Rev (to be published)

BRINKMAN J, KUYPERS HGJM (1972) Slitbrain monkeys: cerebral control of ipsilateral and contralateral arm, hand, and finger movements. Science 176(4034): 536-539

BROOKS VB (1979) Motor programs revisited. In: TALBOT RE, HUMPHREY DR (eds) Posture and movement. Raven, New York, pp 13-49

EVARTS EV (1981) Role of motor cortex in voluntary movement in primates. BROOKS VB (ed) The nervous system. II. Motor control. American Physiological Society, Bethesda, pp 1083-1120 (Handbook of physiology)

GEORGOPOULOS AP, CAMINITI R, KALASKA JF, MASSEY JT (1983a) Spatial coding of movement: a hypothesis concerning the coding of movement direction by motor cortical populations. In: MASSION J, PAILLARD J, SCHULTZ W, WIESENDANGER M (eds) Neural coding of motor performance. Springer, Berlin Heidelberg New York, pp 327-336

GEORGOPOULOS AP, KALASKA JF, CRUTCHER MD, CAMINITI R, MASSEY JT (1983b) The representation of movement direction in the motor cortex: single cell and population studies. In: EDELMAN GM, COWAN WM, GALL WE (eds) Dynamic aspects of neocortical function. Wiley, New York (to be published)

HAAXMA R, KUYPERS HGJM (1974) Role of occipitofrontal cortico-cortical connections in visual guidance of relatively independent hand and finger movements in rhesus monkeys. Brain Res 71(2-3): 361-366

HOWARTH CI, BEGGS WDA (1981) Discrete movements. In: HOLDING D (ed) Human skills. Wiley, New York

HUMPHREY D (1979) On the cortical control of visually directed reaching: contributions by nonprecentral motor areas. In: TALBOT RE, HUMPHREY DR (eds) Posture and movement. Raven, New York, pp 51-112

IBERALL T, LYONS D (1983) A schema-based language for the control of hand movements. Technical report. Department of Computer and Information Science, University of Massachusetts, Amherst (to be published)

JEANNEROD M (1981) Intersegmental coordination during reaching at natural visual objects. In: LONG J, BADDELEY A (eds) Attention and performance, IX. Erlbaum, Hillsdale, pp 153-168

JEANNEROD M (1983) The timing of natural prehension movements. J Motor Behav (to be published)

JEANNEROD M, BIGUER B (1982) Visuomotor mechanisms in reaching within extrapersonal space. In: INGLE DJ, GOODALE MA, MANSFIELD RJW (eds) Analysis of visual behavior. MIT Press, Cambridge, pp 387-409

LYONS D, IBERALL T (1983) A graphics system for hand simulation. Technical report. Department of Computer and Information Science, University of Massachusetts, Amherst (to be published)

McILWAIN JT (1982) Lateral spread of neural excitation during microstimulation in intermediate grey layer of cat's superior colliculus. J Neurophysiol 47(2): 167-178

MUIR RB, LEMON RN (1983) Cortico spinal neurons with a special role in precision grip. Brain Res 261: 312-316

NAPIER J (1956) The prehensile movements in the human hand. J Bone Joint Surg 38B(4): 902-913

NAPIER J (1962) The evolution of the hand. Sci Am 207(6): 56-62

NAVAS F, STARK L (1968) Sampling or intermittency in hand control system dynamics. Biophys J 8: 252-302

NEISSER U (1976) Cognition and reality: principles and implications of cognitive psychology. Freeman, San Francisco

OVERTON KJ (1983) The acquisition, processing, and use of tactile sensor data for robot control. Ph.D. thesis, Department of Computer and Information Science, University of Massachusetts, Amherst

PAILLARD J (1982) The contribution of peripheral and central vision to visually guided reaching. In: INGLE DJ, GOODALE MA, MANSFIELD RJW (eds) Analysis of visual behavior. MIT Press, Cambridge, pp 367-385

PITTS WH, McCULLOCH WS (1947) How we know universals, the preception of auditory and visual forms. Bull Math Biophys 9: 127-147

POLIT A, BIZZI E (1979) Characteristics of motor programs underlying arm movements in monkeys. J Neurophysiol 42(1): 183-194

ROBINSON DA (1972) Eye-movements evoked by collicular stimulation in alert monkey. Vision Res 12(11): 1795-1808

ROBINSON DA (1981) Control of eye movements . In: BROOKS VB (ed) The nervous system. II. Motor control. American Physiological Society, Bethesda, pp 1275-1320 (Handbook of physiology)

SMITH AM, BOURBONNAIS D (1981) Neuronal activity in cerebellar cortex related to control of prehensile force. J Neurophysiol 45(2): 286-303

Transcortical Reflexes:
Their Properties and Functional Significance

E. V. Evarts

Laboratory of Neurophysiology, National Institute of Mental Health, Building 36, Room 2D10, Bethesda, MD 20205, USA

Introduction

Phillips (1969) ushered in much of our current thinking about the functional significance of transcortical reflexes when he proposed that the segmental circuits linking afferent input with motoneuron output might have been ..."overlaid in the course of evolution by a transcortical circuit of which the powerful and 'recent' corticomotoneuronal projection is the efferent limb."

Subsequently, motor cortex neurons were shown to exhibit reflex responses during volitional movements (Evarts 1973) and Conrad et al. (1974, 1975) and Evarts and Tanji (1974, 1976) found that the properties of these reflex responses conformed to those postulated by Phillips (1969) when he hypothesized that corticomotoneuronal (CM) cells that were active prior to and during a centrally initiated movement would be reflexly excited when the movement was retarded. Fetz et al. (1980) used the technique of spike-triggered averaging to identify CM cells, and they found that CM cells whose activity covaried with a given muscle were reflexly excited when the muscle was stretched.

The demonstration that CM cells are reflexly activated by peripheral afferent input has eliminated any question as to the *existence* of transcortical reflexes, but a number of questions concerning transcortical reflexes remain and several of these questions will be discussed in this paper. But before beginning this discussion, it will be worthwhile to repeat Phillips' statement that the transcortical circuit has been *overlaid* on the segmental circuit. Figure 1 shows this *overlay* and makes it clear that motoneurons simultaneously receive feedback signals via a first pathway that is segmental and a

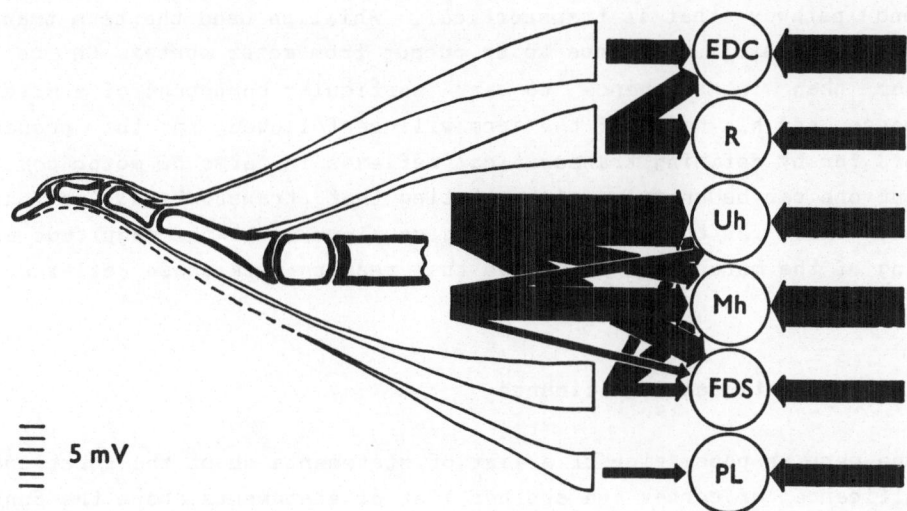

Fig. 1. Distribution of spindle *(left)* and motor cortex *(right)* inputs to motoneurons of different muscles and muscle groups of the baboon's hand. Breadths of *arrow* shafts measure mean quantities of monosynaptic excitatory action (mV scale at left of figure). *Circles* represent motoneuron pools of extensor digitorum communis *(EDC)*, remaining dorsiflexors of wrist *(R)*, hypothenar muscles, interossei, ulnar lumbricals, intrinsic flexor, and adductor of thumb *(Uh)*, remaining intrinsic hand muscles *(Mh)*, flexor digitorum sublimis *(FDS)*, and palmaris longus *(PL)*. (Clough et al. 1968)

second pathway that is transcortical. Phillips used the term transcortical reflex in reference to an output from motor cortex CM cells
rather than in reference to any particular component of a muscle
response, and his usage of the term will be followed in the present
paper, for by defining transcortical reflexes' in terms of motor cortex
output one can be unequivocal in stating that transcortical reflexes
exist, while at the same time being cautious as to the magnitude and
timing of the motoneuron discharge that results from these reflexes.

Assigning Functional Significance

If one were in possession of a list of statements about the functional
significance of cortex and another list of statements about the functional significance of reflexes, one might use these two lists as a
starting point in a discussion of the functional significance of
transcortical reflexes. In actually attempting to prepare such lists,
however, one immediately finds it necessary to narrow the field of
discussion by selecting a particular type of reflex and a particular
cortical field. This paper will focus on proprioceptive reflexes, and
the cortical field to be considered will be the rostral part of primary motor cortex (MI/r), a motor cortex subsector in which neurons
are driven by inputs arising from deep receptors (Strick and Preston
1978; Tanji and Wise 1981). It is to be emphasized that a full consideration of transcortical reflexes would deal not only with MI/r
outputs generated by signals arising in deep receptors but also with
MI responses to inputs arising from cutaneous receptors and with outputs from the postcentral gyrus as well as from MI. The present
treatment of transcortical reflexes must therefore be seen as quite
limited.

Role of Proprioception

In any discussion of the functional significance of proprioceptive reflexes, it is well to begin with Sherrington (1906), who defined
proprioceptors as deep receptors for stimuli that ..."*are traceable to
actions of the organism itself,* and ... since ... the stimuli to the
receptors are deliverd by the organism *itself*, the deep receptors may

be termed *proprio-ceptors*, and the deep field a field of proprioception."

The Latin word *proprius*, meaning own, provided a prefix that called attention to the fact that the organism's *own* acts created the adequate stimuli for these deep receptors. After defining proprioception, Sherrington noted a property of proprioceptive reflexes that is sometimes forgotten: the amount of muscle activity mobilized by proprioceptive inputs is slight. Given this fact, one should not be surprised when it is demonstrated that proprioceptive reflexes are ineffective in achieving load compensation in the face of large external disturbances. (Sherrington pointed out that exteroceptors rather than proprioceptors underlie responses to external disturbances!!) Actually, closed-loop feedback systems are more important when errors are small than when they are large. Under conditions in which errors are large, open-loop systems generate large movements which will reduce error to a value such that closed-loop systems can function effectively. Thus, a priori considerations would lead one to believe that the systems using proprioceptive inputs should be effective for small errors and ineffective for large errors. According to this notion, small errors would be able to elicit effective responses by controlling that proportion of the motoneuronal pool (or the corticomotoneuronal pool) that is near threshold, either having been recruited recently or being on the verge of being recruited. Direct evidence for a role of proprioceptive inputs in precise control has been provided by the work of Goodwin et al. (1978) on control of voluntary jaw movements. The jaw-closing musculature affords a unique opportunity to disrupt the afferent limb of the stretch reflex with minimal damage to the rest of the sensory innervation of the region, and Goodwin et al. found that surgical interruption of the myotatic reflex arc led to instability of voluntary control of steady state force during isometric contraction of jaw muscles in a situation where the monkey had a visual display corresponding to isometric biting force and was required to maintain force at a certain level determined by the visual display. This loss of stability in force control when proprioceptors were lost and when vision was the only feedback mode suggested an important role for muscle spindle afferents in reducing errors of muscle contraction produced by fluctuating levels of motor discharge.

Studies in human subjects have also provided evidence that proprioceptors are especially important in situations involving small amounts of motor output. In studying the behavior of the elbow and thumb joints during passive mechanical oscillations, Joyce et al. (1974) and Brown et al. (1982a, b) determined that reflex responses to passive oscillations were sufficient to modify small oscillations but not large ones: the inherent mechanical properties of the limb overwhelmed the myotatic reflex during large passive oscillations. In other sudies Marsden et al. (1978) found that the short-latency muscle response to stretch (presumably the segmental stretch reflex) increased linearly in size until the force induced by the stretch reached a threshold whereupon longer-latency responses (potentially of supraspinal as well as spinal origin) contributed significantly to load compensation.

Recordings of spindle afferent discharge in man (Vallbo 1973a, b) have provided additional evidence for the high dynamic sensitivity of proprioceptive systems. Vallbo found that spindle inputs during small "spontaneous" irregularities present during normal movement contribute to maintenance of smooth shortening during isotonic muscular contractions. He reasoned that this same exquisitely sensitive reflex system would be relatively ineffective in compensating for large external load disturbances.

Figure 2 shows the activity of proprioceptors under conditions of unperturbed movement in a person whose spindle afferents reflected small movement irregularities produced by his *own* muscles and not by any load change. Here, then, is a clear demonstration that precisely controlled movements are modulated by proprioceptive reflexes in the absence of external disturbances. The fact that external disturbances are useful tools in laboratory investigations has led to undue consideration of the possible role of proprioceptive reflexes in correcting of external disturbances. It is becoming increasingly clear, however, that proprioceptive inputs are of more importance in precisely controlled small movements and in postural stability than in reactions to unpredictable environmental events.

The special role of proprioceptive reflexes in postural stability and precise manual control was noted by Phillips (1969) in his discussion of the functional significance of the transcortical servoloop, and he listed three general approaches that might be used to identify the functional significance of spindle feedback during willed movement of the human hand. Of course, the ideal technique would be one in which a normal subject was deprived of spindle inputs temporarily and reversibly, leaving all other sensory inputs intact. Unfortunately, there is no way of achieving this ideal, and Phillips discussed three compromise approaches: studies of patients who had had dorsal rhizotomies for relief of intractable pain (Foerster 1927), ischemic blockade of *all* sensation from the hand (Merton 1964), and vibration of the tendon in such a way as to generate continuous high-frequency spindle afferent discharge that would block any useful information that spindle afferents might carry to the central nervous system. Citing Foerster's observation that dorsal root section caused severe deficits in finger movements even when visual guidance was present, Phillips commented that

> ...it is hard to conceive that vision could substitute for the flow of precise measurements normally supplied by length receptors associated with motor units in $\alpha\gamma$ linkage. It may be suspected therefore that an important part of the disability is due to loss of the spindle feedback. Foerster (1927) thought that the prompt and correct addressing of the cortical message depended on afferent impulses. Presumably man has a CM projection to the fingers at least as powerful as the baboon's. Normal 'addressing' could well depend in part on spindle input to the CM colonies, and could also be helped at the segmental level by the mutually reinforcing action of CM and spindle impulses at the membranes of the target α motoneurons (c.f. Clough et al. 1968).

The ideas that Phillips put forward on the basis of Foerster's clinical observations and his own neurophysiological studies have recently been supplemented by studies of impairments of motor performance in patients with proprioceptive loss due to large-fiber sensory neuropathy. Rothwell et al. (1982) and Sanes et al. (to be published) found that such patients are impaired in maintenance of steady postures and performance of accurate movements. The patients studied by Sanes et al. had a sensory neuropathy that was clinically characterized by progressive ataxia, absence of vibration and position sense, and moderate decreases in sensitivity to pinprick and light touch. Tendon jerks and stretch reflexes were absent. Muscle biopsy showed little evidence of muscle denervation and muscle strength was normal or near normal. A variety of motor functions was evaluated in these patients, and the most consistent deficits were an inability to maintain a constant position or force and a progressive impairment of movement accuracy as movement size decreased.

The electromyogram and position records of a patient who attempted to maintain a stable hand position against a spring load are illustrated in Fig. 3a. Without the benefit of visual guidance (Fig. 3a, top),

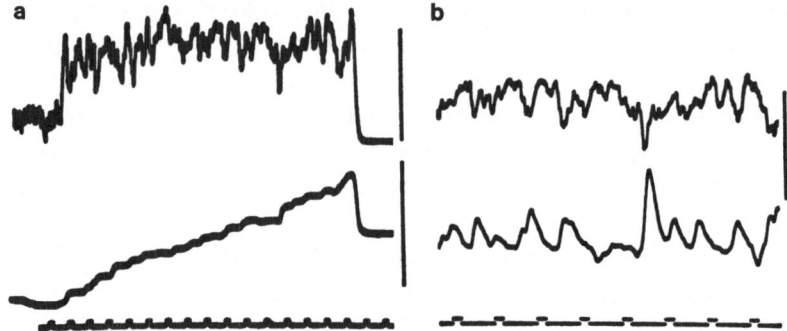

Fig. 2a,b. Response of a spindle primary ending during an isotonic contraction. **a** Relation to joint angle. The *upper* trace represents the impulse frequency of the single unit and the *lower* trace the angle of the metacarpophalangeal joint when the subject slowly flexed his ring finger. The events associated with the second half of this contraction are also illustrated in **b**. Calibrations: 0 and 25 imp/s, 155° *(bottom)* and 145° *(top)*. Time signal: 1 s. **b** Relation to the speed of joint movement. The *upper* trace shows the single unit impulse frequency and the *lower* trace the time derivative of the joint angle signal and, hence, it represents the speed of the joint movement. Calibrations: 0 and 25 imp/s. Time signal: 1 s. (Vallbo 1973b)

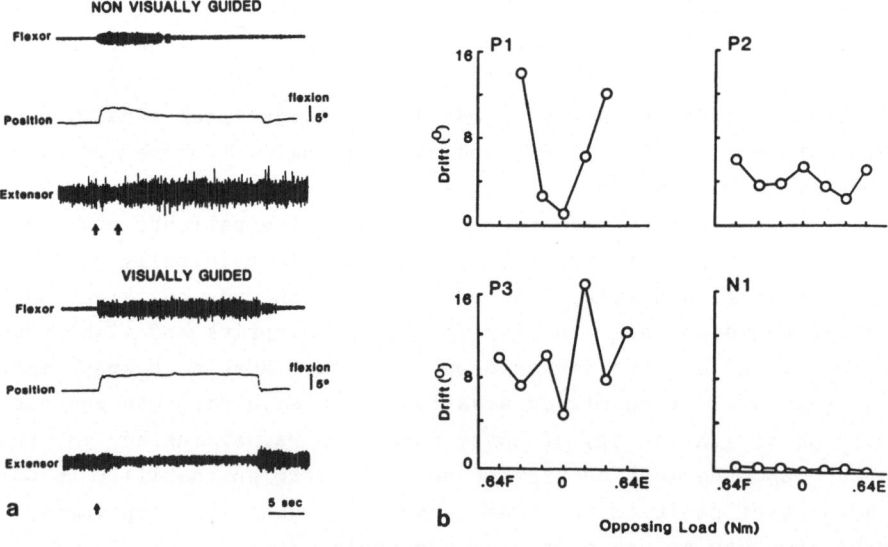

this patient was unable to sustain the levels of flexor and extensor muscle activity necessary for maintaining the 5° flexed position. This postural instability was observed even though the patient had re- peatedly maintained the position with visual guidance (Fig. 3a, bot- tom).

In a second experiment, postural drift was evaluated when patients at- tempted to maintain a steady wrist position against constant torques. When postural maintenance of the hand was visually guided patients maintained hand alignment relatively well, but postural control de- teriorated quickly and markedly (Fig. 3b) when visual guidance was re- moved.

In addition to deficits in maintaining posture without visual gui- dance, patients with sensory loss also showed deficits in making dis- crete movements. The impairments are illustrated in Fig. 4. When visual guidance of hand position was available, positional errors by patients ranged from 1° - 2°, irrespective of movement size. This amount of error was approximately three times that observed in con- trols. Errors increased for both controls and patients when visual guidance of hand position was removed, but the increase in absolute error was substantially greater for patients. The absolute error was greatest for the largest movements, but the error by patients for the smallest movements was *equal to or greater than the intended movement*. Thus, relative error was approximately 100% for the small (3°) move-

Fig. 3a,b. Postural control with and without visual guidance. a Hand position and agonist *(flexor)* and antagonist *(extensor)* EMG during movements against an elastic load of 0.13 Nm deg^{-1}. The leftmost *arrow* under the top and the single *arrow* in the bottom records indi- cate when the target jumped to the new position. The second *arrow* under the top record indicates removal of visual guidance. Visual guidance was restored 5 s after return to the original position. In the lower traces note (a) the ability of the patient to maintain a flexed position against the elastic load and (b) the reciprocal EMG in wrist flexor and extensor muscles. In the upper traces, note the al- most immediate deterioration of positional and muscle control when visual guidance was removed.
b Postural responses of three patients and one control are shown dur- ing maintenance of 10° of flexion without visual guidance. The *abscissa* indicates the constant torques opposing postural maintenance (0.64F = 0.64 Nm opposing flexion; 0.64E = 0.64 Nm opposing exten- sion). The *ordinate* shows the average absolute value of the postural drift for the 20-s period without visual guidance. Patients P1 and P3 drifted the least when no load opposed movement. Only patient P1 showed load dependence on the amount of drift. The normal subject was able to maintain alignment within 0.5° of the wrist rotation. (Sanes et al., to be published)

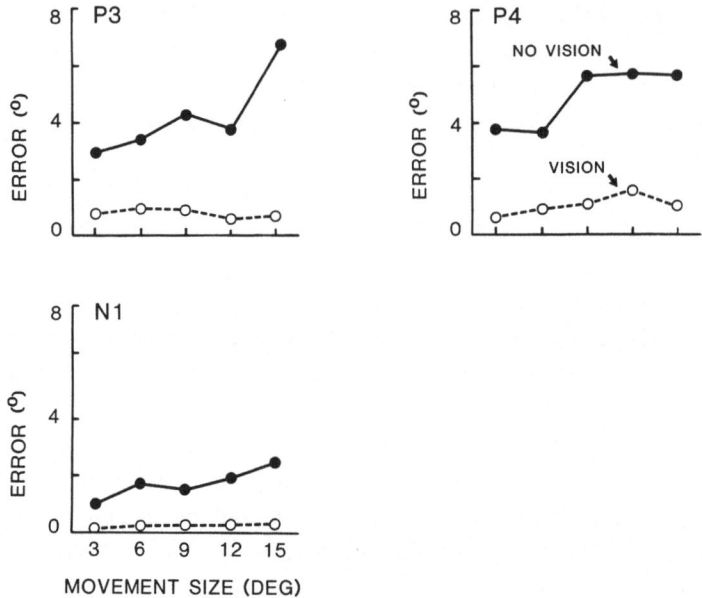

Fig.4 Incremental movements with and without visual guidance. Movements of five sizes are shown for two patients (P3 and P4) and one normal control (N1). Subjects moved with *(open circles, dashed lines)* or without *(closed circles, solid lines)* visual guidance. Movement size on *abscissa* and average absolute value of movement error on the *ordinate*. (Sanes et al., to be published)

ment, while the relative error was approximately 30 - 50% for the largest movement (15°) studied in this experiment.

Patients with the deficits described above had losses of touch as well as of proprioception and, of course, their deficits reflected the roles of afferent loss at all levels of the neuraxis. Thus, while these deficits can provide a point of departure for considering the functional significance of motor cortex outputs elicited by proprioceptive inputs, it is obvious that clinical disturbances resulting from peripheral neuropathy reflect much more than the function of the signals leaving MI as a result of deep inputs. We are thus led to consider the special functional role of MI, and in what follows there will be a discussion of those functions that may especially depend upon the motor cortex.

Role of Primary Motor Cortex (MI)

The large proportion of MI that is devoted to control of speech and manipulation has been immortalized in the *homunculus* of Penfield and Rasmussen (1950). This disproportionate allocation may in part reflect the precision of movements of lips, tongue, and fingers, just as the large macular representation in visual cortex reflects macular visual acuity. But if movement precision alone is the key to space distribution, why is there no representation of eye muscles in MI? There is, of course, a frontal eye field (area 8), but area 8 seems more analogous to premotor cortex (area 6) than to MI. This has suggested that in addition to precision, there may be some other characteristic of those movements controlled by motor cortex. Kuypers (1964) and others (see Phillips and Porter 1977) have proposed that this additional characteristic is individuation and fractionation. Movements of the fingers and of the speech muscles involve endless numbers of different combinations of contraction in individual muscles. Furthermore, the idea that precision alone determines cortical representation is not consistent with the fact that movements of the shoulder girdle have a relatively small representation in motor cortex in spite of the fact that such movements are quite precise. If movement error is defined as the percentage by which the magnitude of a movement deviates from what was intended, then both proximal (shoulder and elbow) and distal (wrist and finger) movements have errors of about 5%. These errors are for experimental subjects (usually students) in laboratory settings. Athletes, musicians, craftsmen, marksmen, etc. can achieve much greater movement accuracy. But for either skilled or unskilled subjects, movement errors (as a percent of movement amplitude) are about the same for proximal and distal joints. It would therefore seem that the innumerable combinations of positions, forces, and temporal sequences involved in speech and manipulation may have necessitated the large motor cortex representation for these movements.

Proprioception and Multicomponent Systems

A clue as to how proprioceptive inputs might participate in the control of the innumerable combinations of different cortically controlled movements is provided by work on the role of afferent inputs in control of speech. Abbs and Cole (1982) found that many of the consequences of afferent inputs delivered in the course of speech depend upon open-loop control mechanisms. Evidence for operation of open-loop control in multiarticular coordination was provided by studies of the effects of unanticipated loads applied to the jaw during coordinated movements involving the upper lip, lower lip, and jaw in the generation of a labial occlusion [p]. It was found that compensation for loads applied to the jaw depended on movements of the upper and lower lips: the reflex response to retardation of jaw closure invloved a greater movement by the lips. Abbs and Cole pointed out that "when the jaw contributed a large displacement to the opening or closure of the oral cavity, the upper lip and lower lip contributed proportionately less, and conversely."

The properties of these orofacial responses were elucidated in experiments involving disturbances injected into ongoing movements, but their true importance must be primarily in connection with normal, undisturbed speech: it is unusual to have one's articulatory movements stopped by external forces. In normal, unperturbed speech the open-loop systems discussed by Abbs and Cole will automatically take into account the status of the neuromuscular system with respect to the goal of the impending movement and generate an output that will acquire the goal in spite of the fact that the individual contributions of the numerous participating muscles will be quite different, depending on a variety of factors. The automatic coordination of different components underlying undisturbed speech is discussed by Hunker and Abbs (1982) in a paper on respiratory movement control during speech. They noted that

> ...because the speech motor mechanism consists of several
> systems (e.g., laryngeal, respiratory, orofacial), coordina-
> tion across these semi-independent systems as well as among
> the components within a system is not trivial. Relevant to
> this problem is the neural control process superimposed over
> several potentially independent movements to ensure their
> complementary contribution in achieving common goals. The
> operation of motor equivalence implies that there are sever-
> al levels of motor programming; the more general "higher"
> levels are not involved with control of detailed subcom-
> ponents. Several studies support motor equivalence coordi-
> nation in the limbs and in orofacial speech movements. The
> control of the respiratory system for speech may operate
> similarly, given its multiple degrees of freedom in produc-
> ing speech alveolar air pressures. That is, alveolar air
> pressure can be produced by many different combinations of
> recoil and active muscle forces. Similarly, a given lung
> volume can be produced by many different combinations of rib
> cage and abdominal movements.

Hunker and Abbs evaluated the automatic coordination of contributions
to expiratory air flow by rib cage (RC) and abdomen (AB). Their data

> ...suggested motor equivalence at several levels in respi-
> ratory motor control during speech: (1) when initial lung
> volumes were equivalent, RC and AB positions varied recipro-
> cally in their common contributions to those lung volumes;
> (2) when initial lung volumes were unconstrained, they var-
> ied considerably, requiring reciprocal muscle adjustments to
> offset the varying recoil forces in achieving equivalent al-
> veolar pressures; (3) when expiratory lung volume trajecto-
> ries for a sentence were equivalent, relative contributions
> of the RC and AB varied considerably throughout these tra-
> jectories.

Limb muscle responses to afferent inputs also exemplify the features
elucidated by Abbs and his colleagues. Such features have been documented by Traub et al. (1980) in an experiment that involved recording muscle activity from the longer flexor of the thumb during a movement in which the thumb was not displaced directly (by external forces
applied directly to the object upon which the thumb was acting), but
indirectly (by moving the hand with the result that the thumb was
pulled away from the object against which it was flexing). To clarify
the distinction between direct and indirect displacements, let us
first consider direct displacements. An example of a direct displacement would be seen in the case of a thumb flexion during which the object being moved by the thumb is pushed backwards so as to reverse the
movement direction and to stretch the thumb flexors while the subject
is attempting to make a smooth movement of fixed velocity. Another
example of a direct displacement is seen for the case in which there
is a reduction of resistance to movement so that movement velocity increases beyond what was intended and the muscle shortens more rapidly
than was intended.

The signs of the reflex responses to direct displacements conform to
the signs of segmental stretch reflexes. In contrast, the *indirect*
displacements used by Traub et al. occurred in situations in which
elimination of the discrepancy between actual and intended movement
required that the sign of the muscle response be *opposite* to that expected on the basis of connectivity of the segmental stretch reflex.
Their experiment on indirect displacements involved externally produced arm movements that pulled the thumb away from the lever against

which it was actively flexing. This indirect displacement "unloaded" the · actively contracting thumb muscle. The segmental reflex response of a muscle to unloading is a decrease of activity. For the *indirect* displacement, however, when the subject's goal was to maintain contact between thumb and lever, a corrective response required an *increase* of activity in the unloaded muscle, and this is exactly what was observed. This response (whose sign was opposite to that of the segmental stretch reflex) occurred at a latency between 50 and 60 ms.

As part of the same series of experiments described above, Traub et al. carried out a "sherry-glass" experiment in which subjects were asked to maintain the thumb and index finger a few millimeters from the rim of a glass full of sherry as if they were about to pick it up. In this situation a muscle response (latency 50 to 60 ms) was evoked either by pulling or releasing the wrist. The authors noted that while the sherry-glass response was automatic and ..."could not be substantially increased by effort of will"..., it was nevertheless ..."dependent on the subject's intent, for if he chose to ignore the fragility of the glass and knock it over, no response was observed."

Properties of Motor Cortex Reflexes

Motor cortex pyramidal tract neurons (PTNs) have been studied in situations where the set of the subject (a monkey) called for a muscle response that was sometimes opposite to that dictated by segmental stretch reflexes. In this work (Evarts and Tanji 1974), monkeys were rewarded for making a movement specified by a prior instruction stimulus (IS). The IS (a red or green lamp) told the monkey *what* to do (red meant that the monkey should prepare to pull and green meant that it should prepare to push), but the monkey was required to withhold movement until the rod in its hand was displaced by a torque motor. There were two possible directions of rod displacement by the motor (toward or away from the monkey). The displacement served as a trigger stimulus (TS) telling the monkey *when* to carry out the movement called for by the prior IS.

Figure 5 shows EMG records from the biceps muscle for the four possible combinations of the two ISs and the two TSs. Of these four IS-TS combinations, the one at the upper left (Fig. 5a) was associated with maximum biceps activity: the TS (a displacement involving biceps

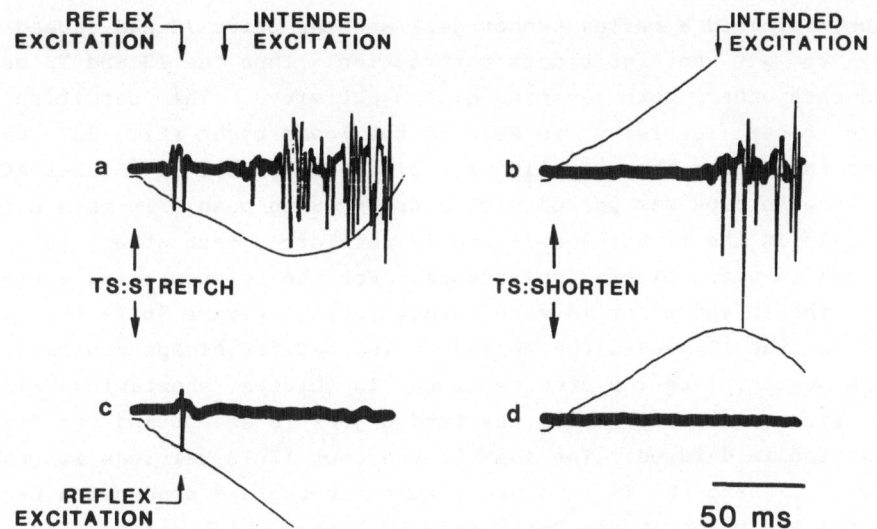

Fig. 5a–d. Biceps activity for different IS-TS combinations. ISs to pull **(a** and **b)** or push **(c** and **d)** and TSs involving handle displacements stretching biceps **(a** and **c)** or shortening biceps **(b** and **d)** could be combined in four possible ways. Each pair of traces shows biceps EMG activity and the output of the potentiometer coupled to the handle, with *upward* deflection of the potentiometer trace indicating movement of the handle toward the monkey and *downward* deflection of the potentiometer trace indicating movement of the handle away from the monkey. In **a** the IS was "get ready to pull" and the TS stretched biceps, with the result that biceps showed two phases of excitation: (a) reflex excitation due to biceps stretch and (b) intended excitation due to the monkey's set to pull. Biceps stretch with the monkey set to push **(c)** elicited the reflex without subsequent intended discharge. Only the intended phase occures **(b)** following a reflexly inhibitory TS with the monkey set to pull. (Evarts and Tanji 1974)

stretch) elicited a reflex tendon jerk and the prior IS had caused the monkey to get set for biceps contraction. Thus the IS and TS reinforced each other, both favoring biceps activity. The condition of minimum biceps activity is seen at the lower right (Fig. 5d), where neither the TS nor the IS called for biceps activity: a TS that shortened the biceps was paired with a prior IS to push. In this situation, like in the first, the TS and IS reinforced each other, but now both called for biceps quiescence. For the two remaining parts of Fig. 5, the TS and prior IS were antagonistic. Figure 5b is the case in which the IS caused the monkey to get set for biceps contraction, but the segmental reflex effects of the TS (biceps shortening) inhibited biceps activity. Here the tendon jerk is absent and the biceps contraction is delayed. The last of the four IS-TS pairings is shown in Fig. 5c, where the TS involved biceps stretch and produced a tendon jerk, but the prior IS was "get ready to push," which called for biceps silence rather than biceps contraction.

EMGs, such as those shown for single trials in Fig. 5, can be depicted for multiple trials using the same sorts of raster displays that are commonly used for single unit discharge. Figure 6 shows such raster displays for a muscle that was tonically active during the hold period prior to the displacing TS, and the presence of this tonic EMG activity provided a background allowing one to see inhibition. There is dissociation of the inhibitory and excitatory components of the muscle responses in Fig. 6b and c: in b, reflex inhibition precedes intended excitation when a TS that shortens (and therefore inhibits) the muscle triggers an intended movement requiring muscle excitation. In contrast, excitation precedes inhibition in Fig. 6c, where the TS stretches the muscle and causes a brief excitation that is then followed by the inhibition necessary for the intended movement.

Figure 7 illustrates MI/r activity during this task and demonstrates a dissociation of the two components of the PTN responses that parallels the dissociation seen in the EMG. The PTN in Fig. 7 was active when the monkey voluntarily pushed and was reflexly excited by externally produced displacements that opposed push. The most intense and prolonged PTN response to TS occurred when an IS meaning "get ready to push" had preceded a reflexly excitatory, displacing TS that opposed push by moving the handle toward the monkey (Fig. 7a). In contrast, the PTN was almost totally silenced when an IS meaning "get ready to pull" preceded a TS that moved the handle away from the monkey (Fig. 7d). Note the analogy between this situation and that observed

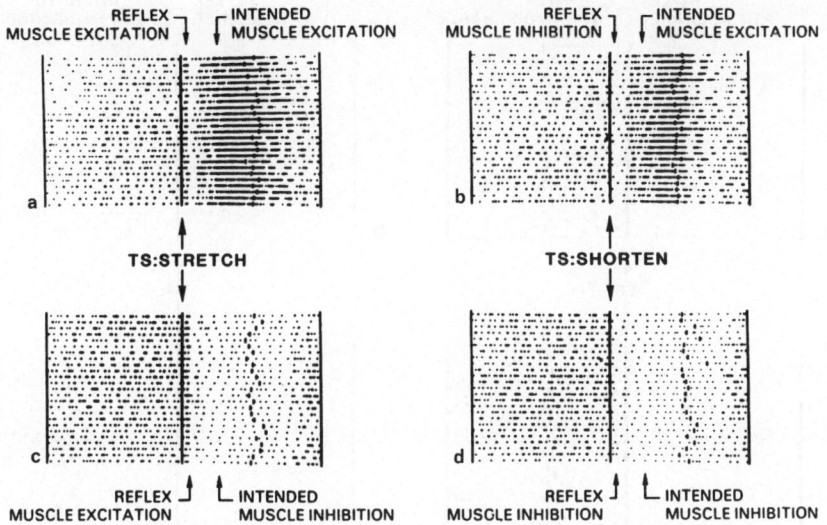

Fig. 6a–d. Response of triceps muscle to stretching and shortening TSs. To create these displays the EMG was rectified and fed to a voltage-to-frequency converter whose pulse outputs generated the rasters. In **a** and **b** the IS was "get ready to push," while for **c** and **d** the IS was "get ready to pull." At the *left*, the TS stretched triceps and evoked reflex muscle excitation, while at the *right* the TS shortened and inhibited triceps activity. In **a** the intended muscle excitation following reflex excitation began at a latency of 60 ms, whereas in **c**, where the TS inhibited triceps, the intended response had a latency of 80 ms. Rasters show a 1-s period of muscle activity, with 500 ms displayed before TS and 500 ms displayed after TS. The single heavy mark in each row to the right of the central line indicates the time at which the intended movement was completed. The central line marks the time of occurrence of the TS delivered by the torque motor. (Evarts and Tanji 1976)

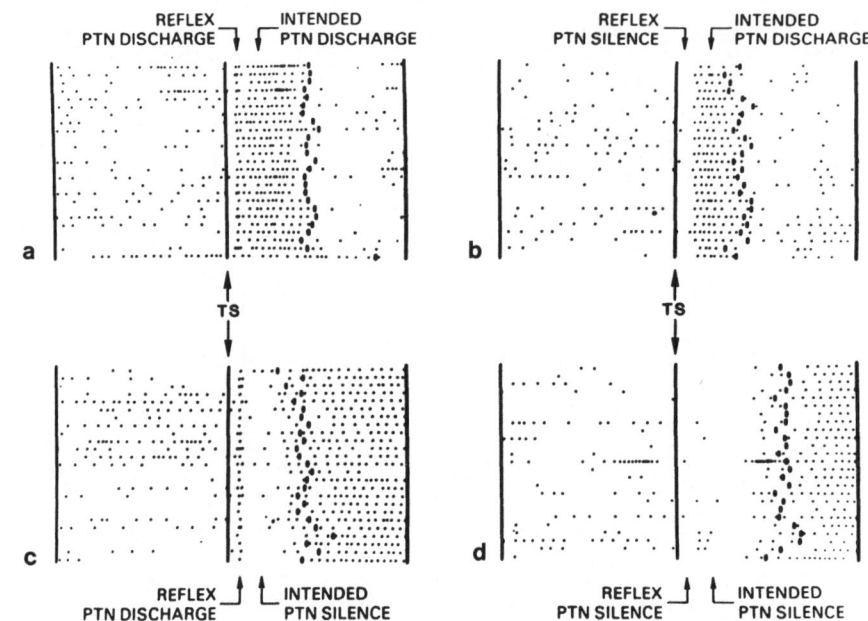

Fig. 7a–d. Dissociation of reflex and intended PTN discharge. These rasters parallel the triceps muscle rasters in Fig. 6. The PTN whose activity is represented discharged when the monkey pushed and was silent when it pulled. TSs that opposed push (**a** and **c**) reflexly excited the PTN, while TSs that assisted push (**b** and **d**) reflexly silenced the PTN. In **a** the reflex PTN discharge resulting from limb displacement merged into the intended PTN discharge, while in **c** the reflex PTN discharge was cut off by the intended PTN silence that accompanied pull. In **b** the TS reflexly silenced the PTN, but this reflex silence was followed by intended discharge.

The single heavy mark in each raster row indicates the time at which the monkey completed the intended movement. These heavy marks occur later in **d** than in **b**, since in **d** the monkey was set to pull and TS opposed and delayed completion of the intended pull by moving the handle away from the monkey, while in **b** the monkey was set to push and the TS displacement of the handle away from the monkey assisted the intended push movement and allowed it to be completed more quickly. Likewise, the heavy marks in **a** are later than in **b**, because in **a** the TS opposes the intended push movement, while in **b** the TS assists the intended push movement.

Each raster row depicts activity for 500 ms before and after the displacing TS. (Evarts and Tanji 1976)

for muscle in Fig. 6, where the muscle was most active when the direction of the TS was *opposite* to the direction of instructed movement
with which the muscle discharged. So, too, with this PTN: the PTN
was active with intended push and responded reflexly to an oppositely
directed external displacement.

The data that have now been presented show that PTNs have two phases
of activity following a limb displacement. The first (20-ms latency)
phase is analogous to a segmental stretch reflex, in that it depends
on the direction of limb displacement, while the second phase (40-ms
latency) depends on motor set. The set-dependent phase of muscle activity (Fig. 5 and 6) occurs at a latency of about 60 ms, sufficiently
long after the 40-ms latency set-dependent phase of motor cortex activity so that the latter can contribute to its occurrence. These two
phases of motor cortex output have different properties and different
functional roles. The first phase follows the rules of a closed-loop
negative feedback control system, whereas the second phase operates
according to the open-loop principles discussed by Abbs and Cole
(1982). It seems likely that the first phase is especially important
in maintenance of postural stability, while the second phase is important in movements that involve reaching a goal from a variety of different starting positions and with a variety of different contributions by the components of a multicomponent coordinated system, such
as the system involved in speech. The properties of motor cortex
responses are quite different, depending on the extent to which one or
the other of these two control modes is dominant, and in what follows
there will be a separate discussion of MI responses in these two different phases of motor control.

Postural Stability and Micromovement

As a volitional change of position becomes smaller and smaller (e.g.,
a hand movement under a dissecting microscope), the associated change
of muscle activity also becomes smaller, and for sufficiently small
movements it requires rather fortuitous placement of EMG electrodes to
pick up the changes of electrical activity in the relevant muscles.
For motor cortex, however, it is not at all difficult to pick up
changes of activity even with movements so small that they are associated with minimal activity of the relevant muscles. The sensory
responses of MI/r neurons were studied in monkeys making very small

movements (Evarts and Fromm 1977), and it was found that precentral neurons whose activity changed with precisely controlled, small, supination-pronation movements almost invariably showed sensory responses to small perturbations which pronated or supinated the forearm. Conversely, most units that were recruited with large, high-velocity movements, but that failed to show changes with small movements, were unaffected by these kinesthetic inputs. Moreover, units showing intense, short-latency responses to perturbations delivered during postural stability usually failed to respond when these same perturbations were delivered immediately prior to a high-velocity movement. These results suggested that proprioceptive reflexes continuously modulate motor cortex neuron discharge during the closed-loop control mode used in accurate positioning and precise fine movement, whereas such modulation is attenuated before and during high-velocity active movements.

Figure 8 shows activity of a PTN prior to and during a small volitional movement and illustrates the intense discharge elicited in this PTN by a displacement that opposed the movement. In this figure, the opposing displacement was 5°, whereas the volitional movements preceded by intense discharge of many MI/r PTNs were often as small as 1°. In order to conclude that the PTNs involved in controlling 1°-movements are continuously modulated by proprioceptive feedback, it was necessary to observe effects of sensory inputs resulting from smaller displacements. Schmidt et al. (1982) obtained data on MI responses to 1° and smaller wrist displacements in monkeys trained to perform a wrist flexion-extension task and found that 44% of task-related PTNs responded to wrist displacements of less than 1°. About two-thirds of the responsive neurons showed the types of responses that were appropriate to subserve wrist stabilization. It was concluded that sensory feedback was effectively controlling motor cortex output during small movements and that the sign of the feedback was such as to stabilize limb position.

Cheney and Fetz (1983) recorded responses of 21 CM cells whose target muscles were stretched or shortened. Of these 21 cells, 20 were excited at short latency (24 ± 9 ms) by stretch, but eight were also excited at short latency (22 ± 7 ms) when their target muscles were shortened. The existence of these significant proportion of bidirectional CM cells means that one effect of the transcortical reflex output is to stiffen muscles (independent of the direction of the displacement), while another effect (based on the directionally specific

149

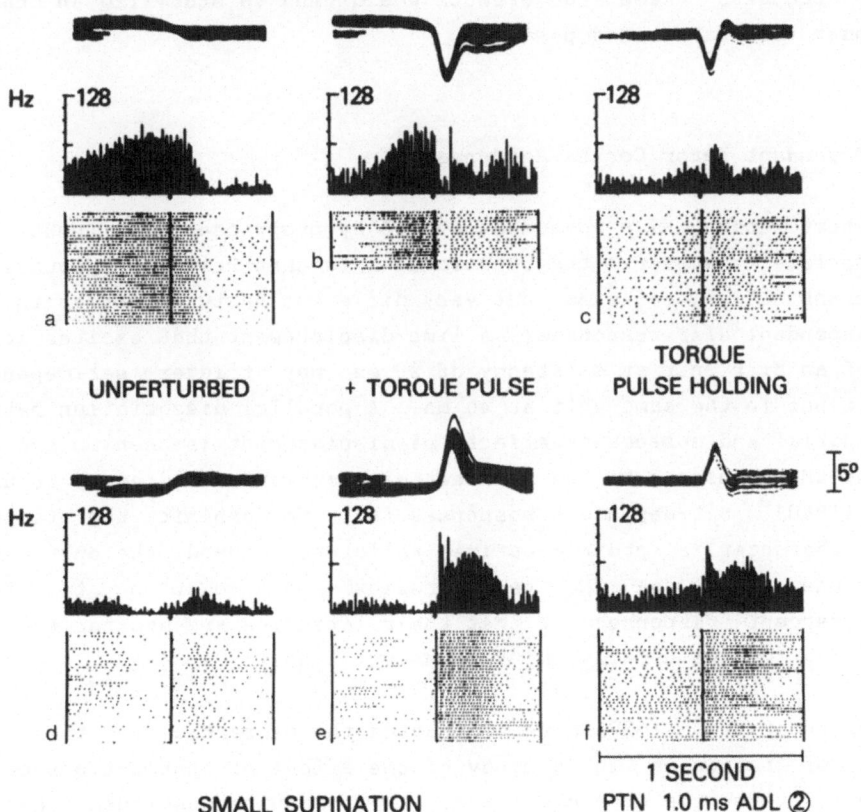

SMALL PRONATION

UNPERTURBED   + TORQUE PULSE   TORQUE
PULSE HOLDING

SMALL SUPINATION

1 SECOND
PTN 1.0 ms ADL ②

Fig. 8a–f. Reflex PTN responses during postural stability and precise
active movement. Three sorts of displays are shown: Superimposed position traces representing the position of the handle are at the *top*
of each set of displays. In the *center* of each set of displays is a
histogram corresponding to unit discharges. At the *bottom* of each
display are rasters of unit discharges where each row corresponds to a
trial and each dot in a row corresponds to a single impulse in a neuron. This PTN was active prior to and during a small active pronating
movement **(a)** and reciprocally silent for a small active supinating
movement **(d)**. In accord with the transcortical servo hypothesis, the
PTN was excited by displacements that opposed the active movement with
which it discharged **(e, f)**, and inhibited by "assisting" displacements
(b, c). The displays of unperturbed small movements **(a, d)** are
aligned on detection of movement onset. In **b** and **e** a torque pulse was
delivered at the time when movement onset was detected, so that both
active movement onset and occurrence of an external displacement occur
at the middle of the displays marked *TORQUE PULSE*. In **c** and **f** torque
pulses were delivered during a period of steady holding. (Fromm and
Evarts 1978)

cells) is to selectively increase discharge in the stretched muscle. Taken togther, these two effects would tend to stabilize an ongoing movement or a maintained posture.

Set-Dependent Motor Cortex Responses

The short-latency MI/r responses evoked by proprioceptive inputs are appropriate to subserve the functional role that Phillips hypothesized for transcortical reflexes, but very different features appear in the set-dependent MI/r responses: a limb displacement that excites activity of an MI/r unit at a latency of 20 ms may trigger set-dependent quiescence in the same unit at 40 ms. A parallel dissociation between the initial and subsequent effects of displacement is seen in the case of the EMG responses to the indirect displacements studied by Traub et al. (1980). Set-dependent responses fail to exhibit the features that characterize proprioceptive reflexes. Indeed, the only reason that one might use the term "reflex" in relation to these set-dependent responses is that their latencies are shorter than the latencies that we usually think of as volitional.

Colebatch et al. (1979) have contrasted the properties of these two sorts of responses in a study of the effect of instructions on EMG responses to muscle stretch in man. Their results confirmed the observations of Hammond (1956) as to the effect of prior instruction on stretch-evoked muscle activity occurring at a latency of 40 - 70 ms. In addition, they showed that the effect of a prior instruction to resist or not resist depended on the interval between occurrence of the instruction and the muscle stretch. Thus, no subject was able to control the 40 - 70 ms EMG activity on the basis of instructions given simultaneously with the initiation of muscle stretch. The results were compatible with the idea that the

> ...information relevant to reflex modification such as we
> have observed must arise before presentation of the load.
> It seems likely that the signals which are important for
> such reflex modification are initiated when the decision is
> made to 'resist' or to 'let go'. It is possibly relevant
> that alterations in the activity of motor cortical cells
> occur in response to instructions regarding a forthcoming
> response, and before the perturbation which evokes that
> response has occurred (Evarts and Tanji 1974; Tanji and
> Evarts 1976). By pre-setting the excitability levels within
> the long-latency pathway according to instruction and in-
> tent, responses to perturbations could be pre-set: whether
> or not perception of those perturbations occurs need not
> then be concerned with the sizes of responses evoked.

One may think of the instruction as having caused the brain to "commit itself" to the *automatic* execution of a particular act upon occurrence of a particular input. Thus, the volitional process, in the sense that we often think of it, has already taken place prior to the limb displacement, and instead of initiating a volitional process at the time the stretch begins, the subject reacts automatically according to a motor-set that is already in place. This formulation fits with the ideas of Traub et al. (1980), who noted that the "grab reflex" in their sherry-glass experiment was not modifiable by volition, but that it could be present or absent depending on the prior decision of the subject.

Perhaps the best way to think of these quick, but set-dependent, responses is in the terms used by J. Hughlings Jackson. One might think that the decision process following an instruction starts in Jackson's highest level, but that once the decision has been made and the motor-set has been adopted, the more automatic middle level of motor control (including MI) can react with remarkable speed.

Additional Questions

What pathways mediate the two phases of MI/r activity evoked by limb displacement? What is the strength of the transcortical reflex? What are the mechanisms for dynamic switching between the closed-loop controls that are so important in maintenance of postural stability and the open-loop controls that characterize the goal-oriented transients seen in speech and manual control? These are but a few of the questions that have not been dealt with in this paper. As to the pathways involved, the recent papers by Asanuma et al. (1983a, b, c) consider the current state of knowledge and summarize evidence showing that there are several routes (both corticocortical and thalamocortical) over which sensory inputs may activate MI/r in the closed-loop mode. There are also alternative pathways for the set-dependent responses of MI/r, as elucidated by Schell and Strick (1983) in work on the projections of globus pallidus via thalamus to area 6 and thence to MI. As to the strength of transcortical reflexes, we know that the impulse frequencies generated by sensory inputs to CM cells can be as great as the frequencies occurring in CM cells prior to volitional movements. To say that reflex CM outputs generated by sensory inputs are unimportant would thus amount to saying that CM outputs in general are unimpor-

tant, and this view would contradict a vast amount of neurophysiological data: motoneuronal excitatory postsynaptic potentials (EPSPs)
generated by CM synapses and segmental synapses (see Fig. 1) are of
similar orders of magnitude. Finally, the dynamic processes that may
switch MI from a closed-loop to an open-loop mode are considered by
Wiesendanger (1981) and Wiesendanger and Miles (1982) in discussion of
cortical gating. All three of these topics (pathways to MI, strength
of MI reflexes, and dynamic gating) are of enormous importance and it
may be expected that the next decade will see exciting new developments in these areas, developments that will greatly enhance our
understanding of the properties and functional significance of transcortical reflexes.

References

ABBS JH, COLE KL (1982) Consideration of bulbar and suprabulbar afferent influences upon speech motor coordination and programming. In:
GRILLNER S, LINDBLOM B, LUBKER J, PERSSON A (eds) Speech motor control. Pergamon, Oxford, pp 159-186

ASANUMA C, THACH WT, JONES EG (1983a) Cytoarchitectonic delineation of
the ventral lateral thalamic region in the monkey. Brain Res Rev 5:
219-235

ASANUMA C, THACH WT, JONES EG (1983b) Distribution of cerebellar terminations and their relation to other afferent terminations in the
ventral lateral thalamic region of the monkey. Brain Res Rev 5:
237-265

ASANUMA C, THACH WT, JONES EG (1983c) Anatomical evidence for segre-
gated focal groupings of efferent cells and their terminal ramifica-
tions in the cerebellothalamic pathway of the monkey. Brain Res Rev
5: 267-297

BROWN TIH, RACK PMH, ROSS HF (1982a) Forces generated at the thumb in-
terphalangeal joint during imposed sinusoidal movements. J Physiol
332: 69-85

BROWN TIH, RACK PMH, ROSS HF (1982b) A range of different stretch reflex responses in the human thumb. J Physiol 332: 101-112

CHENEY PD, FETZ EE (1984) Primate corticomotoneuronal cells contribute
to long latency stretch reflexes. J Physiol (to be published)

CLOUGH JFM, KERNELL D, PHILLIPS CG (1968) The distribution of
monosynaptic excitation from the pyramidal tract and from primary
spindle afferents to motoneurones of the baboon's hand and forearm. J
Physiol 198: 145-166

COLEBATCH JG, GANDEVIA SC, McCLOSKEY DI, POTTER EK (1979) Subject instruction and long latency reflex responses to muscle stretch. J Physiol 292: 527-534

CONRAD B, MATSUNAMI K, MEYER-LOHMANN J, WIESENDANGER M, BROOKS VB (1974) Cortical load compensation during voluntary elbow movements. Brain Res 71: 507-514

CONRAD B, MEYER-LOHMANN J, MATSUNAMI K, BROOKS VB (1975) Precentral unit activity following torque pulse injections into elbow movements. Brain Res 94: 219-236

EVARTS EV (1973) Motor cortex reflexes associated with learned movement. Science 179: 501-503

EVARTS EV, FROMM C (1977) Sensory responses in motor cortex neurons during precise motor control. Neurosci Lett 5: 267-272

EVARTS EV, TANJI J (1974) Gating of motor cortex reflexes by prior instruction. Brain Res 71: 479-494

EVARTS EV, TANJI J (1976) Reflex and intended responses in motor cortex pyramidal tract neurons of monkey. J Neurophysiol 39: 1069-1080

FETZ EE, FINOCCHIO DV, BAKER MA, SOSO MJ (1980) Sensory and motor responses of precentral cortex cells during comparable passive and active joint movements. J Neurophysiol 42: 1070-1089

FOERSTER O (1927) Schlaffe und spastische Lähmung. In: BETHE A, BERGMANN G, EMBDEN G, ELLINGER A (eds) Handbuch der normalen und pathologischen Physiologie. Springer, Berlin, pp 900-901

FROMM C, EVARTS EV (1978) Motor cortex responses to kinesthetic inputs during postural stability, precise fine movement and ballistic movement in the conscious monkey. In: GORDON G (ed) Active touch. Pergamon, Oxford, pp 105-117

GOODWIN GM, HOFFMAN D, LUSCHEI ES (1978) The strength of the reflex response to sinusoidal stretch of monkey jaw closing muscles during voluntary contraction. J Physiol (Lond) 279: 81-111

HAMMOND PH (1956) The influence of prior instruction to the subject on an apparently involuntary neuromuscular response. J Physiol (Lond) 132: 17-18

HUNKER CJ, ABBS JH (1982) Respiratory movement control during speech: evidence for motor equivalence. Soc Neurosci Abstr 8: 946

JOYCE GC, RACK PMH, ROSS HF (1974) The forces generated at the human elbow joint in response to imposed sinusoidal movements of the forearm. J Physiol (Lond) 240: 351-374

KUYPERS HGJM (1964) The descending pathways to the spinal cord, their anatomy and function. In: ECCLES JC, SCHADE JP (eds) Organization of the spinal cord. Elsevier, Amsterdam, pp 178-200

MARSDEN CD, MERTON PA, MORTON HB, ADAM A (1978) The effect of lesions of the central nervous system on long-latency stretch reflexes in the human thumb. In: DESMEDT JE (ed) Long loop mechanisms. Karger, Basel, pp 334-341 (Cerebral motor control in man, vol IV)

MERTON PA (1964) Human position sense and sense of effort. Symp Soc Exp Biol 18: 387-400

PENFIELD W, RASMUSSEN T (1950) The cerebral cortex of man. Macmillan, New York

PHILLIPS CG (1969) Motor apparatus of the baboon's hand. Proc Soc Lond [Biol] 173: 141-174

PHILLIPS CG, PORTER R (1977) Corticospinal neurones: their role in movement. Academic, London

ROTHWELL JC, TRAUB MM, DAY BL, OBESO JA, THOMAS PK, MARSDEN CD (1982) Manual motor performance in a deafferented man. Brain 105: 515-542

SANES JN, MAURITZ KH, EVARTS EV, DALAKAS MC, CHU A (to be published) Impaired motor control in humans without proprioception.

SCHELL GR, STRICK PL (1984) The origin of thalamic inputs to the arcuate premotor and supplementary motor areas. J Neurosci 4: 539-560

SCHMIDT EM, McINTOSH JS, GLENN LL (1982) Sensory response in motor cortex neurons from an unloading perturbation. Soc Neurosci Abstr 8: 539

SHERRINGTON CS (1906) On the proprio-ceptive system, especially in its reflex aspect. Brain 29: 467-482

STRICK PL, PRESTON JB (1978) Multiple representation in the primate motor cortex. Brain Res 154: 366-370

TANJI J, EVARTS EV (1976) Anticipatory activity of motor cortex neurons in relation to direction of an intended movement. J Neurophysiol 39: 1062-1068

TANJI J, WISE SP (1981) Submodality distribution in sensorimotor cortex of the unanesthetized monkey. J Neurophysiol 45: 467-481

TRAUB MM, ROTHWELL JC, MARSDEN CD (1980) A grab reflex in the human hand. Brain 103: 869-884

VALLBO AB (1973a) The significance of intramuscular receptors in load compensation during volunatry contractions in man. In: STEIN RB, PEARSON KB, SMITH RS, REDFORD JB (eds) Control of posture and locomotion. Plenum, London, pp 211-226 (Advances in behavioral biology, vol 7)

VALLBO AB (1973b) Muscle spindle afferent discharge from resting and contracting muscles in normal human subjects. In: DESMEDT JE (ed) New developments in electromyography and clinical neurophysiology, vol 3. Karger, Basel, pp 251-262

WIESENDANGER M (1981) Organization of secondary motor areas of cerebral cortex. In: BROOKS VB (ed) The nervous system, vol II. Motor control. Am Physiol Soc, Bethesda, pp 1121-1147 (Handbook of physiology, part 2)

WIESENDANGER M, MILES TS (1982) Ascending pathway of low-threshold muscle afferents to the cerebral cortex and its possible role in motor control. Physiol Rev 4: 1234-1270

Small Hand Muscles in Percision Grip:
A Corticospinal Prerogative?

R. B. Muir

Department of Physiology, University of Hongkong, Li Shu Fan Building, 5, Sassoon Road, Hongkong

Introduction

It is clear from both clinical and experimental evidence that the lateral corticospinal tract is in some way critically concerned with the
fine control of hand and finger movements. Without an intact corticospinal tract human patients and experimental primates alike have
proved unable to produce movements requiring relatively independent
control of the digits, such as a "precision grip" between the tips of
the index finger and thumb (Horsley 1909; Penfield 1954; Lawrence
and Kuypers 1968; Lawrence and Hopkins 1976; Napier 1956). However,
the precise nature of the corticospinal contribution to the regulation
of finger movements remains uncertain.

In the rhesus macaque, the distribution of corticospinal terminals has
been shown by microanatomical studies to include the most dorsolateral
portion of the ventral horn in the cervical enlargement (Kuypers and
Brinkman 1970), the region occupied especially by motoneurons of the
hand and forearm. The distribution extends, in the chimpanzee, to include the more medially located motoneuron pools belonging to proximal
musculature (Kuypers 1981). But even in the human, in the first tho-
racic spinal segment corticospinal terminals would appear to be most
densely concentrated in the lateral prominence of the anterior horn
(Schoen 1964) where motoneurons of the intrinsic hand muscles are situated.

The microanatomical evidence is complemented by the electrophysiological findings of Phillips and his colleagues (Phillips and Porter 1964;
Clough et al. 1968) who demonstrated that the motoneurons of the distal forelimb muscles received "larger quantities of monosynaptic exci-

tatory action" than those of proximal muscles, in response to electrical stimulation of restricted areas of motor cortex. The corticomotoneuronal synapses possess the special property of temporal facilitation (Porter 1970), and especially when the facilitation is compounded
by the arrival in rapid succession of several impulses within a
high-frequency burst (Muir and Porter 1973), these synapses are capable of producing a large and rapid depolarization of the motoneuronal
membrane. Consequently, the monosynaptic corticomotoneuronal connections have been implicated in the control of fine finger movements.
But until quite recently (Muir and Lemon 1983), there has been no
direct evidence of such an involvement.

Although the techniques pioneered by Evarts (1966) for recording the
natural activity of cortical neurons in the conscious monkey have been
in use now for about 2 decades, surprisingly most of the work has concentrated upon arm and wrist movements not requiring precise control
of the individual fingers. Moreover, until relatively recently (Fetz
et al. 1976; Lemon and Muir 1981a, b), it has not been possible to
identify the target muscles of the excitatory output of motor cortical
neurons from which spike activity could be recorded. Consequently, it
has been necessary to resort to one or other of two methods in attempting to interpret the significance of observed discharge patterns
of cortical neurons: (a) a "black box" approach in which a relationship is sought between the neuronal discharge frequencies and one or
more parameters of the monkey's motor performance (Humphrey 1972),
such as the torque or force produced in an external manipulandum
(Evarts 1968; Schmidt et al. 1975; Smith et al. 1975; Cheney and
Fetz 1980; Evarts et al. 1983; Hepp-Reymond and Diener 1983), or
(b) a prediction of the postsynaptic effects of the neuronal discharges based upon electrophysiological evidence and assuming a possible terminal destination for the axon of the cortical neuron (Porter
and Muir 1971).

While the black box method might ultimately yield results which are
highly valuable for the control of a limb prosthesis, it would seem to
have a limited capacity for providing neurophysiological insight into
the relative contributions of different neuroanatomical subsystems.
The second method, on the other hand, achieves its full significance
when the actual synaptic destination of the cortical neuron can be determined in the conscious animal, and the functional effects of the
target neuron in terms of muscle action can be described.

When an analysis of this kind can be completed it will become possible to provide a definitive answer to questions such as whether or not control of the intrinsic hand muscles during precision grip is a prerogative of corticospinal neurons. Already we can begin to explore the answer to this question, and the time is now ripe to accelerate our efforts toward achieving a detailed analysis of the neurophysiological processes which enable the "higher" primates to manipulate objects in their environment in a precise and intricate manner. In what follows I propose to bring together several preliminary and sometimes fragmentary pieces of evidence, which I believe might help to shed some light on this question and perhaps provide impetus for further investigation.

First some observations are presented concerning the activity of hand muscles during precision grip, then evidence is summarized which argues directly for corticospinal control of such hand muscle activity. However, in subsequent sections findings are reported which could be seen to question the exclusive right of corticospinal neurons to command the hand muscles during relatively independent finger movements.

Muscle Activity During Precision Grip

Our knowledge of the functional anatomy and peripheral neuroanatomy of the human hand has recently been set on firmer foundations by the pains-taking work of Landsmeer (1976) and Sunderland (1978). Electromyographic analysis of the muscular contributions to a number of different elementary finger movements (Long and Brown 1964; Landsmeer and Long 1965) leaves no doubt about the complexity of muscular action upon the fingers, and the astute clinical observations of Sunderland (1944) reveal something of the plasticity of muscle contributions to a given movement as patients employ different strategies to compensate for innervative deficits in particular muscle groups.

Because of the complexity of the hand machinery, it has been necessary in our laboratory to begin the analysis of hand muscle action during precision grip using an empirical approach; a combination of electromyographic recording and frame-by-frame inspection of videotape recordings has been used to examine the average contribution of different hand and forearm muscles during repetitive performance by a human subject of a stereotyped precision grip squeeze between index

158

finger and thumb. For comparison, a power grip (Napier 1956) squeeze of a cylinder by the whole hand was also investigated. These particular motor tasks were selected for their similarity with the movements performed by monkeys in the experiments of Muir and Lemon (1983).

For the precision grip, the subject flexed his three ulnar digits close to the palm while extending the index finger and thumb through a slot in a horizontal acrylic (Perspex) top plate to reach the tips of two vertically positioned Perspex levers spaced 5 cm apart. The levers were attached to the shafts of potentiometers which gave an electrical record of lever movement as the subject squeezed against the restraining force of springs fixed between the levers and the frame of the manipulandum. The stationary end of the springs could be adjusted by a wingnut to provide various amounts of pretensioning. The subject was required to displace each lever tip from its resting position through 4 mm into a target zone which was a further 4 mm wide. For the records shown in Fig. 1, this required a force of approximately 60 N. Whenever the finger lever and the thumb lever were both within their target zones a tone sounded. Maintenance of both target positions, uninterrupted, for 1-2 s caused a lamp to be switched on, an electrical trigger pulse to be generated, and a counter to be incremented, indicating successful completion of the trial. The subject then released the levers and removed his hand from the manipulandum. After a delay of at least 1 s a new trial was begun.

The records shown in Fig. 1 are computer averages of eight segments of data, each consisting of one complete trial and the start of the next following squeeze. The computer was triggered by the pulses indicating the end of the 1.2-s period of target maintenance. Since all signals were delayed by 2.5 s relative to the trigger pulses, the averaged include 2.5 s of pretrigger data; an arrowhead indicates the time of the trigger in the figure.

The manipulandum used for the power grip consisted of a brass cylinder, 4.4 cm in diameter, in which a longitudinal slot was cut from end to end. The wall of the cylinder opposite the slot was thinned by filing to increase the compliance between the two halves of the cylinder when the cylinder was grasped and squeezed between the fingers on one side and the thumb on the other. A thin rubber sleeve was fitted snugly over the outside of the cylinder. A microswitch mounted inside the cylinder detected the approximation of the two halves of

159

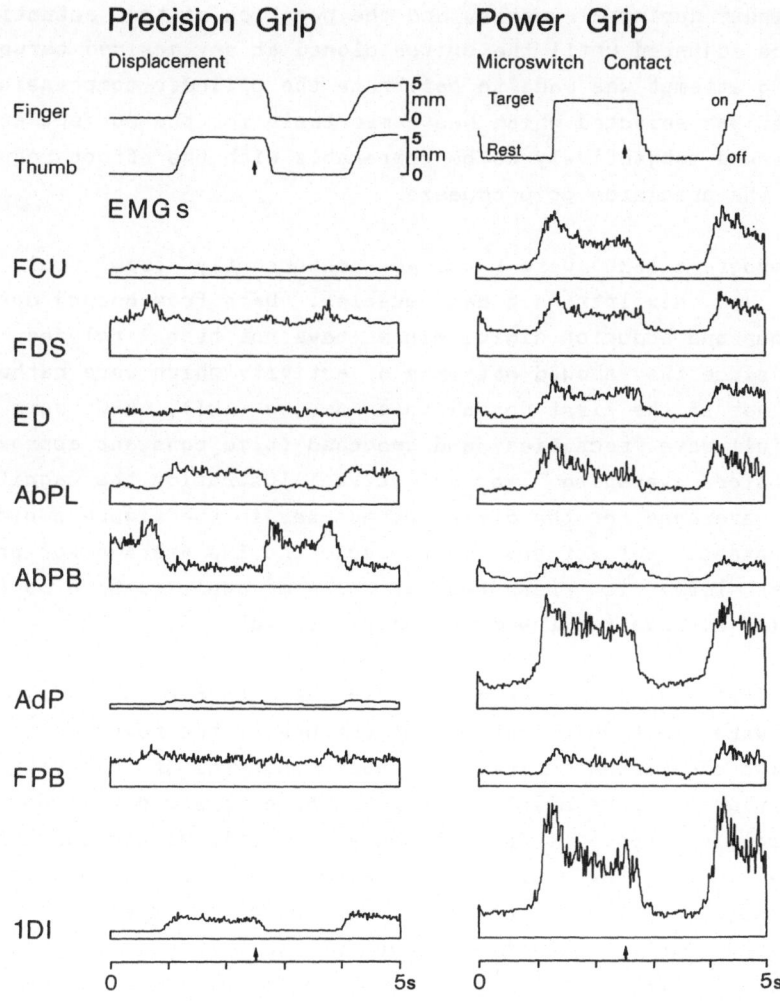

Fig. 1. Rectified EMGs from forearm and hand muscles of a human subject, computer averaged (8 sweeps) with respect to precision grip and
power grip movements; the uppermost traces are corresponding averages
of the displacement of the finger and thumb levers of precision grip
(each measured from its resting position toward the other lever) and
of the closure of a microswitch which signaled the target force for
the power grip. The averaging computer was synchronized by a trigger
after 1.2 s of target maintenance, as indicated by the *arrowheads*.
The absolute magnitudes of the EMG averages are uncalibrated and the
amplification is different from one muscle to another. However, for a
given muscle the precision grip and power grip traces are plotted with
the same vertical calibration. Muscle identification: *FCU*, flexor
carpi ulnaris; *FDS*, flexor digitorum superficialis; *ED*, extensor digitorum; *AbPL*, *AbPB*, abductor pollicis longus and brevis; *AdP*, adductor pollicis; *FPB*, flexor pollicis brevis; *1DI*, first dorsal interosseous. Note that the minimum values (baseline levels) in the EMG
averages are exaggerated to undefined extent by the contributions of
noise and nonlinearities in the apparatus

the cylinder during a squeeze, and the position of the actuating lug could be adjusted until the switch closed at any desired target pressure. No attempt was made to calibrate the cylinder compression, but a target was selected which was comfortable for the subject and which was assessed subjectively to be comparable with the effort required to perform the precision grip squeeze.

Electromyograms (EMG) were recorded concurrently from four forearm muscles and six intrinsic hand muscles. Data from second dorsal interosseous and abductor digiti minimi have not been included in the figure, since they showed patterns of activity which were rather similar to that of the first dorsal interosseous (1DI). EMGs were amplified, full-wave rectified, and smoothed (time constant approximately 20 ms) before averaging. For effective illustration the magnification of the averages for the different muscles in the figure could not be kept constant. But for any given muscle the EMG average for precision grip (at left) is presented at the same amplification as the corresponding average for the power grip (at right).

What is striking from these results is the extent to which all of the muscles were coactivated during performance of the power grip, in contrast with the marked variation in their patterns of activity during precision grip. Some muscles, such as the extensor digitorum (ED) and flexor pollicis brevis (FPB), showed very little modulation of their activity during the different phases of the precision grip trials, while others, such as 1DI and the thumb abductors, are clearly related to the movements; note the antagonistlike behavior of abductor pollicis brevis (AbPB) and abductor pollicis longus (AbPL) during precision grip.

When similar recordings are made from a monkey trained to perform precision grip and power grip squeezes, the results are usually less stereotyped than those from human subjects. Monkeys tend to show far greater variability, from trial to trial and from one recording session to another, in the posture of the hand relative to the manipulandum and the strategy they use to perform the movements. Even the differences between precision grip and power grip are sometimes not so strikingly obvious, although clear differences are always to be found, such as is evident in the records of Fig. 2 where averages from three forearm muscles and three small hand muscles are presented using the same format as in Fig.1. Note, for example, the difference in the contribution of adductor pollicis (AdP) to the two types of grip and

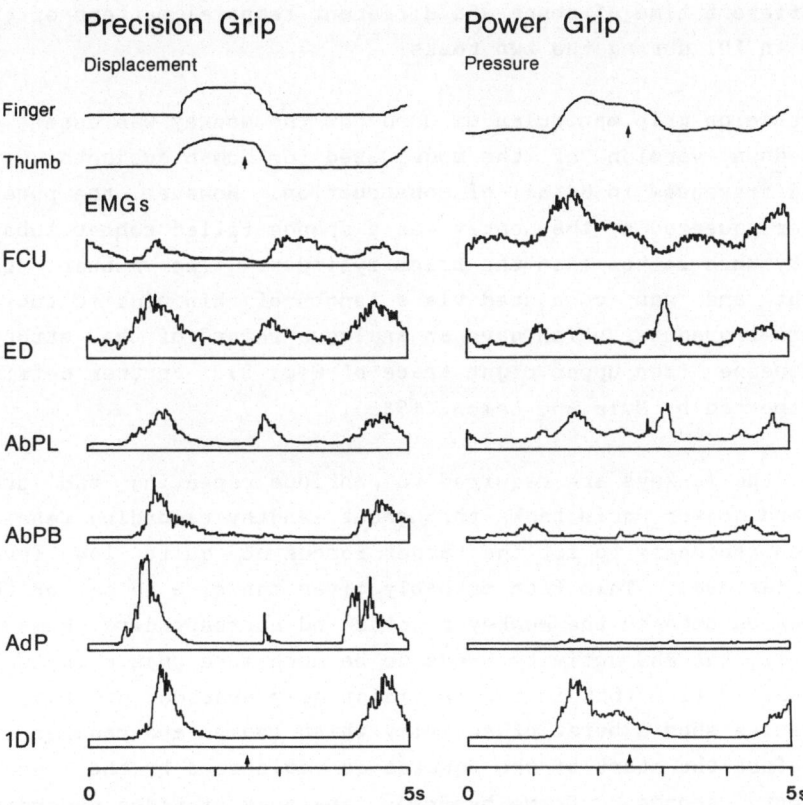

Fig. 2. Averaged (n =8), rectified EMGs recorded from a monkey performing precision grip and power grip movements. The format of the figure is the same as for Fig. 1, except that the uppermost trace at the right is the averaged analogue pressure record. Muscle identities and triggering like in Fig. 1. The displacement and pressure records are uncalibrated, but the finger and thumb each moved about 3 mm for the precision grip

the different time of onset and different temporal pattern of the activity in 1DI during the two tasks.

The precision grip manipulandum used for the monkey was essentially a scaled-down version of the model used for human subjects, with some small differences in detail of construction. However, the power grip cylinder squeezed by the monkey was a sponge-filled rubber tube, which was very much softer than the brass cylinder. The rubber tube was airtight and was connected via a length of thin plastic tubing to a pressure transducer which gave an analogue record of the strength of the squeeze (see upper right trace of Fig. 2). Further details have been reported by Muir and Lemon (1983).

Because the monkeys are required to continue repeating the precision grip and power grip tasks throughout lengthy recording sessions, it has been necessary to set the target forces at quite low levels to avoid fatigue. This fact probably gives the clue to one of the main differences between the monkey records and corresponding human averages, viz, the EMG activity tends to be much more phasic in the monkey averages. Notice that in the precision grip averages of Fig. 2, all six muscles show a burst of activity which begins and reaches its peak well before the start of the squeeze as registered by the lever displacement records. Frame-by-frame analysis of video recordings reveals that for about 500 ms before the start of the squeeze the monkey is positioning the hand above the manipulandum and shaping the hand and the fingers appropriately to place the index finger and thumb in contact with the levers. It thus seems that with the small forces required to squeeze the levers (less than 1 N), relatively more muscle activity is necessary to form the correct posture of the hand than is needed actually to produce the squeeze. This contrasts with the power grip records where the dominant muscle activity tends to be closely associated with the squeeze, as indicated by the averaged pressure record. For both types of grip in Fig. 2, the ED and AbPL muscles show a further burst of activity in association with the release of the squeeze.

Despite the general similarity of the activity patterns of the different muscles in relation to the formation of the precision grip posture, there are clear differences from one muscle to another in the precise time of onset and time course of the activity. Clear fractionation of hand muscle activation is evident in both the monkey and the human records during performance of precision grip sequences. It is this fractionation of muscle excitation in space and time that has been suggested to be the special contribution of the corticomotoneuronal neurons of the precentral gyrus (Phillips and Porter 1964, Muir and Lemon 1983).

Corticospinal Control of Precision Grip

The method of spike-triggered averaging was first applied for the identification of the target muscles of motor cortical neurons in the conscious monkey by Fetz and his colleagues (Fetz et al. 1976), and the technique is now well established (Fetz and Cheney 1980; Fetz 1981; Lemon and Muir 1981a, b), if not widely used. Cheney (1980) has applied it also to the identification of rubromotoneuronal connections.

The technique is thought to reveal the transient increase of firing probability in motoneurons receiving excitatory postsynaptic potentials via corticomotoneuronal synapses. When the rectified EMG from a given muscle is subjected to computer averaging, with the averaging computer triggered by the natural discharges of a cortical cell, a brief increase in EMG activity is sometimes seen in the average several milliseconds after the neuronal spike, a so-called postspike facilitation (PSF). The timing of the PSF is usually consistent with its generation by monosynaptic corticomotoneuronal excitation acting upon motoneurons of that muscle. Probabilistic arguments imply that such an effect would be unlikely if the excitation occurred via a polysynaptic pathway, although it is now accepted that inhibitory responses can sometimes be seen in the spike-triggered averages, and the shortest route for corticomotoneuronal inhibition is thought to be disynaptic (Jankowska et al. 1976). Cortical neurons for which a PSF is demonstrated in one or more muscles are thus very probably corticomotoneuronal neurons (CMN).

Muir and Lemon (1983) have recorded the activity of single cortical neurons identified as pyramidal tract neurons (PTN) by antidromic stimulation for the medullary pyramid. Some of these PTNs have yielded a PSF when tested against the rectified EMGs of intrinsic hand muscles and, in some cases, forearm muscles. Example spike-triggered averages are presented in the left half of Fig. 3 for a cortical neuron with PSFs in ED and AbPB. EMGs from AbPL, AbDM, 1DI, and FPB were

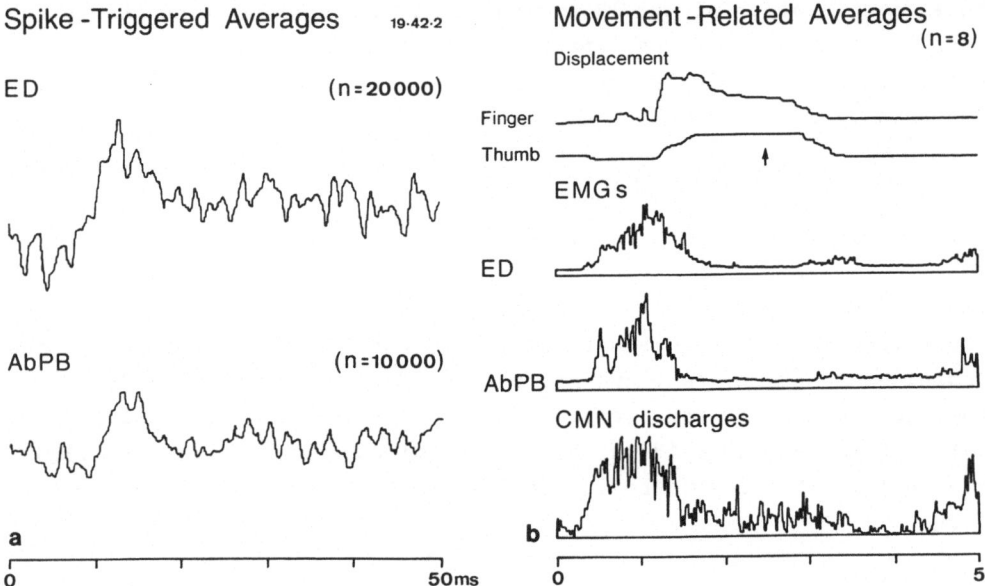

Spike-Triggered Averages   19·42·2

ED   (n=20000)

AbPB   (n=10000)

a

0   50ms

Movement-Related Averages
                              (n=8)
Displacement

Finger

Thumb

EMGs

ED

AbPB

CMN discharges

b

0   5s

Fig. 3. a Spike-triggered averages of the rectified EMGs of extensor digitorum and abductor pollicis brevis with respect to the discharges of a corticomotoneuronal neuron (CMN) in a conscious monkey, showing a post-spike facilitation in each record. **b** Averaged EMGs of the same two muscles in relation to the performance of a precision grip squeeze, with averaged lever displacement records above. The lowest trace shows the corresponding average of the firing of the CMN; the discharges were converted to standard pulses and then filtered (time constant 8.5 ms) before averaging

also averaged with respect to the discharges of this PTN but did not show a PSF. The EMGs of the target muscles of this CMN were then rectified, smoothed by a filter with a time constant of 8.5 ms, and averaged with respect to the precision grip squeeze movement, as in Figs. 1 and 2, for eight repetitions. These averages are given in the right half of Fig. 3, together with the corresponding averaged position traces for the finger and thumb levers. Also shown is the averaged firing of the CMN for the same eight repetitions of the task; each spike of the neuron triggered the generation of a brief rectangular pulse of constant amplitude and duration, and the resulting pulse train was subjected to the same filtering as for the EMGs and averaged in the same manner.

Especially during the onset of activity and the squeezing together of the levers, there is an obvious relationship between the temporal pattern of CMN activity and the activity in the target muscles. This particular identified corticospinal neuron was most probably playing an active role in the production of the precision grip movement.

In an attempt to determine to what extent such CMNs are especially or exclusively concerned with precision grip, Muir and Lemon (1983) made a quantitative comparison of the activity of each identified CMN and the activity of its facilitated muscle during the target-maintenance phase of the two different tasks. They found that all such neurons investigated discharged less in relation to target muscle activity for the power grip than for the precision grip; many CMNs showed markedly less activity during the power grip, even though the target muscles were in every case the more active for the power grip. This they interpreted as clear evidence for a preferential involvement of these hand muscle CMNs in the "fractionation of muscle action required for discrete movements of individual digits."

Pyramidal Tract Stimulation During Precision Grip

Electrical stimuli delivered to the medullary pyramid of a conscious monkey for antidromic identification of cortical neurons can also produce responses in the EMGs of hand muscles (Lemon and Muir 1983). The stimulus-evoked responses appear to be particularly strong when the motoneuron pools are facilitated by voluntary activity, and the stimuli sometimes produce flick movements of the digits during the squeeze phase of the precision grip task.

Figure 4 shows some averaged EMG responses to pyramidal tract stimulation during a period of repetitive performance by the monkey of the precision grip task. As illustrated in the figure, the most common response was an early brief excitation followed by a period of reduced activity. That inhibitory connections contribute to this reduction of EMG activity is argued by the fact that the "inhibition" was sometimes much more striking than the preceding excitation and occasionally a purely inhibitory response has been seen. In the present context I wish to draw attention to just three aspects of the results:

1. The latencies of the earliest hand muscle responses to pyramidal stimulation were up to 3 ms longer than the shortest latencies of PSFs in the hand muscle EMGs, even after applying the smallest feasible correction for the difference in conduction distances (cortex to pyramid) in the two cases.
2. The latencies and durations of the inhibitory components of the responses were too long for the inhibition to have been caused solely by a simple disynaptic pathway. A polysynaptic contribution must have been involved.
3. In some cases (not illustrated in the figure) a di- or triphasic excitatory response was seen, with the latter components of excitation occurring so late that they could easily have involved intracranial loops or other descending pathways.

Hand Muscle Activity After Motor Cortical Ablation

Since the direct corticomotoneuronal projection to the lateral motoneuronal cell groups in the monkey arises almost entirely from the precentral gyrus, including the rostral bank of the central sulcus (Kuypers 1981), ablation of the precentral gyrus can be considered effectively to have destroyd the corticomotoneuronal pathway and removed the cortical connections from that area to subcortical centers of the brain. The opportunity arose recently to make some preliminary EMG recordings from the hand muscles of three monkeys which, 6 - 10 months earlier, had received extensive surgical ablation of the precentral gyrus in connection with behavioral studies. The lesion extended from the longitudinal fissure almost to the superior bank of the lateral sulcus and included the caudal bank of the arcuate sulcus. These monkeys never regained the ability to perform fine movements with the

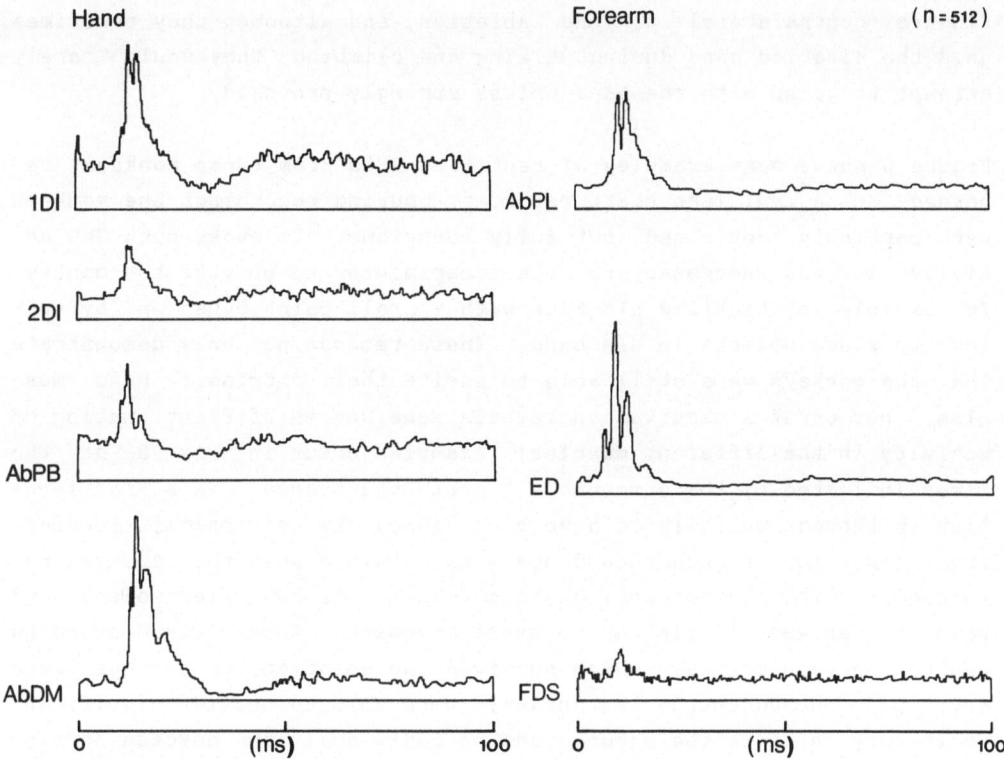

Fig. 4. Computer averages of rectified EMGs showing responses to single-pulse electrical stimulation (200 μA, 0.2 ms) of the medullary pyramid contralateral to the recorded muscles in a conscious monkey during repetitive performance of a precision grip squeeze. Stimuli were delivered between a tungsten microelectrode implanted in the pyramidal tract and a remote anode in the cranium. The timing of the stimuli was unrelated to the monkey's self-paced performance of the motor task. Muscle identities: *2DI*, second dorsal interosseous; *AbDM*, abductor digiti minimi; see also the legend to Fig. 1

fingers contralateral to the ablation, and although they sometimes used the disabled hand during walking and climbing, they would rarely attempt to grasp with the hand unless strongly provoked.

Figure 5 shows some examples of rectified EMGs from these monkeys recorded on a multipen chart recorder. During recordings the monkeys were partially restrained, but fully conscious. To evoke such EMG activity it was necessary for the experimenter to provoke the monkey, for example, by tickling his face with a small paint brush and by trying to place objects in his hand. These records not only demonstrate that the monkeys were still able to excite their intrinsic hand muscles, but careful observation reveals some degree of fractionation of activity in the different muscles; examples occur in Fig. 5 at the times indicated by the arrowheads beneath the traces. This fractionation is thought unlikely to have been caused by peripheral feedback from the limb, as such recordings were obtained when the hand was not in contact with any external objects (except the EMG electrodes) and when it showed little or no overt movement. Apparently descending central nervous pathways which survived the ablation, or perhaps were developed subsequent to the surgery, were able to provide significant excitatory input to the motoneurons of individual hand muscles or restricted combinations of muscles.

It is also noteworthy that in one of the lesioned monkeys electrical stimuli were delivered to the skin of the proximal phalynx of the index finger and these stimuli evoked both long-latency (33 - 41 ms) and short-latency (17 - 24 ms) reflex responses in some intrinsic hand muscles. The excitability of these reflexes seemed to be under the strong influence of other, possibly voluntary, central nervous activity.

Conclusions

EMG recordings from human subjects and monkeys confirm that a considerable degree of independence is required in the activation of the various hand and forearm muscles during the production of precision grip movements and reveal something of the intricate temporal patterns of activation needed, especially during the formation of the appropriate hand posture at the beginning of a squeeze between index finger and thumb. The power grip, on the other hand, when performed by a

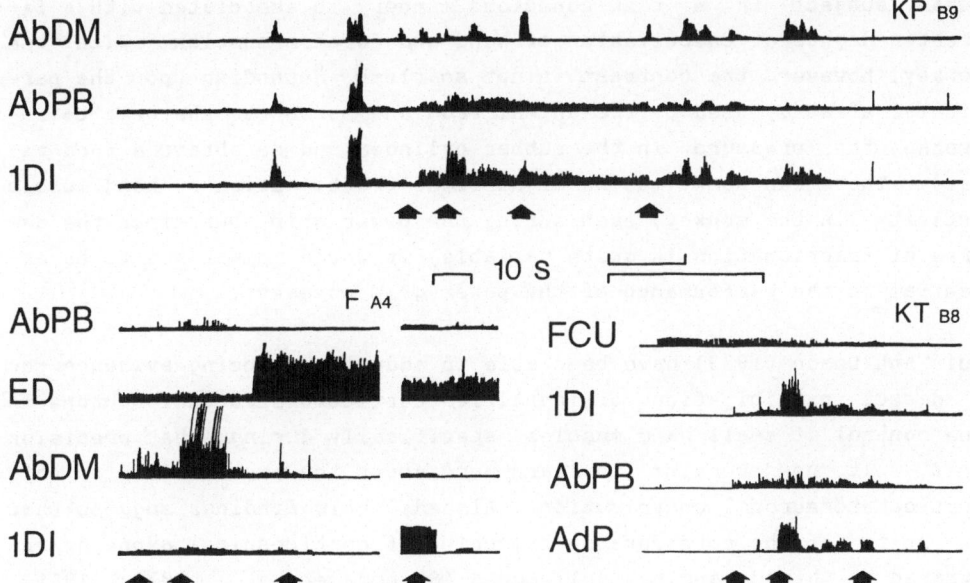

Fig. 5. Rectified EMGs, plotted on a chart recorder from hand and forearm muscles of three conscious monkeys *(KP, F,* and *KT)* during voluntary activity. The monkeys had received an extensive ablation of the motor cortex contralateral to the recorded muscles 6 - 10 month earlier. The *arrowheads* indicate times when a clear change has occurred in the level of activity of one or two hand muscles without a comparable change in the other(s). Muscle identities are the same as in Figs. 1 and 4

human subject in a task-conscious manner, is associated with a far greater degree of coactivation of hand and forearm muscles. With the monkey, however, the contrast is not so clear, depending upon the particular strategy used by the animal from one moment to the next to increase the pressure in the rubber cylinder and so obtain a food reward. There can sometimes be significant fractionation of hand muscle activity in the monkey, even during the power grip, but since the degree of fractionation is quite variable, it would appear not to be essential to the performance of the power grip movement.

Muir and Lemon (1983) have been able to adduce convincing evidence for a direct participation of identified corticomotoneuronal neurons in the control of small hand muscles, specifically during the precision grip. It now remains to learn more about the precise form of this corticomotoneuronal contribution. Already their findings suggest that in spite of the considerable branching of corticospinal axons demonstrated by Shinoda and his colleagues (Shinoda et al. 1979, 1981), divergence of the corticospinal axons to contact different pools of hand muscle motoneurons is apparently quite restricted. However, it would be surprising if the corticospinal neurons did not represent one step in the process of selecting particular *combinations* of finger muscles needed specifically in particular phases of intricate movements; both excitatory and inhibitory connections could be expected to participate in such a selection.

The findings of Muir and Lemon (1983) also emphasize the limitations of the black box approach to the analysis of neuronal discharge behavior. Neurons have been recorded which although clearly related to the production of a precision grip squeeze, discharge scarcely at all during the power grip, even though their identified target muscles are more active for the power grip. Cheney and Fetz (1980) have described identified CMNs which fired at high frequency in relation to the generation of ramp-and-hold wrist movements but which discharged only an occasional spike when the monkey made ballistic movements of the wrist, again even though the target forearm muscles were more active during the ballistic movements. Such neurons cannot be expected to show a consistent relationship with the force in a manipulandum except under very restricted conditions when the target muscles themselves are active in proportion to the external force. If the nature of the movement or the strategy used by the monkey should change, then the relationship will break down.

The preliminary observations presented in the preceding sections provide suggestive evidence that both polysynaptic corticofugal connections and other descending motor pathways might be capable of contributing to the control of hand muscles during precision grip, i.e., that the CMNs do not necessarily have exclusive rights of command. Polysynaptic effects are unlikely to be revealed by the spike-triggered averaging method; it is therefore quite possible that in addition to the probably monosynaptic effects disclosed by the PSF, collaterals of CMNs might also have influence via polysynaptic routes. The same possibility must exist for other PTNs for which a PSF cannot be revealed. This provides a possible explanation for the longer latencies of stimulus-evoked responses compared with PSFs; perhaps transmission through polysynaptic connections becomes much more secure when a synchronized descending volley occurs following stimulation. The stimulus-evoked inhibition which often follows the excitatory response probably also occurs via polysynaptic routes which might also play a role in the voluntary control of precision grip. It is noteworthy that a similar excitatory-inhibitory sequence has sometimes been seen in response to intracortical microstimulation (Lemon and Muir, unpublished observations).

The long-latency excitatory responses to pyramidal stimulation and the voluntary activation of intrinsic hand muscles contralateral to an ablated motor cortex both suggest that other descending tracts might also have ready access to motoneurons of hand muscles, perhaps even with a specificity of connection which enables some degree of fractionation of muscle activity. In the case of the lesioned animals, it is possible that such connections have arisen, or at least increased their effectiveness, by terminal sprouting to occupy synaptic sites vacated by degenerating corticomotoneuronal fibers.

Corticomotoneuronal neurons undoubtedly play an important part in the selection and sequencing of muscles during precision grip movements, but they did not seem to have an exclusive right of command of the hand muscle motoneurons.

Acknowledgements. I greatefully acknowledge the collaborative assistance of the following colleagues: A. Beishuizen and F. Bijl with the work on muscle activity during precision grip; Dr. R.N. Lemon with the work on the corticospinal control of and pyramidal tract stimulation during precision grip, and hand muscle activity after cortical ablation; G.W.H. Mantel and E.S. Buijs with the work on corticospinal control of precision grip; and Prof. Dr. H.G.J.M. Kuypers and Dr. A.M. Huisman with the work on hand muscle activity after cortical ablation. J. van der Burg, P. van Alphen, and E. Klink are thanked for their valuable technical, photographic, and secretarial assistance respectively. The work was supported by Grant 13-46-91 from the FUNGO/ZWO (Dutch Organization for Fundamental Research in Medicine).

References

CHENEY PD (1980) Response rubromotoneuronal cells identified by spike-triggered averaging of EMG activity in awake monkeys. Neurosci Lett 17: 137-142

CHENEY PD, FETZ EE (1980) Funtional classes of primate corticomotoneuronal cells and their relation to active force. J Neurophysiol 44: 773-791

CLOUGH JFM, KERNELL D, PHILLIPS CG (1968) The distribution of monosynaptic excitation from the pyramidal tract and from primary spindle afferents to motoneurones of the baboon's hand and forearm. J Physiol 198: 145-166

EVARTS EV (1966) Pyramidal tract activity accociated with a conditioned hand movement in the monkey. J Neurophysiol 29: 1011-1027

EVARTS EV (1968) Relation of pyramidal tract activity to force exerted during voluntary movement. J Neurophysiol 31: 14-27

EVARTS EV, FROMM C, KRÖLLER J, JENNINGS VA (1983) Motor cortex control of finely greated forces. J Neurophysiol 49: 1199-1215

FETZ EE (1981) Neuronal activity associated with conditioned limb movements. In: TOWE AL, LUSCHEI ES (eds) Handbook of behavioral neurobiology, vol 5. Plenum, New York

FETZ EE, CHENEY PD (1980) Postspike facilitation of forelimb muscle activity by primate corticomotoneuronal cells. J Neurophysiol 44: 751-772

FETZ EE, CHENEY PD, GERMAN DC (1976) Corticomotoneuronal connections of precentral cells detected by post-spike arrages of EMG activity in behaving monkeys. Brain Res 114: 505-510

HEPP-REYMOND M-C, DIENER R (1983) Neural coding of force and rate of force change in the precentral finger region of the monkey. In: MASSION J, PAILLARD J, SCHULTZ W, WIESENDANGER M (eds) Neural codingof motor performance. Springer, Berlin Heidelberg New York, pp 315-326 (Exp Brain Res, Suppl. 7)

HORSLEY V (1909) The Linacre lecture on the function on the so-called motor area of the brain. Br Med J 2: 125-132

HUMPHREY DR (1972) Relating motor cortex spike trains to measures of motor performance. Brain Res 40: 7-18

JANKOWSKA E, PADEL J, TANAKA R (1976) Disynaptic inhibition of spinal motoneurones from the motor cortex in the monkey. J Physiol 258: 467-487

KUYPERS HGJM (1981) Anatomy of the descending pathways. In: BROOKS VB (ed) The nervous system. Am Physiol Soc, Bethesda (Handbook of physiology, section I, vol II, part 1)

KUYPERS HGJM, BRINKMAN J (1970) Precentral projections to different parts of the spinal intermediate zone in the rhesus monkey. Brain Res 24: 29-48

LANDSMEER JMF (1976) Atlas of anatomy of the hand. Livingstone, Edin-burgh

LANDSMEER JMF, LONG C (1965) The mechanisms of finger control, based on electromyograms and location analysis. Acta Anat 60: 330-347

LAWRENCE DG, HOPKINS DA (1976) THe development of motor control in the rhesus monkey: evidence concerning the role of corticomotoneuronal connections. Brain 99: 235-254

LAWRENCE DG, KUYPERS HGJM (1968) The functional organization of the motor system in the monkey. I. The effects of bilateral pyramidal lesions. Brain 91: 1-14

LEMON RN, MUIR RB (1981a) Functional connections between motorcortex neurones and small hand muscles in the conscious monkey. Neurosci Lett [Suppl] 7: S345

LEMON RN, MUIR RB (1981b) Direct facilitation of intrinsic hand muscle activity by motor cortex neurones in the conscious monkey. J Physiol 320: 47P

LEMON RN, MUIR RB (1983) Responses of hand and forearm muscles to py-ramidal tract stimulation during voluntary hand movements in the mon-key. J Physiol 338: 31P

LONG C, BROWN ME (1964) Electromyographic kinesiology of the hand: muscle moving the long finger. J Bone Joint Surg 46A: 1683-1706

MUIR RB, LEMON RN (1983) Corticospinal neurons with a special role in precision grip. Brain Res 261: 312-316

MUIR RB, PORTER R (1973) The effect of a preceding stimulus on tempo-ral facilitation at corticomotoneuronal synapses. J Physiol 228: 749-763

NAPIER JR (1956) The prehensile movements of the human hand. J Bone Joint Surg 38B: 902-913

PENFIELD W (1954) Mechanisms of voluntary movements. Brain 77: 1-17

PHILLIPS CG, PORTER R (1964) The pyramidal projection to motoneurones of some muscle groups of the baboon's forelimb. In: ECCLES JC,

SCHADÉ JP (eds) Physiology of spinal neurons: Elsevier, Amsterdam, pp
222-245 (Progress in brain research, vol 12)

PORTER R (1970) Early facilitation at corticomotoneuronal synapses. J
Physiol 207: 733-745

PORTER R, MUIR RB (1971) The meaning of motoneurones of the temporal
pattern of natural activity in pyramidal tract neurones of conscious
monkeys. Brain Res 34: 127-142

SCHMIDT EM, JOST RG, DAVIS KK (1975) Reexamination of the force relationship of cortical cell discharge patterns with conditioned wrist
movements. Brain Res 83: 213-223

SCHOEN JHR (1964) Comparative aspects of the descending fibre systems
in the spinal cord. In: ECCLES JC, SCHADÉ JP (eds) Physiology of
spinal neurons: Elsevier, Amsterdam, pp 222-245 (Progress in brain
research, vol. 11)

SHINODA Y, ZARZECKI P, ASANUMA H (1979) Spinal branching of pyramidal
tract neurons in the monkey. Exp Brain Res 34: 59-72

SHINODA Y, YOKOTA J-I, FUTAMI T (1981) Divergent projection of individual corticospinal axons to motoneurons of multiple muscles in the
monkey. Neurosci Lett 23: 7-12

SMITH AM, HEPP-REYMOND M-C, WYSS UR (1975) Relation of activity in
precentral cortical neurons to force and rate of force change during
isometric contractions of finger muscles. Exp Brain Res 23: 315-332

SUNDERLAND S (1944) Volutary movements and the deceptive action of
muscles in peripheral nerve lesions. Aus N Z J Surg 13: 160-183

SUNDERLAND S (1978) Nerves and nerve injuries, 2nd edn.   Livingstone,
Edinburgh

Relations Between Two-Dimensional Arm Movements and Single-Cell Discharge in Motor Cortex and Area 5: Movement Direction Versus Movement End Point

A.P.Georgopoulos, J.F.Kalaska, and R.Caminiti

The Philip Bard Laboratories of Neurophysiology, Department of Neuroscience, and Department of Physiolgy, The Johns Hopkins University, School of Medicine, Baltimore, MD 21205, USA

An important question regarding the CNS mechanisms subserving the control of movement in space concerns the variable to which changes in cell discharge preceding movement onset might be related, that is, the behavioral variable predicted by these changes. In the case of aimed movements this question assumes a special significance, for specific hypotheses have been put forth on how the motor system might produce them. According to one idea, what matters is the final position of the arm (Polit and Bizzi 1979); according to another, it is the particular parameters of movement (e.g., direction) that are programmed (Rosenbaum 1980). Apparently, both factors are of importance. For example, information about the location of the target, and therefore the end point of the movement, has to be taken into account, even if movement parameters have to be calculated. On the other hand, movements with certain specifications (e.g., to be made in a desired direction) can be produced at will without reference to a target. It is likely that in CNS motor structures both kinds of relations, that is, to the target and/or to the parameters of the upcoming movement, can be found, depending on the structure under study and the conditions of the task. We explored these relations in two cortical areas involved in the central control of movement, namely, the motor cortex and area 5 of the posterior parietal cortex, under conditions that dissociated the target of the movement from its direction.

Experimental Procedures

Three rhesus monkeys (4-5 kg body weight) were trained, using operant conditioning techniques, to move a manipulandum to a common end point, starting from different positions on a plane. The two-dimensional tracking apparatus used has been described before (Geogopoulos et al. 1981). It consists of a planar working surface and an articulated manipulandum that can move freely over the surface in front of the animal. Nine light-emitting diodes were placed on that plane in a circular pattern, with one at the center and eight on the circumference of a circle of 8-cm radius. An angle of 45° was subtended between the center and any two adjacent peripheral lights. The monkeys were trained to capture the lights, as they came on, within a transparent plexiglass circle attached to the distant end of the manipulandum, close to the point where the animal grasped it. In one of the tasks used in the present study (out → center task), one of the peripheral lights came on first, and the animal had to capture it and hold the manipulandum there for a variable period of time (1.2 - 4.8 s). Then the center light came on, and the animal had to move to that light, capture it, and hold there for a period of time to receive a liquid reward. Thus the end point of the movements was always the same, but the movement directions differed. In another task, the center light came on first and the peripheral light afterward, so that movements were made from a common starting point to eight different targets on the circle (center → out task). These tasks are shown schematically in Fig. 1. The peripheral lights in either task were presented in a randomized block design. Individual movements exhibited a dome-shaped velocity profile with a peak value of 200 - 400 mm/s in different movements.

The discharge of single cells was recorded extracellularly. The surgical procedures, electrophysiological methods, and data collection and analysis techniques used have been described elsewhere (Georgopoulos et al. 1982).

The objective of the data analysis was to evaluate the hypotheses that cell discharge preceding the onset of movement relates to the end point or to the direction of the movement. We proceeded as follows. First, the frequency of cell discharge (spikes/s) during the reaction time (RT) was calculated for every trial. The RT was the time that elapsed from the onset of the target to the beginning of movement. The latter was calculated from the velocity record (obtained by numer-

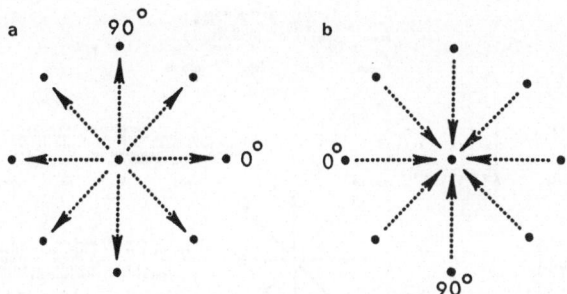

Fig. 1a,b. Schematic diagram of the tasks used. a Center → out task;
b out → center task

ical differentiation of the position data) using a threshold cross and
an iteration procedure that checked for aberrant velocity values.   A
comparison with acceleration records revealed that this threshold was
very close to the moment of abrupt change in the acceleration.   Then,
the   presence   of   a significant change in cell discharge, as compared
with the background activity preceding the target onset, was   assessed
and its latency determined, as described before (see Fig. 1 in Georgopoulos et al. 1982).   The cells that showed changes in discharge during the RT were selected for further study.   Finally, variation in the
frequency of discharge in these cells with the direction of   movements
was   evaluated   using   an   analysis   of   variance.   In the out → center
task, a relation to the movement end point would   be   indicated   by   a
lack of significant variation of cell discharge with the eight directions of movement, for all the movements ended in the same point;   on
the   other   hand, the presence of such variation would suggest a relation to the direction of movement.

Results

A total of 185 cells in the motor cortex (three monkeys) and 128 cells
in area 5 (two monkeys) were studied.   These cells were active in the
tasks used and were related to movements of the contralateral arm,   as
determined by examination of the animals outside the behavioral task.
They belong to a larger population of cells whose relations to the
direction of movement were described in previous papers (Georgopoulos
et al. 1982; Kalaska et al. 1983).   The entry points of the penetrations into the motor cortex are shown in Fig. 3 of Geogopoulos et

178

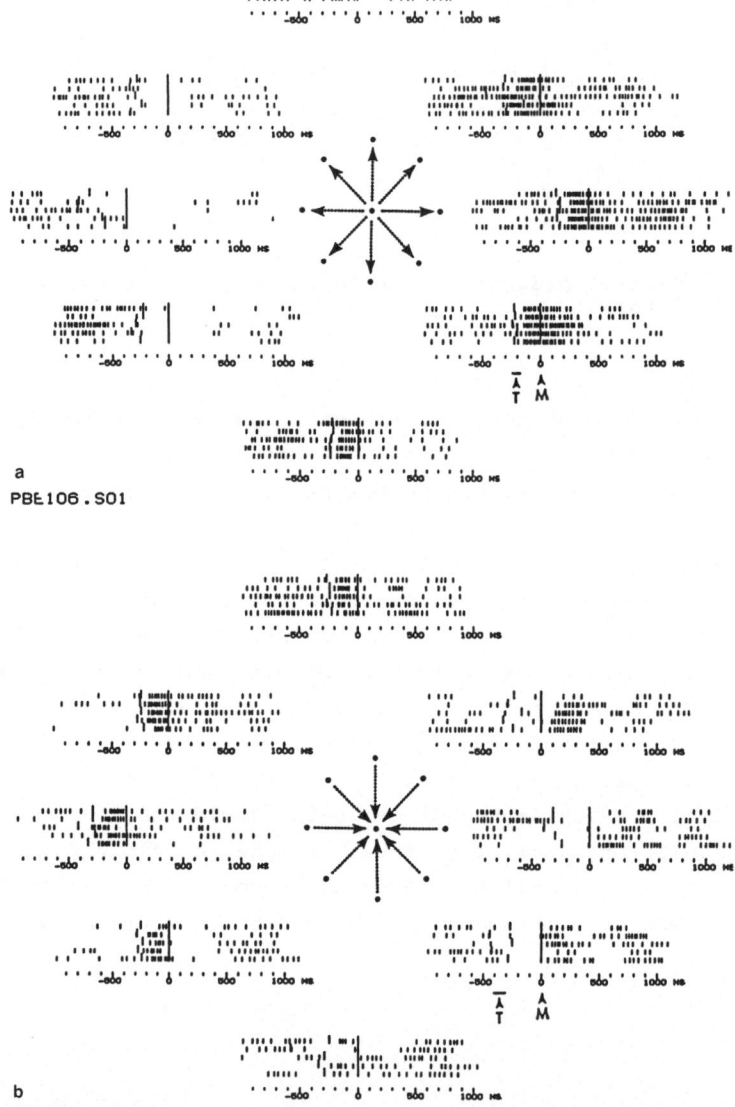

a
PBE106.S01

b
PBE106.S02

Fig. 2. Impulse activity of a cell in motor cortex during the center → out **(a)** and out → center **(b)** task. The direction of movement is indicated in the diagrams in the center. Rasters are repetitions of 5 trials in each direction and are aligned to movement onset *(M)*. *T with bar*, approximate time of target onset. This stimulus onset is shown for every trial as a *longer bar* preceding movement onset. Movements ended at about 500 ms after their onset

al. (1982) (the three right hemispheres); those of penetrations into
area 5 are shown in Fig. 2 of Kalaska et al. (1983).

Of the cells studied, 154/185 (83%) in the motor cortex and 55/128
(43%) in area 5 showed significant changes during the RT, as compared
with their activity during the prestimulus period. In 151/154 (98%)
of those cells in the motor cortex and 54/55 (98%) cells in area 5 the
discharge rate during the RT varied significantly with the direction
of movement, as determined by the analysis of variance (F test,
$p < 0.05$). Examples from the two areas studied are shown in Figs. 2
and 3. In b of both figures, raster activity in the out → center task
is shown from five trials per each movement direction. It can be seen
that cell discharge varied in an orderly fashion with movement direction, so that discharge rate was highest with movements in a preferred
direction and decreased gradually with movement made farther and
farther away from that preferred one, as described before (Georgopoulos et al. 1982; Kalaska et al. 1983). This directionality was
preserved in the center → out task, shown in Figs. 2a and 3a.
Figure 4 shows data from another cell in the form of directional tuning curves. The frequency of discharge during the RT is plotted
against the movement direction following the conventions of Fig. 1.
Filled circles and solid lines are from the center → out task; open
circles and interrupted lines are from the out → center task. Notice
the orderly variation of cell discharge during the RT with the direction of the upcoming movement and the similarity of the directionality
in both tasks used.

In three cells in motor cortex and in one cell in area 5, an increase
of discharge rate was observed of the same strength for all movement
directions in the out → center task. This could suggest an invariance
possibly related to the movement end point. However, these cells
showed a uniform increase in discharge in the center → out task as
well, in which the movement end points differed. This indicates that
these nondirectional changes in cell activity did not relate to the
end point of movement, but to some other factor.

Discussion

A salient finding of this study is that the frequency of cell discharge in the motor cortex and area 5 preceding the movement relates

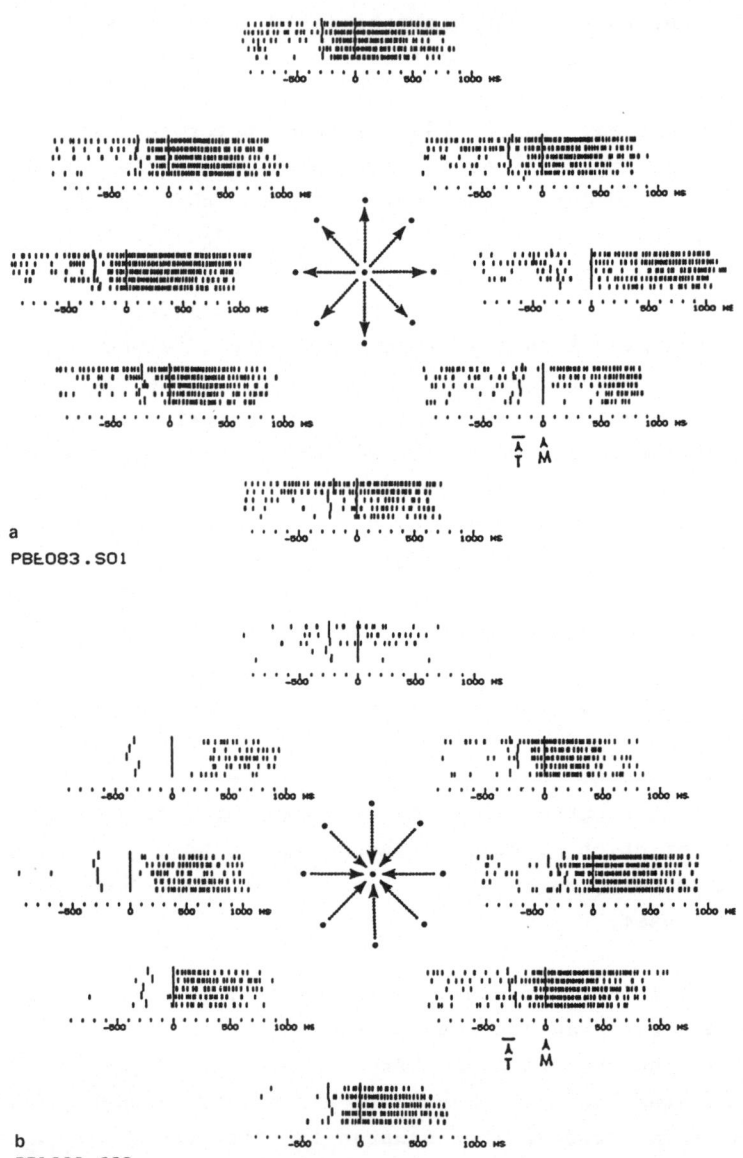

a

PBŁO83.SO1

b

PBŁO83.SO2

Fig. 3a,b. Impulse activity of cell in area 5 during the tasks used. Conventions as in Fig. 2

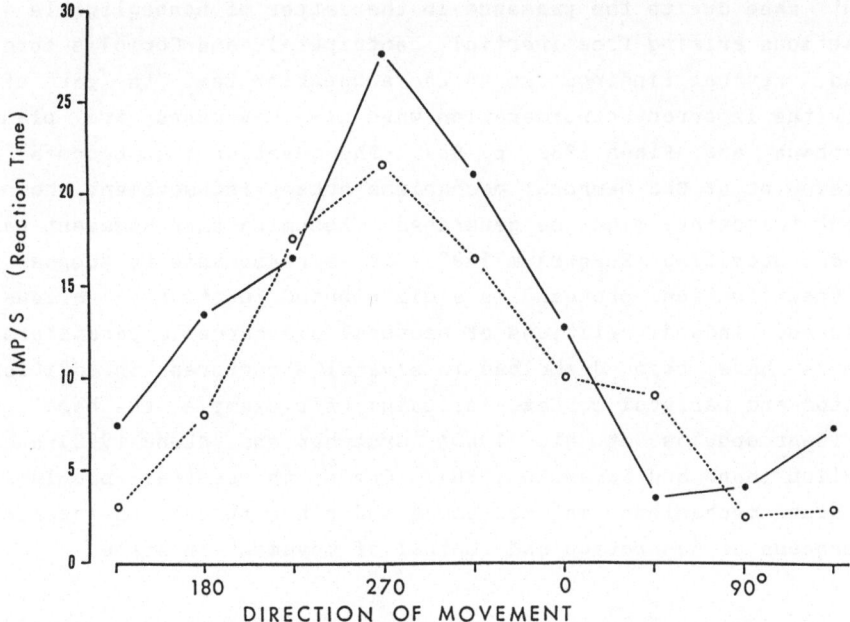

Fig. 4. Directional tuning curves during the reaction time for a different motor cortical cell. *Filled circles* and *solid lines*, data from the center → out task; *open circles* and *interrupted lines*, data from the out → center task. The variation in cell discharge with the direction of movement was similar in both tasks

to the direction of the upcoming movement and not to its end point. This strengthens the hypothesis proposed on the basis of experiments in which the direction of movement and its end point were confounded, namely, that these cortical areas are concerned with the control of movement direction (Georgopoulos et al. 1982; Kalaska et al. 1983). Moreover, the similarity of findings in motor cortex and area 5 suggests that these areas may process directional information in a similar way (Kalaska et al. 1983).

Overall, the present findings support the idea that central motor mechanisms may program the upcoming movement as a set of spatial parameters, as proposed by Rosenbaum (1980). This, in turn, suggests a direct control of the movement trajectory in space and supports a recent notion put forth by Hollerbach and Flash (1982) on theoretical grounds, namely, that the trajectory of movement may indeed be the controlled variable in multijoint movements. These workers pointed out that strategies developed for single-joint movements (e.g., end point control) (see Sakitt 1980) may not be generalizable to the multijoint case due to the presence in the latter of nonnegligible joint interactions arising from inertial, centripetal, and Coriolis torques. Instead, several findings led to the suggestion that "the path of the hand is the important consideration when arm movements are planned" (Hollerbach and Flash 1982, p. 68). The question then becomes, from the viewpoint of CNS neuronal mechanisms subserving movement, how the movement trajectory might be generated. Assuming that movement parameters are specified (Rosenbaum 1980), it is reasonable to suppose that this specification process is a distributed function of various CNS structures. Indeed, relations of neuronal discharge to parameters of movements have been described in several structures, in addition to the motor and parietal cortex, including, for example, the basal ganglia (Georgopoulos et al. 1983; Crutcher and DeLong 1983) and the cerebellum (Mano and Yamamoto 1980). One of the central problems in CNS motor mechanisms is how these and other structures interact in this process of generation and control of movement in space.

Acknowledgements. This research was supported by United States Public Health Service Grants NS17413 and NS0226, which we gratefully acknowlegde. J.F. Kalaska was a Postdoctoral Fellow of the Medical Research Council of Canada, 1978-1981. R. Caminiti was a CNR-NATO Fellow, 1980-1981 and a Howell-Cannon Foreign Scholar, 1981-1982.

References

CRUTCHER MD, DeLONG MR (1983) Single cell studies of the primate putamen. II. Relations to direction of movement and pattern of muscular activity. Exp Brain Res (to be published)

GEORGOPOULOS AP, KALASKA JF, MASSEY JT (1981) Spatial trajectories and reaction times of aimed movements: effects of practice, uncertainty, and change in target location. J Neurophysiol 46: 725-743

GEORGOPOULOS AP, KALASKA JF, CAMINITI R, MASSEY JT (1982) On the relations between the direction of two-dimensional arm movements and cell discharge in primate motor cortex. J Neurosci 2: 1527-1537

GEORGOPOULOS AP, DeLONG MR, CRUTCHER MD (1983) Relations between parameters of step-tracking movements and single cell discharge in the globus pallidus and subthalamic nucleus of the behaving monkey. J Neurosci 3: 1586-1598

HOLLERBACH JM, FLASH T (1982) Dynamic interactions between limb segments during planar arm movements. Biol Cybern 44: 67-77

KALASKA JF, CAMINITI R, GEORGOPOULOS AP (1983) Cortical mechanisms related to the direction of two-dimensional arm movements: relations in area 5 and comparison with motor cortex. Exp Brain Res 51: 247-260

MANO NJ, YAMAMOTO KI (1980) Simple-spike activity of cerebellar Purkinje cells related to visually guided wrist tracking movement in the monkey. J Neurophysiol 43: 713-728

POLIT A, BIZZI E (1979) Characteristics of motor programs underlying arm movements in monkeys. J Neurophysiol 42: 183-194

ROSENBAUM DA (1980) Human movement initiation: specification of arm, direction and extent. J Exp Psychol [Gen] 109: 444-474

SAKITT B (1980) A spring model and equivalent neural network for arm posture control. Biol Cybern 37: 227-234

Differences of Neuronal Responses in Two Cortical Motor Areas in Primates

J. Tanji

Department of Physiology, Hokkaido University, School of Medicine, Sapporo 060, Japan

Although the presence of motor areas rostral to the precentral motor cortex of primates has lang been known from the work of Horseley and Schaefer (1888) and the Vogts (Vogt and Vogt 1919), Penfield and Welch (1949, 1951) were the first to distinguish a secondary or supplementary motor area (SMA), situated in the superior frontal gyrus on the medial wall of the hemisphere and in the upper bank of the cingulate sulcus, from the primary area in the precentral gyrus. Their studies and subsequent reports suggest that SMA is involved in a different way in neural motor control mechanisms than the precentral motor cortex (PCM). Electrical stimulation of SMA produced complex, synergistic movements of the contralateral limbs and body, with occasional ipsilateral limb movements (Bancaud and Talairach 1967; Penfield and Welch 1951). Cessation of ongoing movements or speech was also observed on stimulating this area. This contrasted with the results found for the primary motor area in the precentral gyrus, where focal stimulation elicited movements localized in limited portions of the contralateral body parts. Studies of movement disturbance after ablation of SMA have shown that the effects are subtle, often transient, with no apparent paralysis or paresis, again contrasting greatly with effects of PCM lesions. In man, unilateral large lesions produced little or no lasting deficits (Laplane et al. 1977; Penfield and Jasper 1954; Waltregny 1972); Erickson and Woolsey (1951) observed a transient grasp reflex. Penfield and Jasper (1954) also noted a tendency for forced grasping in their patients along with slowness of movements of the contralateral limbs. In the monkey, Travis (1955) found that unilateral ablation of SMA gave rise to a transient grasp reflex and a bilateral hypertonia. Coxe and Landau (1975), however, did not observe persistent changes in muscle tone or posture.

Experimental Brain Research, Suppl. 10
© Springer-Verlag Berlin · Heidelberg 1985

In order to establish how the two cortical motor areas take part in various aspects of motor performance, single-cell recordings were made from both SMA and PCM of individual monkeys; and neuronal responses were compared during the performance of both simple and somewhat complex motor tasks.

Comparison of Responses Related to Simple Wrist Movements

A number of reports have demonstrated that a group of SMA neurons are active in association with limb movements and that some of these cells change their activity prior to the onset of movement (Brinkman and Porter 1979; Tanji and Kurata 1979; Smith 1979; Wise and Tanji 1981), just like PCM neurons. Thus, even with simple movements, it is a mistake to conclude that SMA takes no part in the execution of distal limb movements. These findings necessitated an investigation of how neuronal activity in the two motor areas differs in relation to the execution of simple movements. A series of experiments were designed in our laboratory to quantitatively compare the neuronal responses in the two areas. Monkeys performing wrist movements in response to sensory signals of three different modalities were used.

Monkeys were trained to perform alternate flexions and extensions at the wrist in response to a visual, an auditory, or a somatosensory triggering signal, which was presented to the animal in a randomized sequence. The visual signal was a red light-emitting diode (LED) placed in front of the animal. The auditory signal was a tone burst of 1 kHz, with an intensity of 30 dB above background noise level. The tactile signal was a vibratory stimulus applied to the monkey's hand. The vibration was generated by a servo-controlled DC torque motor driven by a burst of square wave oscillation at 500 Hz. The reward was given when the correct response was initiated within 400 ms after the onset of the triggering signals.

A representative example of neuronal activity in the precentral motor cortex is shown in Fig. 1a. The neuron, identified as a pyramidal tract neuron, was active with wrist extension, but not active with flexion. In the left column of the figure, discharges are aligned at the onset of the signals calling for wrist extension. Neuronal response latency was calculated on the basis of the peristimulus time histogram. The latencies to the visual, auditory, and tactile signals

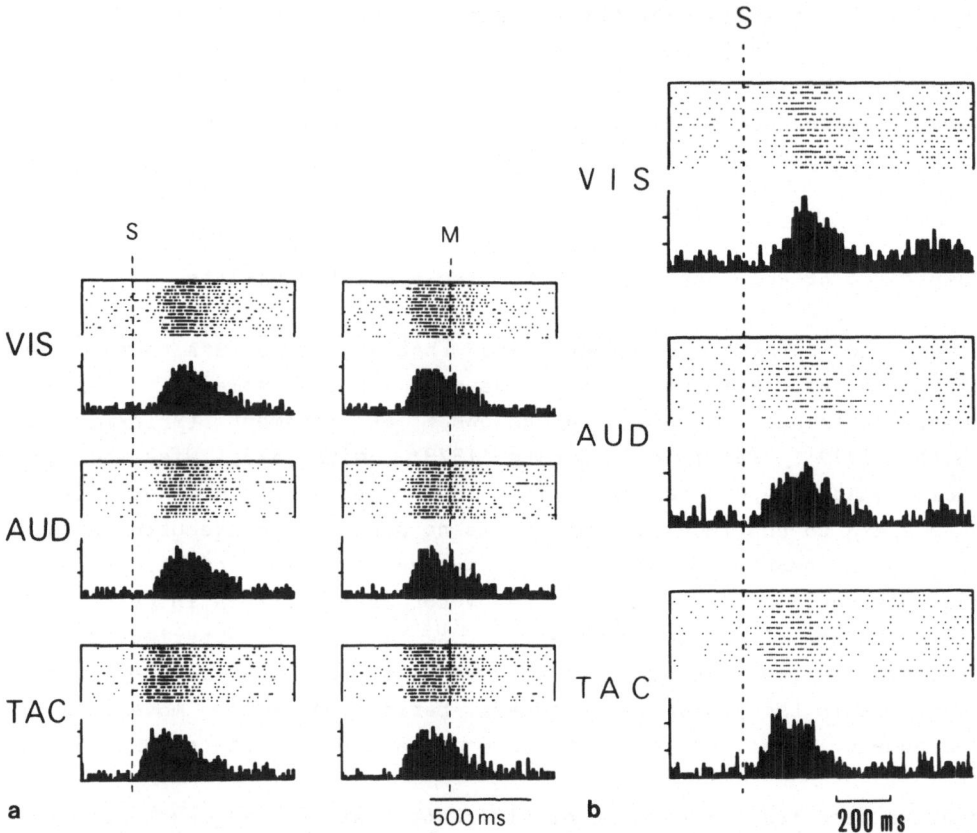

Fig. 1. a Discharges of a neuron in the precentral motor cortex, whose activity increased with wrist extension. In the *left* column, discharges are aligned at the onset of visual, auditory, and tactile signals *(dotted line* labeled *S)*. In the *right* column, discharges are aligned at the start of the wrist extension movement *(M)*. A step on the ordinate scale of the perievent histogram denotes 64 impulses/s. **b** Discharges of an SMA neuron, whose activity increased with wrist extension. A step in the ordinate scale of the histogram denotes 40 impulses/s. *VIS, AUD,* and *TAC* indicate responses following visual, auditory, and tactile trigger signals, respectively

were 122, 100, and 40 ms, respectively. These values correspond well to those obtained in previous reports (Evarts 1966, · 1973). In the right half of Fig. 1a, discharges are aligned using the onset of wrist extension detected by the potentiometer. The discharge pattern is representative of PCM neurons in that the discharge increase is simi- lar, regardless of the modality of the triggering sensory signal. The activity of the neuron resembled that of EMG activity recorded in the forearm, except for earlier onset of activity changes.

A representative example of SMA neurons which showed distinct movement-related activity is shown in Fig. 1b. The activity of this SMA neuron is indistinguishable from that of precentral neurons, ex- cept that the magnitude of activity change is smaller, and the response latencies to auditory and visual signals are less. Out of 1050 neurons recorded from SMA, 399 (38%) responded in association with the wrist movements. This percentage is much smaller than in PCM, where 212 out of 262 neurons (81%) responded with the movements. In 124 out of the 399 movement-related SMA neurons, the magnitude of activity was different, depending upon the modality of the signal to which the animal responded. Such an example is shown in Fig. 2. The neuron was more active in relation to wrist extension triggered by an auditory signal and less responsive to a visual signal. It was almost unresponsive when the triggering signal was tactile.

Another property of SMA neurons differentiating them from precentral motor cortex neurons is the occurrence of responses time-locked to the visual or auditory signal. In the PCM, no neurons were found to respond with a close time relation to the visual or auditory signals. The finding is in accordance with previous reports, although the pres- ence of precentral neurons driven by somatosensory input is well known (Evarts 1973; Lamarre et al. 1978). However, some SMA neurons, like the one shown in Fig. 3, responded with a discharge time-locked to the onset of the sensory signal. The discharge in this neuron increased 84 ms after the onset of the visual signal. There were 68 SMA neurons exhibiting discharges time-locked to the visual and/or auditory sig- nal. A class of SMA neurons was activated by the tactile signal, but the magnitude of such somatosensory-evoked responses was smaller than that observed in the motor cortex.

The magnitude of movement-related activity in the two motor areas was compared quantitatively as follows. In individual neurons, the change of discharge frequency from the background level of discharge was cal-

188

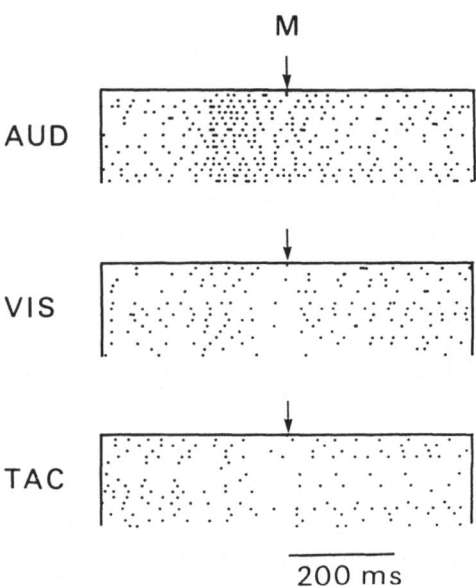

M

AUD

VIS

TAC

200 ms

Fig. 2. Discharges of an SMA neuron whose activity is different, depending on the modality of the signals calling for wrist extention. *M* indicates movement onset

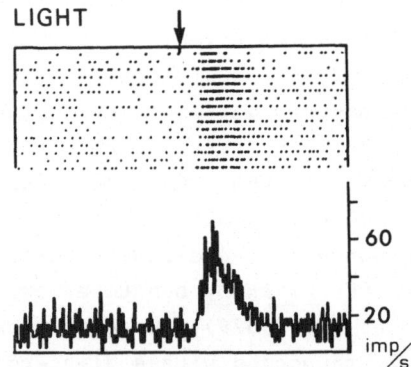

LIGHT

60

20
imp/s

Fig. 3. Discharges of an SMA neuron exhibiting temporal locking of activity increase to the onset of the visual signal *(arrow)*

culated during the 200-ms premovement period for each movement triggered by the three different signals. In visually triggered movement,
the mean magnitudes for SMA neurons and precentral neurons were 32 and
53 impulses/s, respectively. The difference was significant with
$p < 0.00001$. Similar differences were observed in movements triggered
by auditory and tactile signals. The latencies of neuronal responses
to the three signals were also compared. Mean latencies of SMA neurons and PCM neurons are shown in Table 1. The latencies of SMA neurons to visual and auditory signals were significantly shorter than
those of precentral neurons. For the tactile signal the reverse was
true, though the difference was much smaller. Finally, the extent to
which the neuronal activity is coupled to motor output was compared.
The correlation between the onset of neuronal response to a visual or
auditory signal and the onset of wrist movement was calculated from
the data obtained in individual trials. The relationships for most of
the precentral neurons had Pearson correlation coefficients of more
than 0.6. Only 24% of SMA neurons had coefficients of more than 0.6.
Of the 23 precentral neurons thus analyzed, 20 had a correlation coef-
ficient of statistical significance, but only 12 of 39 SMA neurons had
significant coefficients. Apparently, the onset of neuronal activity

Table 1. Neuronal response times to sensory signals triggering wrist
movements

	Modality of triggering signal		
	Visual	Auditory	Tactile
Supplementary motor area	128	108	68
Precentral motor cortex	157	136	59

Values are means in milliseconds of the intervals between the onset of
sensory signals and the onset of neuronal activity changes

in the SMA is not as well correlated with movement onset as that in the PCM.

Judging from the smaller magnitude of the movement-related activity, and the lack of strong correlations of neuronal activity with movement onset, SMA seems to be more remote from the peripheral motor apparatus than PCM. Thus, for simple movements like the one performed in the present experiments, SMA plays some subsidiary role in motor control. The functional significance of the close relationship of SMA neurons to visual or auditory signals remains to be studied.

Comparison of Neuronal Responses to Motor Instructions

Since the SMA has only a minor part in the execution of simple movements, we must look for aspects of motor performance in which the area plays a more significant role. Recent studies in man and monkey suggest a role for SMA in generating a preparatory state for, or even in programming, forthcoming movements. The readiness potential, a potential preceding the initiation of movement by several hundred milliseconds, is known to have its maximum at the vertex (Kornhuber and Deecke 1965). The vertex maximum of the potential persisted without reduction even in patients with parkinsonism who had no such potential in the precentral region (Deecke and Kornhuber 1978). Since the location of the vertex electrode was presumed to be on SMA, this finding was taken to suggest a role for SMA in the process of developing a preparatory state for the impending movement. A technique of measuring regional blood flow in the cortex provided evidence that SMA was active when subjects were internally simulating, but not performing, a motor sequence (Roland et al. 1980). Orgogozo and Larsen (1979) also found that SMA exhibited much higher blood flow during the execution of complex movement sequences than during simple muscle contractions. These observations led them to suggest a role for SMA in programming the motor sequence.

In a subsequent study (Tanji et al. 1980) an attempt was made to observe how neuronal activity in SMA altered with development of a preparatory state for an impending movement in primates. The motor task required the animal to respond properly to a sudden perturbing stimulus delivered to the forearm in two different directions. An instruction as to the direction of the animal's movement, pushing or pulling the forelimb, was given several seconds before the occurrence of the stimulus. A number of SMA neurons exhibited instruction-induced changes of activity during the period intervening between the instruction and the perturbation-triggered movement. These particular neurons were not active in association with the movement itself. Such neuronal activity to instructions specifying the direction of the forthcoming movement, however, is commonly found in PCM (Tanji and Evarts 1976) and in the premotor area (Weinrich and Wise 1983) as well.

In what behavioral paradigm, then, is SMA more active than PCM? Our recent findings provide, at least in part, answers to this question. In a preliminary experiment, a monkey was trained to press a key in two different modes. In one mode it was instructed to respond to a tactile vibratory signal given to its right hand and to press a key with the same hand. In the second mode the animal was instructed not to respond to the same vibratory stimulus, but to remain motionless until a tone burst of 1000 Hz was given a few seconds later. Only then was the monkey required to respond with the key press. The first mode was indicated by an auditory signal of 100 Hz and the second mode by one of 300 Hz. The interval between the instruction onset and the vibration onset was varied from 2 to 5 s. Out of 122 task-related neurons in SMA, 70 responded to instructions. Of these 70, 40 exhibited different responses to the two different instructions, like the examples shown in Fig. 4. The SMA neuron displayed in Fig.4a increased its discharge and the other neuron (Fig. 4b) decreased its discharge when the animal was instructed to ignore the tactile signal and wait for a subsequent signal. When the instruction was to respond to the forthcoming tactile signal and press the key, their discharge remained unaltered. No changes of activity were detected for these neurons in response to triggering signals or in relation to the key press movement. When instructed to ignore the tactile signal, 20 neurons changed their activity preferentially, and 13 neurons responded when instructed to react to the tactile signal. Seven SMA neurons showed reciprocal responses. Such selective responses were rare in PCM, where only two neurons showed slightly different responses to instructions. The findings indicate that the two motor areas respond differently to instructions telling the animal how to start the forthcoming movement.

In order to investigate whether SMA is more related than PCM to a process where the animal is prepared to selectively respond to a particu-

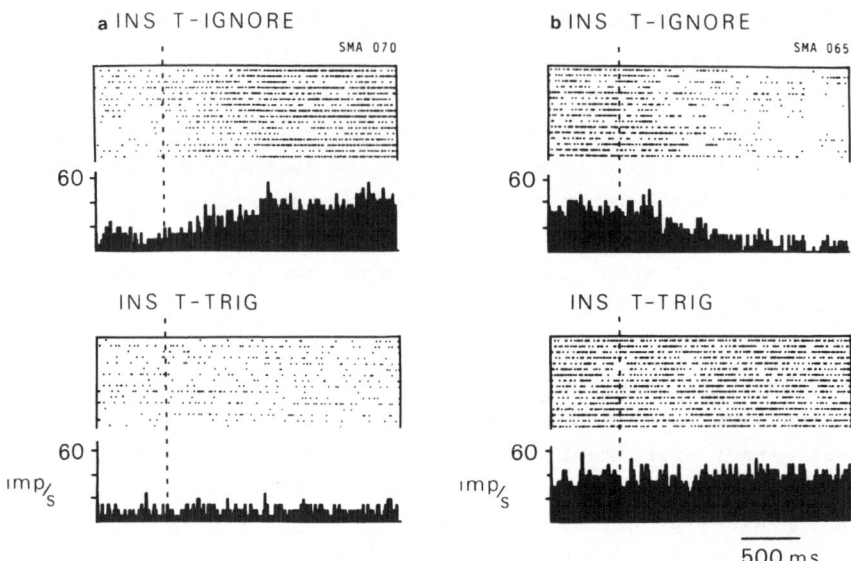

Fig. 4a,b. Examples of two SMA neurons exhibiting instruction-induced activity changes. These neurons responded selectively to the instruction not to start the movement in response to the tactile signal. *INS T-IGNORE,* instruction to ignore tactile signal; *INS T-TRIG,* instruction to respond to tactile signal

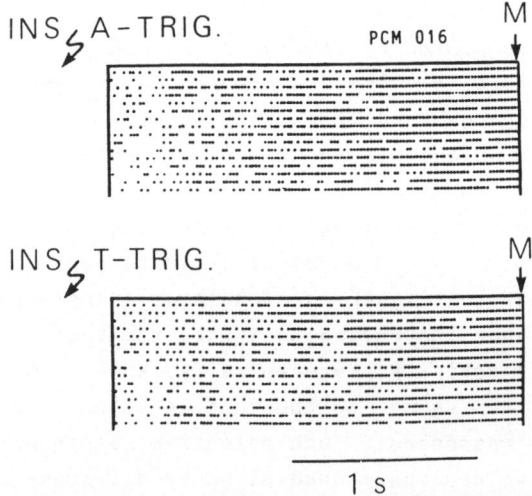

Fig. 5. An example commonly observed in the precentral motor cortex of nonselective neuronal responses to two different instructions. Gradual increase of activity developed toward the onset of movement indicated by *arrows M. INS A-TRIG,* instruction to respond to auditory signal; *INS T-TRIG,* instruction to respond to tactile signal

lar modality of sensory signals, the following series of experiments was performed using a more refined behavioral paradigm. Three monkeys were trained as follows. When an instruction was given by a 100 Hz sound signal, the monkey was required to press a pedal in response to a forthcoming vibratory signal of 40 Hz, but not to respond to a high-frequency tone burst of 1 kHZ. When the instruction was a 300-Hz sound signal, the reward contingency was reversed, requiring a response to the 1 kHz auditory signal, but not to the vibratory signal. The motor response was required to be commenced within 400 ms after the triggering signal. When required not to start the movement, the monkey had to remain perfectly motionless. If any hand movements were detected with a tension transducer, the trial was canceled and the next sequence started. The occurrence of movement-triggering signals and nontriggering signals was randomized.

A striking difference between instruction-induced responses in PCM and in SMA was observed. In PCM, only 22% of task-related neurons responded to instructions, and most of the responses to the two instructions were similar, regardless of whether the instruction called for a response triggered by a tactile or an auditory signal. Such an example is shown in Fig. 5. This type of nonselective instruction response may reflect a preparatory process for the key press movement or an elevated arousal state for the performance of the motor task. In contrast, 49% of task-related SMA neurons responded to instructions. Among them, 69% exhibited differential responses to two instructions. The example in Fig. 6 shows a neuron whose activity increased after the instruction to respond to the auditory signal and not to the tactile signal. When the instruction was the reverse, the neuronal activity remained unaltered.

Since in this paradigm the movement triggered by the auditory or tactile signal was exactly the same, the selective instruction response was not related to preparation for different movements. Instead, such a neuronal response was specific to the state where animals were prepared to respond selectively to signals of different modality. The finding seems to indicate that one of the motor functions in which SMA is more involved than PCM is the modality-specific preparatory process for sensory-triggered motor responses. SMA may be involved in many other motor functions, which are likely to be revealed with appropriate experimental approaches.

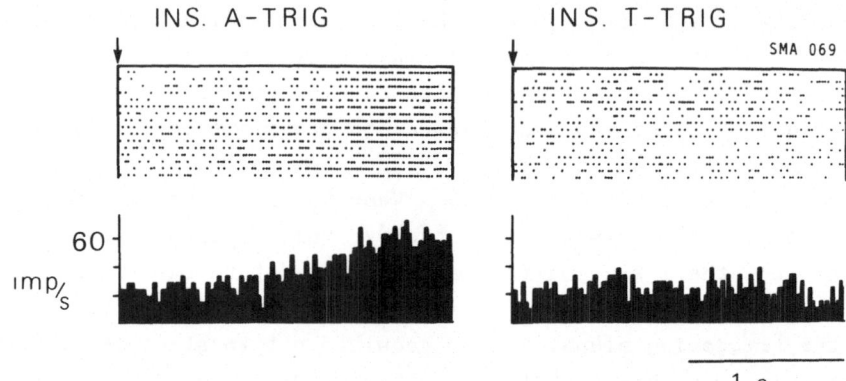

Fig. 6. An example often observed in SMA of selective neuronal responses to instructions. This neuron increased its discharge only when instructed to react to the auditory signal *(INS A-TRIG)* and not to the tactile signal *(INS T-TRIG)*

Acknowledgements. This work was supported in part by a grant-in-aid from the Scientific Research Fund from the Ministry of Education, Science and Culture, Japan.

References

BANCAUD J, TALAIRACH J (1967) Organisation fonctionelle de l'aire motrice supplémentaire. Endeignements apportés par la stereo EEG. Neurochirurgie 13: 343-356

BRINKMAN C, PORTER R (1979) Supplementary motor area in the monkey: activity of neurons during performance of a learned motor task. J Neurophysiol 42: 681-709

COXE WS, LANDAU WM (1975) Observation upon the effects of supplementary motor cortex ablation in the monkey. Brain 88: 763-773

DEECKE L, KORNHUBER HH (1978) An electrical sign of participation of the mesial 'supplementary' motor cortex in human voluntary finger movement. Brain Res 159: 473-476

ERICKSON TC, WOOLSEY CN (1951) Observation on the supplementary motor area of man. Trans Am Neurol Assoc 76: 50-56

EVARTS EV (1966) Pyramidal tract activity associated with a conditioned hand movement in the monkey. J Neurophysiol 29: 1011-1027

EVARTS EV (1973) Motor cortex reflexes associated with learned movement. Science 179: 501-503

HORSELEY V, SCHAEFER EA (1888) A record of experiments upon the functions of the cerebral cortex. Philos Trans R Soc Lond 179: 1-45

KORNHUBER HH, DEECKE L (1965) Hirnpotentialänderungen bei Wilkürbewegungen und passiven Bewegungen des Menschen: Bereitschaftspotential und reafferente Potentiale. Pflugers Arch Ges Physiol 284: 1-17

LAMARRE Y, BIOULAC B, JACKS B (1978) Activity of precentral neurons in conscious monkeys: effects of deafferentation and cerebellar ablation. J. Physiol (Paris) 74: 253-264

LAPLANE E, TALAIRACH S, MEININGER V, BANCAUD J, ORGOGOZO JM (1977) Clinical consequences of corticectomies involving the supplementary motor area in man. J Neurol Sci 34: 301-314

ORGOGOZO JM, LARSEN B (1979) Activation of the supplementary motor area during voluntary movement in man suggests it works as a supra-motor area. Science 206: 847-850

PENFIELD W, JASPER H (1954) Epilepsy and the functional anatomy of the human brain. Little, Brown, Boston, pp 88-102

PENFIELD W, WELCH K (1949) The supplementary motor area in the cerebral cortex of man. Trans Am Neurol Assoc 74: 179-184

PENFIELD W, WELCH K (1951) The supplementary motor area of the cerebral cortex. Arch Neurol Psychiatr 66: 289-317

ROLAND PE, LARSEN B, LASSEN NA, SKINHØJ E (1980) Supplementary motor area and other cortical areas in organization of voluntary movements in man. J Neurophysiol 43: 118-136

SMITH AM (1979) The activity of supplementary motor area neurons during a maintained precision grip. Brain Res 172: 315-327

TANJI J, EVARTS EV (1976) Anticipatory activity of motor cortex neurons in relation to direction of an intended movement. J Neurophysiol 39: 1062-1068

TANJI J, KURATA K (1979) Neuronal activity in the cortical supplementary motor area related with distal and proximal forelimb movements. Neurosci Lett 12: 201-206

TANJI J, TANIGUCHI K, SAGA T (1980) The supplementary motor area: neuronal responses to motor instructions. J Neurophysiol 43: 60-68

TRAVIS AM (1955) Neurological deficiencies following supplementary motor area lesions in *Macaca mulatta*. Brain 78: 174-201

VOGT C, VOGT O (1919) Allgemeinere Ergebnisse unserer Hirnforschung. J Psychol Neurol (Leipzig) 25: 279-462

WALTREGNY A (1972) L'épilepsie de l'aire motorice supplémentaire. Med Hyg 31: 815-816

WEIRICH M, WISE SP (1983) The premotor cortex of the monkey. J Neurosci 9: 1329-1344

WISE SP, TANJI J (1981) Supplementary and precentral motor cortex: contrast in responsiveness to peripheral input in the hindlimb area of the unanesthetized monkey. J Comp Neurol 195: 433-451

Activity of Areas 4 and 7 Neurons During Movements Triggered by Visual, Auditory, and Somesthetic Stimuli in the Monkey: Movement-Related Versus Stimulus-Related Responses

Y. Lamarre, G. Spidalieri, and C. E. Chapman

Université de Montréal, Faculté de médicine, Centre de recherche en sciences neurologiques, Départment de physiologie, Case postale 6128, Succursale A, Montréal, Québec H3C 3J7, Canada

Introduction

In this report, we wish to examine specifically the timing of unitary discharge in the motor cortex (area 4) in relation to elbow movement and to conditioning stimuli of different sensory modalities used to cue the movement. Single-unit recordings in monkeys have firmly established that the discharge frequency of a majority of area 4 neurons changes before the onset of a voluntary movement and that it can be correlated with some parameters of movement (see review by Evarts (1981). These observations strongly suggest that area 4 neurons are causally involved in initiating movements. However, the time at which area 4 neurons come into play during the period between the presentation of a triggering stimulus and the release of a command to move is still a matter of discussion. It seems reasonable to assume that neurons in which the activity is strongly dependent on the triggering stimulus are involved in the early stages of planning the movement, while those whose activity is mainly dependent on the movement are more involved in the final execution stage.

To study this we have trained monkeys to perform identical movements of the arm in response to three different conditioning stimuli (visual, auditory, and somesthetic) while recording from neurons of the contralateral precentral cortex. In analyzing the responses of these neurons, we have attempted to determine if their activity was mainly related to the performance of the movement or mainly related to the sensory cues. We have also investigated the responses of motor cortex

Experimental Brain Research, Suppl. 10
© Springer-Verlag Berlin · Heidelberg 1985

neurons when the reaction time (RT) was lengthened following a lesion of the contralateral dentate nucleus of the cerebellum. For comparison, neurons were also recorded in area 7 in the same experimental conditions. It was found that neurons in area 4 showed movement-related responses, in contrast with the stimulus-related responses displayed by neurons in area 7.

Methods

The experiments were performed on two adult monkeys (*Macaca mulatta*) seated in a primate chair. Single-unit recordings were obtained in the cerebral cortex (areas 4 and 7). The upper arm was extended horizontally from the shoulder at an angle of about 25° in front of the coronal plane and lay comfortably in a shallow trough which was hinged about the elbow joint. During training sessions the animal learned to perform flexion and extension movements of the forearm in response to each of the stimuli. Three conditioning stimuli were randomly alternated: a light (light-emitting diode (LED) unless otherwise indicated), a tone, and small torque applied at the elbow, in either the flexion or extension direction, via a torque motor. The displacement produced by this pulse was usually less than 1°. Fruit juice rewards were delivered when movements of sufficient amplitude (15° - 30°) in the required direction were initiated within 500 ms. There were no reference points for starting or stopping.

After training, the animals were anesthetized with pentobarbital and a cylindrical plastic chamber was attached to the skull under aseptic conditions to allow microelectrode recordings to be made (Lamarre et al. 1970). Multistranded stainless steel wires insulated with Teflon were implanted into selected arm and neck muscles and passed under the skin to a multichannel connector attached to the skull. Electrooculograms (EOG) were also recorded. Glass-coated tungsten microelectrodes were used to record neuronal activity during the performance of the tasks. The outputs of the angular displacement transducer, EMG electrodes, and microelectrode were amplified in a conventional manner and displayed on an oscilloscope. Selected data were recorded on multichannel magnetic tape or directly on photographic paper. On-line data acquisition was performed by a PDP-9 computer, which was also programmed to control the task. Neural spike intervals, pulse replicas of the EMG activity (Evarts 1974), and the angu-

lar displacement (digitized at 200 Hz) were stored on Dec tapes.
Routinely, the onset of changes in neural and muscle activity was calculated both with respect to the onset of the stimulus (RS) and to the
onset of elbow displacement (RM). Figure 1 illustrates these measurements. In the case of units that change their activity before the
onset of movement, the sum of RS and RM is equal to the RT, i.e., the
time from stimulus onset to movement onset.

At the conclusion of the last recording session, the monkeys were
killed with an overdose of pentobarbital. The dura mater was removed
from the depth of the recording chamber and reference points, at known
stereotaxic coordinates, were marked on the surface of the cortex by
penetrations of a microelectrode dipped in India ink. A photograph
was made of the surface of the brain and the coordinates of the point
of entry of each microelectrode track were marked upon the photograph.
A more detailed description of the methods can be found elsewhere
(Lamarre et al. 1983).

Results

Detailed analyses were performed on a total of 365 task-related neurons recorded in the arm area of the precentral cortex (area 4) in the
two monkeys and 72 neurons recorded in area 7 of one animal before and
after lesioning the contralateral dentate nucleus. The effects of the
dentate lesion on the movement and precentral activity have been described in detail elsewhere (Spidalieri et al. 1983).

Figure 2a illustrates the discharge of a motor cortical neuron in relation to flexion of the contralateral elbow. Ninety movements were

Fig. 2a. Rasters of the activity of a motor cortex neuron aligned with
respect to onset of flexion *(left)* and to stimulus onset *(right)*.
Movements with identical average parameters were evoked by light *(L)*,
sound *(S)*, and somesthetic *(M)* inputs. Trials are arranged in order
with respect to reaction time. On the *right*, the *irregular lines* indicate the time of onset of movement. *Arrows* mark the mean reaction
time for each stimulus, given in ms. The two *dots* mark the latency of
the sensory responses to sound (25 ms) and somesthetic stimuli
(35 ms). Note the temporal dissociation between the sensory and
"voluntary" motor responses.
b Plots of onset of activity and reaction time for the unit shown in
a. On the *left*, the time of onset of unit response after each
stimulus *(RS)* is plotted on the *ordinate* as a function of the reaction
time *(RT)*. On the *right*, the time between onset of unit response and
onset of movement *(RM)* is plotted as a function of RT. For both
plots, the corresponding linear regression is also illustrated

199

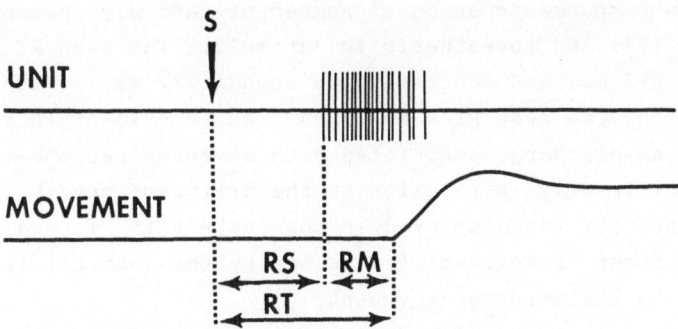

Fig. 1. Time relationships between conditioning stimulus *(S)*, onset of firing change in motor cortical neuron, and onset of movement. Reaction time of movement *(RT)* corresponds to the sum of the latency of the cell response *(RS)* and the time between the onset of firing change in the cell and the onset of movement *(RM)*

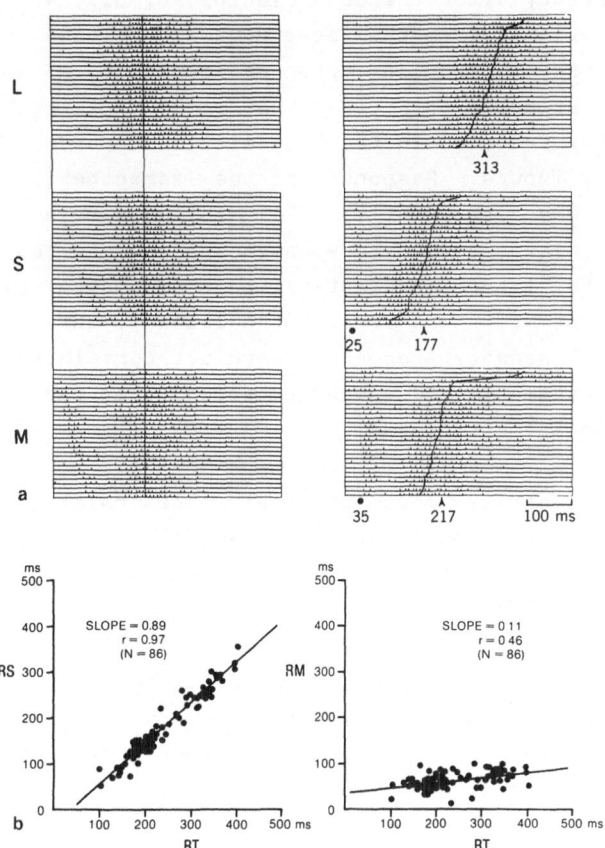

initiated in response to an equal number of randomly presented light (L), sound (S), and somesthetic (M) stimuli. The mean RT was longest with light (313 ms) and shortest with sound (177 ms). With somesthetic stimulation, the mean RT was 217 ms. As is evident in this figure, the pattern of discharge associated with movement responses does not change significantly in relation to the different conditioning stimuli. This activity (as displayed in the rasters in Fig. 2a, left) preceded the onset of movement (indicated by the vertical line) and was time-locked to the onset of movement.

In addition to changes in firing frequency related to the motor responses, this neuron also responded directly to the sensory triggering stimuli. There was a sensory response to sound with a mean latency of 25 ms and a stronger response to somatic stimulation of the elbow after a mean latency of 35 ms. These sensory responses occurred at a fixed latency after the conditioning stimuli, and they were most easily differentiated from the later movement-related responses when the trials were ranked in order of increasing RT and aligned with respect to the stimulus onset (Fig. 2a, right). Altogether, about 40% of the neurons recorded in the two animals responded directly to one or more sensory inputs. Responses to the somesthetic stimulus were observed the most frequently (37%), followed by auditory (11%) and visual (3.5%). These sensory responses appeared to be totally independent of the later movement-related responses, except when they merged at a short RT. Furthermore, the latency, duration, and intensity of these sensory responses were not correlated with RT or any parameters of movement, and whenever tested, they persited after extinction of the motor response.

In order to quantify more precisely the relation between the movement-related responses and the reaction time, we have measured, for each individual trial, the onset of changes in neural activity, both with respect to the onset of the stimulus (RS) and to the onset of elbow displacement (RM), and plotted these values against the RT. Results obtained in this way for the unit of Fig. 2a are shown in Fig. 2b. On such graphs, a unitary response which is closely related with the motor response (i.e., onset latency time-locked to the motor response and varying with the RT) would show a linear correlation with a slope of 1.0 when RS is plotted against RT and a horizontal line (slope of 0) when RM is plotted against RT. The discharge of the unit illustrated in Fig. 2a is clearly related to the motor response, since

the RS-RT regression line has a slope of 0.89 with correlation coeffi-
cient of 0.97, while the RM-RT plot remains uncorrelated with a slope
of 0.11. This unit led the motor responses by a constant time of
about 60 ms. Similar relations have been observed for all cells analyzed in this way. Figure 3a shows the distribution of the RS-RT regression line slopes for 152 units with clearly excitatory responses
and low spontaneous activity. The histogram shows a prominent mode
equal to the expected theoretical value of 1 and a mean of 0.94. All
of these cells had significant correlation coefficients, the distribution of which is shown in Fig. 3b. Similar results were also obtained
with units showing inhibitory responses. The shaded portion of the
histograms represents values for units recorded after lesioning the
dentate nucleus, which produced a delay of movement initiation in
response to visual and auditory signals (Spidalieri et al. 1983).
These values are essentially identical to those obtained from units
recorded before lesion. The question remains, however, whether the
increase in RT observed after dentate lesion can be explained by a
lengthening of RS, RM, or both. We therefore compared the RM and RS
values of the two neuronal populations recorded in the same cortical
regions before and after dentate lesion. Figure 4 illustrates the
distribution of these two sets of values for sound-triggered flexions
in one animal. The hatched histograms represent values obtained before the lesion and the dotted ones, after the lesion. A statistically significant increase ($P < 0.001$) was obtained for the RS values
(Fig. 4c) after the lesion of the dentate, while there was no significant change in the RM values (Fig. 4b). In Fig. 4d we have plotted
the sum of individual RS and RM values (ordinate) as a function of the
corresponding RT. If the delay in RT was totally accounted for by a
delay in the latency of the movement-related responses (RS), the sum
of RS and RM being equal to RT, all of the points would be expected to
fall on a single regression line having a slope of 1 and 0 intercept.
This is, in fact, what was observed when the data obtained before
(filled circles) and after (open circles) the lesion of the dentate
were plotted together.

Further evidence in favor of the view that area 4 neurons in the monkey are better related to the motor responses that to the conditioning
sensory inputs comes from the following experiments. When visual and
auditory stimuli were presented together, the intensity and the duration of both motor and unitary responses remained unchanged, as if
only one stimulus had been given. Figure 5 compares the discharge of

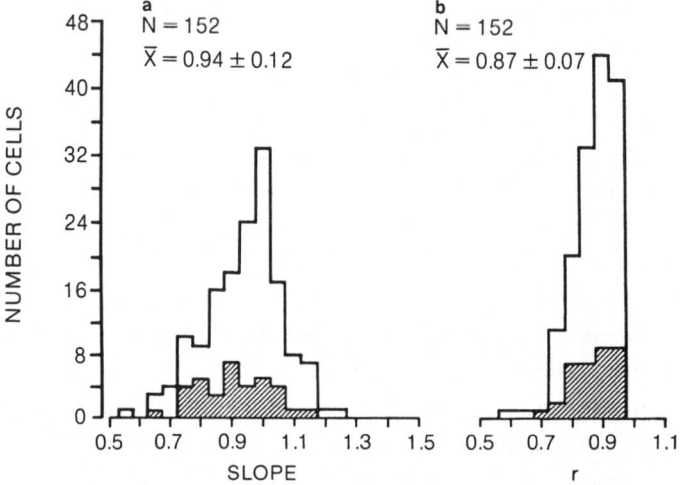

Fig. 3. Frequency distribution of slopes for RS-RT linear regressions
(a) and correlation coefficients **(b)** for 152 area 4 units recorded before *(white)* and after *(shaded)* a contralateral lesion of the dentate

SOUND

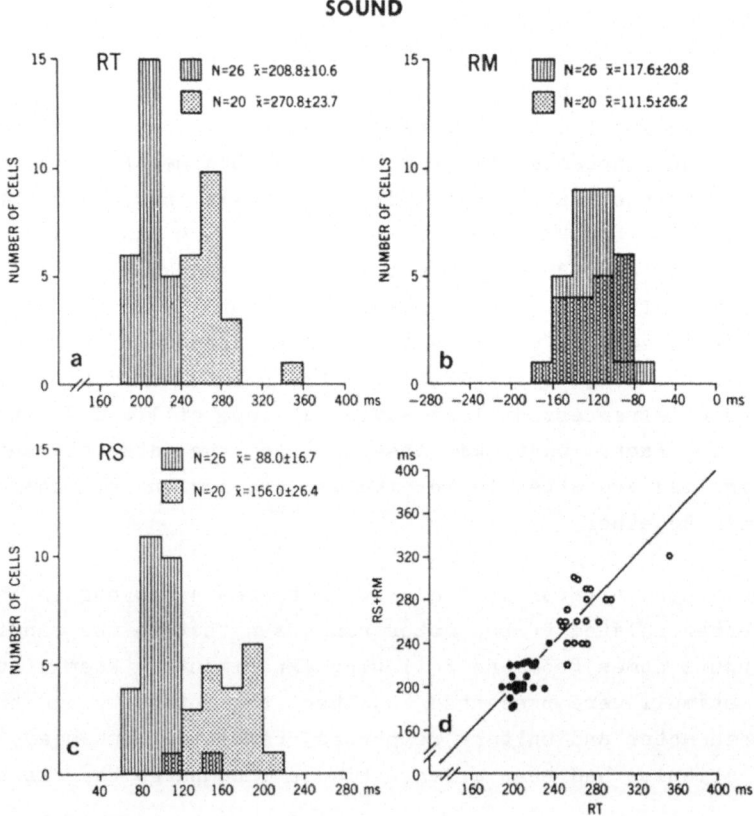

a neuron in response to randomly presented stimuli: light (L), sound (S), and light and sound (L+S) together. With the light, the RT and the latency of the cell discharge were, on the average, 80 ms longer than with sound (Fig. 5a,b). When both stimuli were presented simultaneously, the mean RT and the timing of the cell discharge were about the same as that observed with sound alone (Fig. 5c). This can be compared with the stronger and longer neuronal burst discharge in the theoretical histogram constructed by simple summation of L and S responses (Fig. 5d). It was of interest to compare these data obtained from neurons in area 4 with data obtained in the same experimental conditions from neurons in area 7. In this area, neurons showed responses which were dependent on both the sensory conditioning stimulus and the performance of the movement. A number were activated prior to arm movement about 100 ms after the light signal placed in the right contralateral (side of the moving arm), but not homolateral (left), visual field (Fig. 6a). Such responses were only present if a subsequent arm movement was made in response to the light stimulus and appeared to be independent of the eye movements made towards the light source (Fig. 6b). Figure 7 shows another area 7 neuron with response to both the visual and the auditory stimuli, which were clearly unrelated to the onset of movement (indicated by the oblique line). When both stimuli were delivered together, we observed a simple summation of the two responses (Fig. 7c). Finally, after lesioning the dentate nucleus, the latencies of the responses to light in area 7 were not modified, as was the case for neurons in area 4 (Fig. 8). All of these observations indicate that in area 7, the movement-dependent responses are stimulus-related, rather than movement-related, as in area 4.

Fig. 4. Histograms relating the corresponding mean RTs **(a)**, RMs **(b)**, and RSs **(c)** obtained from 26 *(striped)* and 20 *(dotted)* neurons recorded in the motor cortex contralateral to the operant arm of one monkey performing elbow flexions in response to sound stimuli before and after dentate lesion, respectively. *Abscissas*, time in ms after the onset of the conditioning signal **(a, c)** and with respect to onset of movement **(b)** referred to as time 0. *Ordinates*, number of cells. Mean values of the two neuronal samples are reported at the top of each histogram. As a control **(d)**, the sum of RS and RM is plotted on the ordinate as a function of the corresponding RT reported on the abscissa. As a reference, a line having a slope of 1 and 0 intercept has been traced. *Filled circles*, data from neurons recorded in the intact preparation; *open circles*, data from units recorded after dentate lesion

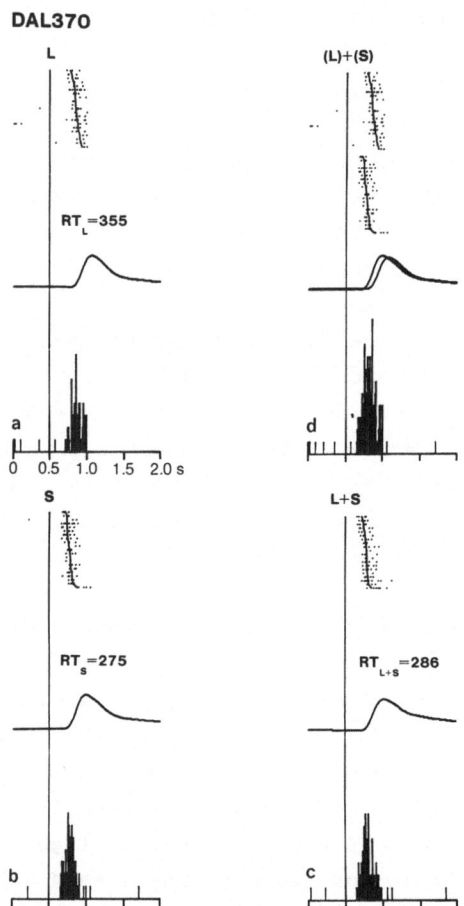

Fig. 5a,d. Rasters and histograms of an area 4 neuron recorded after dentate lesion during flexion of the contralateral arm. Trials are aligned with respect to stimulus presentation and rearranged in order of increasing reaction time. The conditioning stimuli, which were randomly presented, were **(a)** light *(L)*, **(b)** sound *(S)*, and **(c)** light and sound together *(L+S)*. With sound, the reaction time and the latency of the unit discharge were 80 ms shorter than with light **(a, b)**. When both stimuli were presented simultaneously, the reaction time and unit discharge were about the same as with sound alone **(c)**. **d** The simple sum of the L and S responses. One division of the time scale is equivalent to 50° on the movement trace

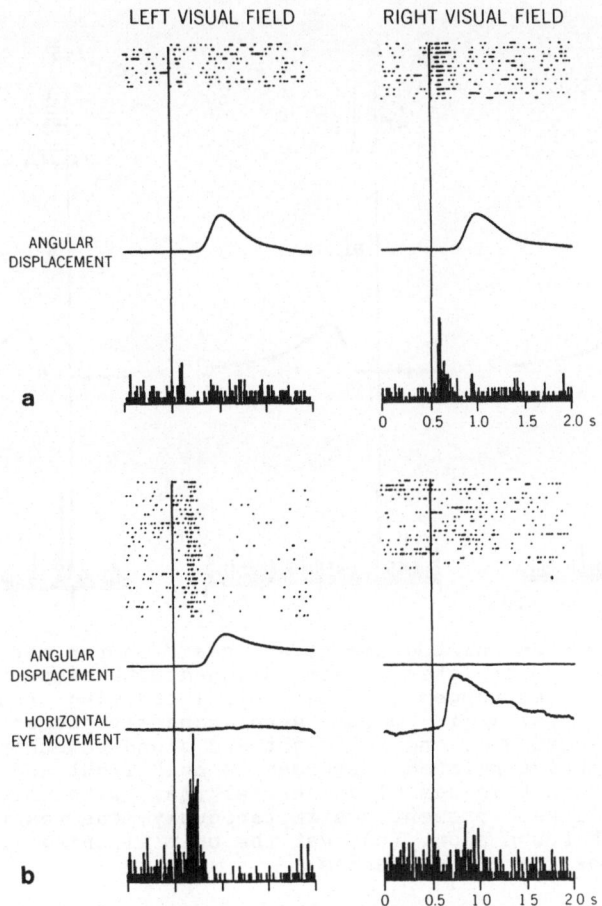

Fig. 6a. Records from a neuron in area 7 which responded only when flexion movement of the right arm was initiated in response to a visual stimulus presented in the right visual field.
b Records from area 7 neuron aligned with respect to light onset. Trials in which no eye movement occurred during arm flexion are shown on the *left*. Displacement and horizontal EOG have been averaged. After extinction of the conditioned response withholding the reward, records were accumulated for trials in which the animal looked toward the light in his contralateral visual field, but did not move the arm *(right)*. The neuron fires 100 ms after the light onset only when the stimulus is followed by the motor response, and this activity appears to be independent of the eye movements

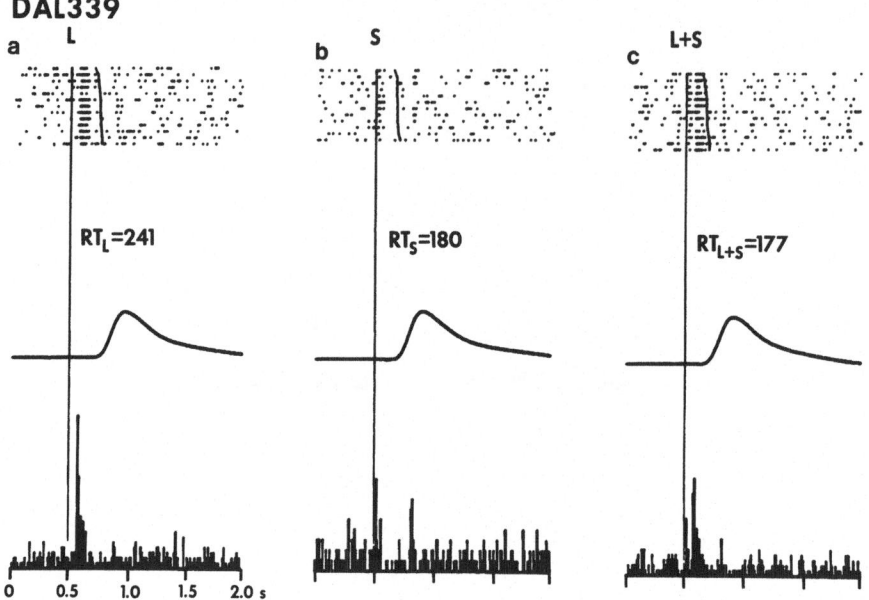

Fig. 7a-c. Rasters and histograms of an area 7 neuron recorded during flexion of the contralateral arm, aligned with respect to stimulus presentation, and rearranged in order of increasing reaction time. The conditioning stimuli, which were randomly presented, were **(a)** light *(L)*, **(b)** sound *(S)*, and **(c)** light and sound together *(L+S)*. The unit showed stimulus-related responses to both light and sound. With sound, the reaction time was 61 ms shorter than with light **(a, b)**. When both stimuli were presented simultaneously, the reaction time was the same as with sound alone **(c)**, yet the unit discharge represented a simple sum of the L and S responses

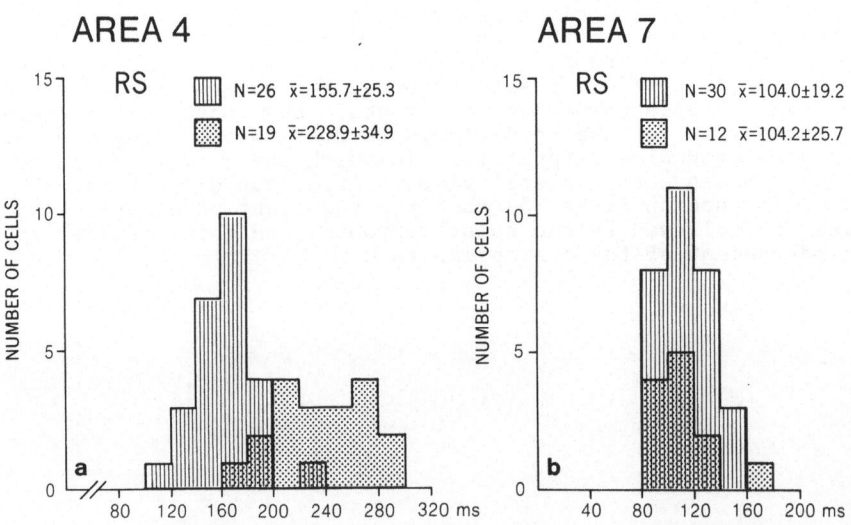

Discussion

In the present study, we were specifically concerned with the timing of neuronal responses relative to triggering stimuli of different sensory modalities and motor responses. We found that all task-related neurons recorded in area 4 showed movement-related responses and that these responses were the same whether the movement was triggered by visual, auditory, or somesthetic stimuli. This is also in accordance with the recent work of Tanji and Kurata (1982). The degree to which neural responses are temporally locked to the arm movement was quantified by plotting the unitary response latency of individual trials as a function of RT. In all instances, progressively longer RTs were associated with progressively longer latencies of the unitary responses (RS). Furthermore, the amount of time by which the unitary responses led the onset of movement (RM) was constant for any one unit, irrespective of the RT and of the modality of the triggering stimuli. After lesioning the dentate the increase in RT was associated with a corresponding increase of RS, but not of RM values. When two triggering stimuli were delivered together (light and sound), the movement-related responses of area 4 neurons remained the same as those observed with one stimulus alone.

In addition to the movement-related responses, some area 4 neurons responded directly to the conditioning stimulus. The most frequent stimulus-related responses occurred with the somesthetic stimulus. In our experimental situation there was no warning stimulus, and we did not observe significant modification of the somatosensory responses in the motor cortex when movement was extinguished. This is in contrast with the modulation of the spontaneous and evoked activity of area 4 neurons which has been observed with paradigms involving a preparatory signal (Evarts and Tanji 1974; Kubota and Hamada 1979; Lecas et al. 1983; Tanji and Evarts 1976). That area 4 neurons in the monkey show responses time-locked to the motor responses is also in agreement with a number of earlier studies (Cheney and Fetz 1980; Evarts 1966; Georgopoulos et al. 1982; Lemon and Porter 1976; Matsumura 1979;

Fig. 8. Histograms comparing mean RS obtained from neurons recorded before *(striped)* and after *(dotted)* dentate lesion in area 4 **(a)** and area 7 **(b)** in response to light stimuli that triggered elbow flexion. *Abscissas,* time in ms after the onset of the conditioning signal. *Ordinates,* number of cells. Mean values of the two neuronal samples are reported at the top of each histogram

Meyer-Lohmann et al. 1977; Smith et al. 1975). However, this is at variance with the results of Murphy et al. (1982) who found that the majority of task-related neurons in the motor cortex of the monkey showed visual cue responses with a mean latency of 150 ms, the earliest onset latency being 70 ms. We found only a small proportion of units (3.5%) with real "photic" responses, and these occurred at short latencies comparable to the latencies observed in animals anesthetized with chloralose (Buser and Ascher 1960; Wall et al. 1953). Recent work by Ghez et al. (1983) indicates that the majority of "lead" cells in area 4 of the cat also show stimulus-related responses to a visual cue. The difference between the results in these studies and ours may, in part, reflect the fact that our experimental paradigm did not involve a visuomotor tracking task.

In contrast with the findings in area 4, neurons in area 7 display stimulus-related responses which are, nevertherless, contingent upon the performance of the conditioned movement, as shown by others [see review by Lynch (1980) and Hyvärinen (1982)]. Stimulus related responses dependent, to varying degrees, on the subsequent movement are also prominent in the premotor (Godschalk and Lemon 1983; Kubota and Hamada 1979; Weinrich and Wise 1982) and supplementary motor cortex (Tanji and Kurata 1982). These findings suggest that such structures are involved in some higher level integrative function which cannot be regarded as either purely motor or purely sensory in nature. Mountcastle et al. (1975) have proposed that associative parietal cortex has a "command function" for movement initiation. The dentate nucelus is also viewed as a source of influence for the initiation of movement [see review by Brooks and Thach (1981)]. Since lesioning of the dentate led to an increase in RT associated with a corresponding increase in the latency of the responses of area 4 neurons, but not of area 7 neurons, one must conclude that the parietal associative cortex is functionally upstream to the neocerebellum and precentral cortex in the process of movement initiation.

In summary, the observations presented in this paper indicate that neurons in area 4 of the monkey display movement-related responses with little if any evidence that they may be involved in the early processing of information required for planning and executing the fast, triggered, ballistic movements that we have studied. In contrast, neurons in area 7 show stimulus-related responses that are movement-dependent, but not directly related to the subsequent motor

output. These findings suggest that the precentral cortex, like the
spinal motoneurons, lies at the end of the chain of events which
transforms a conditioning sensory cue into a motor command, while the
posterior parietal cortex is involved at a very early stage of this
transformation.

Acknowledgements. This research was supported by a Group Grant from
the Medical Research Council of Canada. The authors wish to thank
M.T. Parent and R. Bouchoux for technical assistance.

References

BROOKS VB, THACH WT (1981) Cerebellar control of posture and movement.
In: BROOKS VB (ed) The nervous system. Am Physiol Soc, Bethesda, pp
877-946 (Handbook of physiology, section 1, vol II, part 2)

BUSER P, ASCHER PL (1960) Mise en jeu réflexe du système pyramidal
chez le chat. Arch Ital Biol 98: 123-164

CHENEY PD, FETZ EE (1980) Functional classes of primate corticoneuron-
al cells and their relation to active force. J Neurophysiol 44:
773-791

EVARTS EV (1966) Pyramidal tract activity associated with a conditioned hand movement in the monkey. J Neurophysiol 29: 1011-1027

EVARTS EV (1974) Precentral and postcentral cortical activity in association with visually triggered movement. J Neurophysiol 37: 373-381

EVARTS EV (1981) Role of motor cortex in voluntary movements in primates. In: BROOKS VB (ed) The nervous system. Am Physiol Soc,
Bethesda, pp 1083-1120 (Handbook of physiology, section 1, vol II,
part 2)

EVARTS EV, TANJI J (1974) Gating of motor cortex reflexes by prior instruction. Brain Res 71: 479-494

GEORGOPOULOS AP, KALASKA JF, CAMINITI R, MASSEY JT (1982) On the relation between the direction of two-dimensional arm movements and cell
discharge in primate motor cortex. J Neurosci 2: 1527-1537

GHEZ C, VICARIO D, MARTIN J, YUMIYA H (1983) Sensory motor processing
of targeted movements in motor cortex. In: DESMEDT JE (ed) Motor
control mechanisms in health and disease. Raven, New York, (to be
published)

GODSCHALK M, LEMON RN (1983) Involvement of monkey premotor cortex in
the preparation of arm movements. Exp Brain Res [Suppl] 7: 114-119

HYVÄRINEN J (1982) The parietal cortex of moneky and man. Springer,
Berlin Heidelberg New York, p 202

KUBOTA K, HAMADA I (1979) I. Preparatory activity of monkey pyramidal tract neurons related to quick movement onset during visual tracking performance. Brain Res 168: 435-439

LAMARRE Y, JOFFROY J, FILION M, BOUCHOUX R (1970) A stereotaxic method for repeated sessions of central unit recording in the paralysed or moving animal. Rev Can Biol 29: 371-376

LAMARRE Y, BUSBY L, SPIDALIERI G (1983) Fast ballistic arm movements triggered by visual, auditory and somesthetic stimuli in the monkey. I. Activity of precentral cortical neurons. J Neurophysiol (to be published)

LECAS JC, REQUIN J, VITTON N (1983) Anticipatory neuronal activity in the monkey precentral cortex during reaction time foreperiod: preliminary results. Exp Brain Res [Suppl] 7: 120-129

LEMON RN, PORTER R (1976) Afferent input to movement-related precentral neurones in conscious monkeys. Proc R Soc Lond [Biol] 194: 313-339

LYNCH JC (1980) The functional organization of posterior parietal association cortex. Behav Brain Sci 3: 485-534

MATSUMURA M (1979) Intracellular synaptic potentials of primate motor cortex neurons during voluntary movement. Brain Res 163: 33-48

MEYER-LOHMANN J, HORE J, BROOKS VB (1977) Cerebellar participation in generation of prompt arm movements. J Neurophysiol 40: 1038-1050

MOUNTCASTLE VB, LYNCH JC, GEORGOPOULOS AP, SAKATA H, ACUNA C (1975) Posterior parietal association cortex of the monkey: command functions for operations within extrapersonal space. J Neurophysiol 38: 871-908

MURPHY JT, KWAN HC, MAC KAY WA, WONG YC (1982) Activity of primate precentral neurons during voluntary movements triggered by visual signals. Brain Res 236: 429-449

SMITH AM, HEPP-REYMOND MC, WYSS UR (1975) Relation of activity in precentral cortical neurons to force and rate of force change during isometric contractions of finger muscles. Exp Brain Res 23: 315-332

SPIDALIERI G, BUSBY L, LAMARRE Y (1983) Fast ballistic arm movements triggered by visual, auditory and somesthetic stimuli in the monkey. II. Effects of unilateral dentate lesion on the discharge of precentral cortical neurons and reaction time. J Neurophysiol (to be published)

TANJI J, EVARTS EV (1976) Anticipatory activity of motor cortex neurons in relation to direction of an intended movement. J Neurophysiol 39: 1062-1068

TANJI J, KURATA K (1982) Comparison of movement-related activity in two cortical motor areas of primates. J Neurophysiol 48: 633-653

WALL PD, REMOND AG, DOBSON RL (1953) Studies of the mechanism of the action of visual afferents on motor cortex excitability. Electroencephalogr Clin Neurophysiol 5: 385-393

WEINRICH M, WISE SP (1982) The premotor cortex of the monkey. J Neurosci 2: 1329-1345

Motor and Sensory Properties of Primate Corticomotoneuronal Cells

P. D. Cheney[1], R. J. Kasser[3], and E. E. Fetz[2]

1 Department of Physiology, University of Kansas Medical Center, Kansas City, KS 66103, USA
2 Department of Physiology and Biophysics, Regional Primate Research Center, University of Washington, School of Medicine, Seattle, WA 98195, USA
3 Department of Anatomy, University of Kansas Medical Center, Kansas City, KS 66103, USA

Introduction

The functional role of motor cortex in volitional movement has received much attention since Evarts' first recordings from pyramidal tract neurons in trained monkeys (1966, 1967). However, the interpretation of single-unit recordings in relation to movement has been hindered by lack of evidence for causal relations between the recorded cells and muscle activity. Recently, Fetz and Cheney (1980) used spike-triggered averaging of rectified EMG activity to reveal postspike facilitation (PSF) of muscle activity from single corticomotoneuronal (CM) cells. The purpose of this paper is to review the technique of spike-triggered averaging of EMG activity and some of the functional properties of primate motor cortex cells with functional linkages to forelimb muscles. We will discuss three major issues concerning the motor and sensory properties of CM cells: (a) the output organization of single CM cells, (b) the encoding of movement parameters by CM cells, and (c) the role of CM cells in long-latency stretch-evoked responses of muscle.

Methodology of Spike-Triggered Averaging of EMG Activity

The rationale for spike-triggered averaging of EMG activity as a means of identifying CM cells is that the individual EPSPs produced by these cells in target motoneurons should increase the firing probability of motor units, albeit weakly, at a fixed latency following the occur-

Experimental Brain Research, Suppl. 10
© Springer-Verlag Berlin · Heidelberg 1985

rence of the CM cell spike. This increase in firing probability, time-locked to the occurrence of the CM cell spike, may be detected by averaging the segments of EMG activity associated with many spikes. To illustrate this procedure, Fig. 1 shows the spike discharge of an extension-related CM cell and the EMG activity of a representative extensor muscle. EMG activity is full-wave rectified to avoid possible cancellation of opposite phases of facilitated motor unit potentials occurring at varying latencies. Rectification also distinguishes postspike facilitation from postspike suppression. The middle column in Fig. 1 shows the perispike EMG activity, on an expanded time scale, associated with each of the first five spikes in the record at the left. These segments of the analog EMG waveform, extending from 5 ms before to 25 ms after the cortical cell spike, are selected by the computer, digitized at 4 kHz, and averaged. Cumulative averages of these five EMG records are shown in the right-hand column of Fig. 1. Record 1 is simply the digitized form of the analog EMG waveform and demonstrates that the sampling rate is adequate to resolve even the smallest EMG peaks. The fortuitous postspike peaks in this first EMG record quickly become submerged in noise as additional EMG segments are averaged. However, the average rectified EMG activity associated with 2000 cell spikes (bottom record) shows a well-defined postspike facilitation at a latency of 9 ms following the cortical spike. Overall, PSF had a mean onset latency of 6.7 ms ($n = 346$), a mean peak latency of 10.2 ms ($n = 343$), and a mean amplitude of 9.0% ($n = 164$) above baseline. Several factors combine to determine the actual shape and latency of a particular PSF, including the shape of facilitated motor unit potentials, the conduction velocity of the facilitated motoneurons, and the shape of the underlying CM-EPSP waveforms. Nevertheless, a correlation peak, such as that in Fig. 1, is evidence that the trigger cell is synaptically linked, probably monosynaptically, to motoneurons innervating the muscle whose activity was averaged. We refer to cells generating clear PSF as CM cells. However, it must be remembered that PSF is a direct measure of a cell's correlational linklage to motoneurons, not proof of its anatomical linkage.

The method of spike-triggered averaging is capable of identifying CM cells and their facilitated target muscles, but its application to testing a cell's effect on antagonists of the target muscles is limited by the fact that CM cells are normally inactive during the antagonist phase of alternating movement. To overcome this limitation we de-

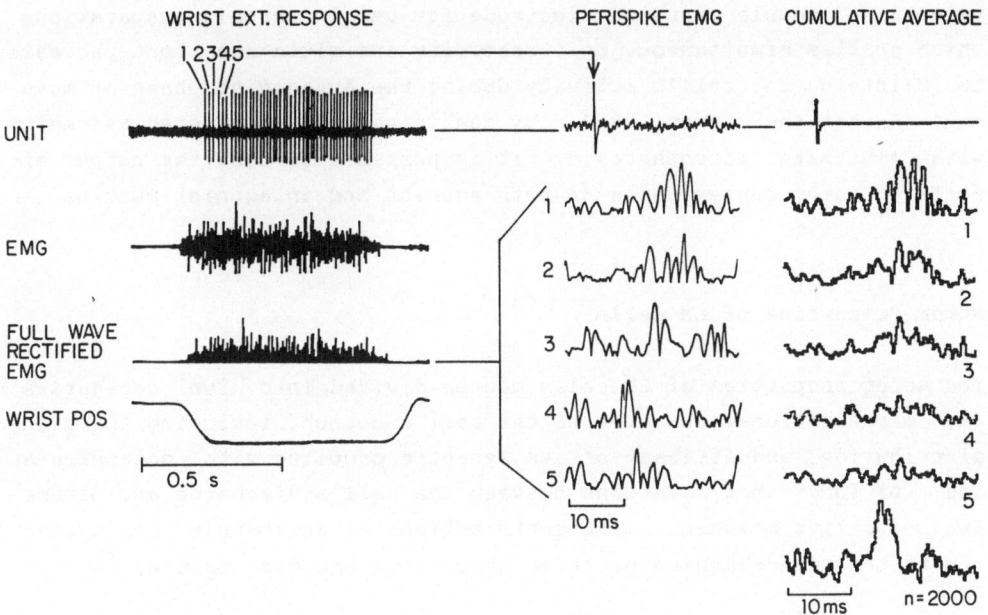

Fig. 1. Spike-triggered averaging procedure used to detect postspike effects. Single extension response of a corticomotoneuronal (CM) cell is shown at *left* with normal and rectified EMG activity. *Middle column* shows the rectified EMG associated with each of the first five spikes in the response at left. *Right column* shows the cumulative averages for the first *n* spikes where *n* = 1,2...5, and 2000. (Fetz and Cheney 1980)

veloped a double-barreled electrode for use in chronic preparations, which enables simultaneous unit recording and glutamate iontophoresis to maintain the cell's activity during the antagonist phase of movement (Kasser and Cheney 1982). By combining spike-triggered averaging with glutamate iontophoresis, it is possible to test the output effects of motor cortex cells on both agonist and antagonist muscles.

Motor Properties of CM Cells

The motor properties of CM cells can be divided into two categories: (a) organizational features of the cell's output, including the sign, distribution, and efficacy of its synaptic coupling with motoneurons, and (b) functional relations between the cell's discharge and parameters of active movement. The contributions of spike-triggered averaging to the understanding of these properties are discussed below.

Output Organization of Corticospinal Neurons

The excitatory or inhibitory nature of a cell's effect on motoneurons of agonist and antagonist muscles can be revealed using spike-triggered averaging together with glutamate iontophoresis to maintain cell activity during the antagonist phase of alternating movement. We have identified three basic patterns of synaptic influence on agonist and antagonist muscles, as illustrated in Fig. 2. Agonist muscles are defined as those with which the cell coactivates during motor tasks. The agonists of each of the cells illustrated in

Fig. 2a-i. Types of CM cell output organization. Examples of response averages and spike-triggered averages of both agonist and antagonist muscles for each cell type are shown. All cells were extension related. Firing rates of the pure facilitation and reciprocal cell examples are abnormally high during the antagonist phase of movement because of glutamate excitation. Number of events averaged in this and all subsequent figures is given in parentheses in the lower right corner of each panel. *Asterisks* in a, b, and c indicate muscles showing either postspike facilitation or suppression. Abbreviations for muscles in this and following figures are: *ECR-L*, extensor carpi radialis longus; *ECU*, extensor carpi ulnaris; *ECR-B*, extensor carpi radialis brevis; *ED*, extensor digitorum; *EDC*, extensor digitorum communis; *FCR*, flexor carpi radialis; *FCU*, flexor carpi ulnaris; *FDP*, flexor digitorum profundus; *FDS*, flexor digitorum superficialis; *PL*, palmaris longus; *PT*, pronator teres

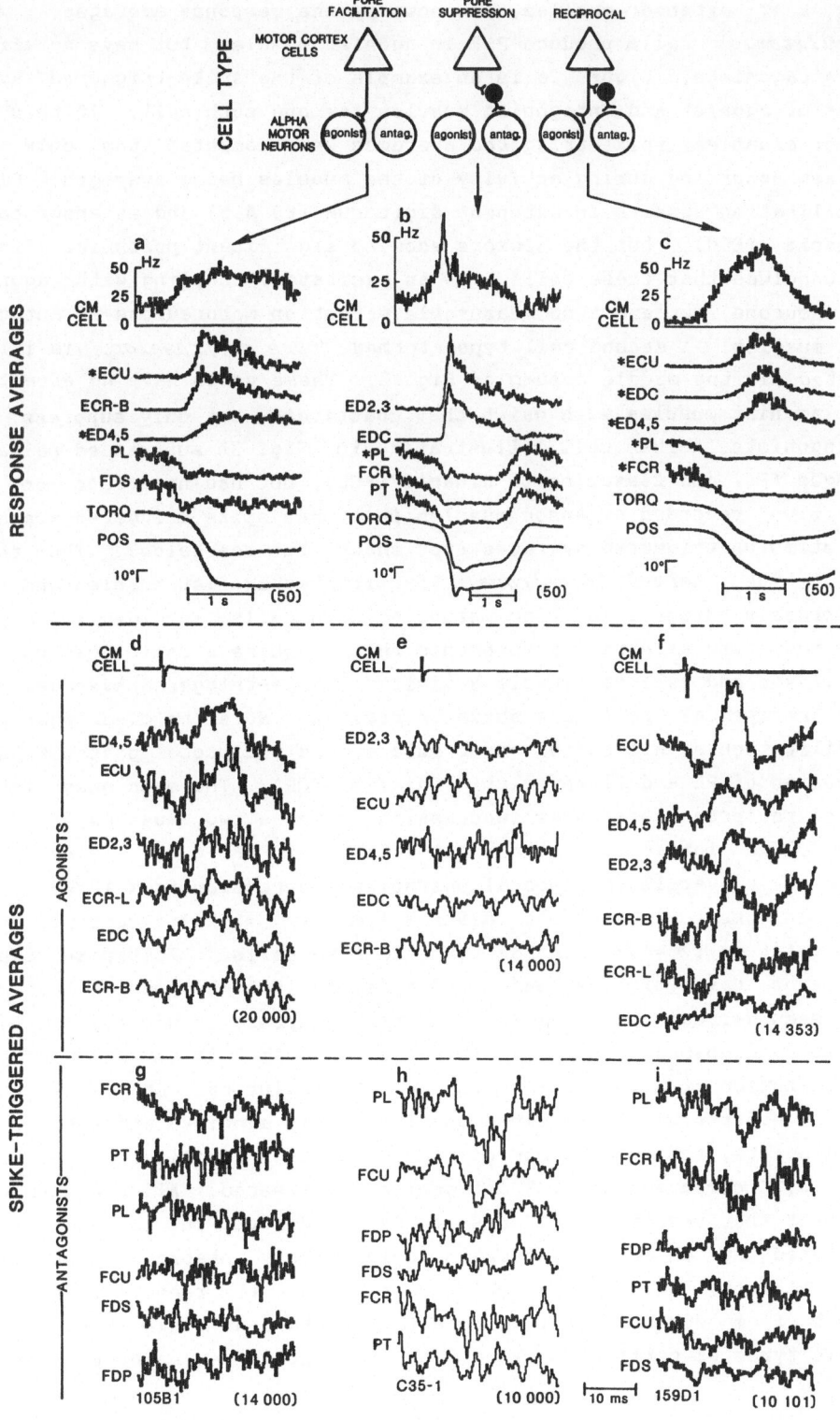

Fig. 2 are extensor muscles, as shown by the response averages. *Pure facilitation* cells produce PSF in agonist muscles, but have no effect on antagonists. Figure 2a is an example of the spike-triggered averages of agonist and antagonist muscles for one such cell. In this and other examples, spike-triggered averages were computed from only the spikes occurring during activity of the muscles being averaged. Clear facilitation appears in extensor digitorum (ED 4,5) and extensor carpi ulnaris (ECU), but the flexors show no significant postspike effect. We conclude that these cells have an excitatory coupling with agonist motoneurons, but exert no measurable effect on motoneurons of antagonist muscles. A second cell type, termed *"pure suppression,"* is illustrated in the middle column of Fig. 2. These cells have no effect on the agonist muscles with which they coactivate, but only suppress the antagonists. The cell illustrated in Fig. 2b suppressed palmaris longus (PL) and flexor carpi ulnaris (FCU), but had no effect on any of five recorded extensor muscles in either spike-triggered averages or stimulus-triggered averages (not shown, but see below). The third cell type, termed *"reciprocal,"* facilitates agonist muscles and also suppresses antagonists. The output of such cells is ideally suited for mediating alternating movements which require a reciprocal pattern of flexor and extensor muscle activity. Spike-triggered averages for one reciprocal cell are shown in Fig. 2c. Note the clear postspike facilitation of all the extensor muscles and reciprocal postspike suppression of PL and flexor carpi radialis (FCR). The mean onset latency of reciprocal postspike suppression in these two muscles (9.7 ms) is 3.7 ms longer than the onset latency of PSF in the six extensors (6.0 ms). Overall, reciprocal postspike suppression from 12 CM cells had an onset latency of 8.9 ± 3.1 ms (n = 20), compared with 5.3 ± 1.6 ms (n = 37) for PSF from the same cells. Postspike suppression was typically weaker and appeared in fewer muscles than PSF. The mean decrease below baseline of peak suppression was $4.1 \pm 1.6\%$ (n = 20), compared with a mean increase of $8.0 \pm 7.0\%$ (n = 37) for peak facilitation from the same cells. Of 11 reciprocal CM cells whose output effects were determined on five or six agonists and five or six antagonists, the mean number of agonist muscles facilitated per cell was 3.1, compared with 1.7 antagonists suppressed. All these factors suggest that reciprocal suppression is not direct, but is most likely mediated by CM axon collaterals to inhibitory interneurons, probably Ia inhibitory interneurons, which are known to receive convergent input from corticospinal neurons (Jankowska and Tanaka 1974). These basic types of corticospinal output organization represent fundamental organizational units available to motor cortex for the control of muscle activity. A particular movement would require selection by the central motor program of cells whose output is appropriate in terms of their basic output organization and in terms of their specific target muscles. As yet, no examples of cells which clearly facilitate both wrist flexor and extensor muscles have been encountered.

A further property of the output organization of CM cells is the extent of divergence of their effects on motoneurons of multiple agonist muscles. By computing simultaneous spike-triggered averages of multiple synergist muscles, we determined that 50% of wrist-related CM cells produced clear PSF in only a single muscle, consistent with a high degree of specificity in the CM control of muscles; however, the remaining cells clearly facilitated two or more synergistic muscles (Fetz and Cheney 1980). These findings do not support the notion that motor cortex output is organized solely in terms of specific control of single muscles, at least for cells related to wrist movements which involve coactivation of many synergists. However, using a more discrete task, precision finger grip, Lemon and Muir (1983) found that all of the seven CM cells they tested facilitated only one of five recorded muscles of the hand and digits.

Another question concerns the distribution of PSF to different motor units within a muscle. This question could in principle be answered by cross-correlating the CM cell spike train with the spike discharges of many single motor units sampled from a single muscle. In practice this experiment has proved to be technically difficult, although the effects of microstimuli on single motor units have been studied (Sawyer and Fetz 1981). Cross-correlating trains of microstimuli applied to cortical output sites (see Fig. 3) with the spike trains of motor units from a single facilitated muscle has shown that minimal single microstimuli typically facilitated all the motor units of a muscle (95%). These findings on the effects of cortical output zones, coupled with others suggesting that neighboring CM cells have similar patterns of terminations with motoneurons of synergist muscles (see below), support the hypothesis than individual CM cells, like Ia afferents, may influence a large fraction of the motor units within a muscle.

SPIKE TRIGGERED AVERAGE   STIMULUS TRIGGERED AVERAGE

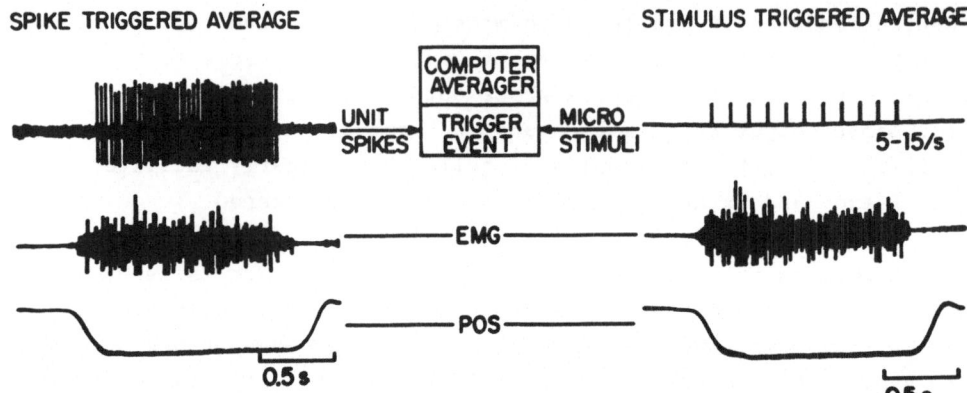

Fig. 3. Comparison of spike- and stimulus-triggered averaging procedures. For stimulus-triggered averaging, intracortical microstimuli
(5 - 10 μA) were applied at low frequency (5 - 15 Hz) to the recording
microelectrode during the phase of movement which engaged the activity
of the cell recorded at that site

Output Effects Revealed by Stimulus-Triggered Averaging

Further insights into the cortical organization of CM cells were obtained by comparing the output effects of a single CM cell with those
of single intracortical microstimuli. Stimulus-triggered averages
(Fig. 3) were computed by applying microstimuli at low intensity
(5 - 15μA) and low frequency (5 - 15 Hz, to avoid temporal summation)
to the site of a recorded CM cell. In tests at 22 CM cell sites,
poststimulus facilitation was observed in the same muscles which
showed PSF in the spike-triggered average; moreover, the relative amplitude of poststimulus facilitation across muscles usually matched
that of postspike facilitation. The absolute amplitude of
poststimulus facilitation, however, was much greater than postspike
facilitation; single 5-μA stimuli evoked facilitation that was six
times stronger than PSF. Figure 4 shows an example of postspike and
poststimulus facilitation for a single cortical site. The greater amplitude of poststimulus facilitation suggests that it is mediated by
several cells located near the electrode and activated by the stimulus
(Rank 1975), whereas postspike facilitation is mediated by only one
cell. However, the fact that the profile of poststimulus facilitation
(across muscles) remains similar to the profile of postspike facilitation, despite a contribution from additional CM cells, suggests that
the output effects of each cell activated by the stimulus are similar
to that of the single cell used to compile the spike-triggered aver-

AVERAGES

SPIKE-TRIGGERED   STIM-TRIGGERED

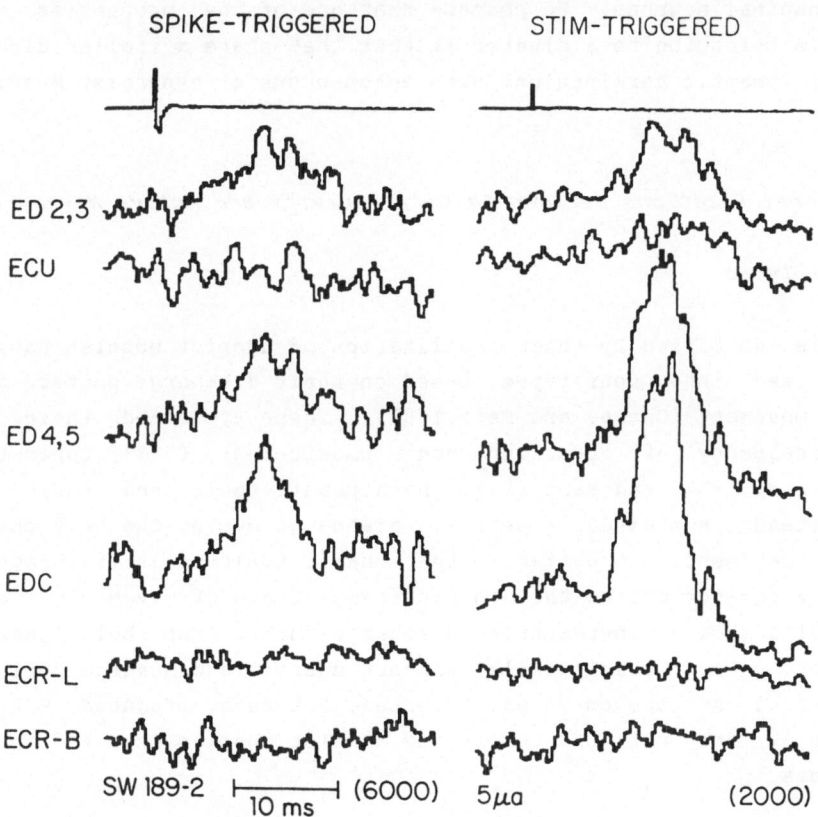

Fig. 4. Example of spike- and stimulus-triggered averages obtained from a single cortical site. Note that the relative amplitude of facilitation is the same in both sets of averages, but poststimulus facilitation is stronger than postspike facilitation. Spike-triggered averages in this case are based on 6000 trigger events, stimulus-triggered averages on 2000 events

age. This conclusion is supported by the observation that neighboring CM cells typically facilitated the same target muscles.

The average size of cortical output zones producing a uniform pattern of poststimulus facilitation in synergist muscles was estimated to be about 800 μm, from measurements of the amplitude of poststimulus facilitation evoked from sites along the bank of the precentral gyrus (Sawyer et al. 1979). These optimal output sites often coincided with the location of clusters of horseradish peroxidase (HRP)-labeled corticospinal neurons identified after cervical injection of HRP.

Jones and Wise (1977) also reported such clustering of HRP-labeled corticospinal neurons. We propose that one of the properties common to cells belonging to a cluster is that they share a similar distribution of synaptic terminations with motoneurons of synergist muscles.

Functional Relations Between CM Cell Activity and Active Movement

CM Cell Types

CM cells identified by their facilitation of agonist muscles have been categorized into four types, based on their discharge pattern during active movement (Cheney and Fetz 1980). These types and their relative frequency of occurrence are: phasic-tonic (59%), tonic (28%), phasic-ramp (8%), and ramp (5%). Both phasic-tonic and tonic cells show steady, sustained, repetitive discharge during the hold phase of agonist movement, but differ in that phasic-tonic cells discharge at a higher frequency during the dynamic (ramp) phase of movement (Fig. 5). Ramp cells show an incrementing discharge during the hold phase of movement. Pure phasic cells, without sustained discharge during the hold period, are common in motor cortex, but never produced PSF, and apparently are not CM cells for agonist muscles involved in wrist movements.

Single motor units in agonist forearm muscles commonly exhibited similar discharge patterns in relation to ramp-and-hold wrist movement (Sawyer and Fetz 1981). Fifty-nine percent of motor units were either phasic-tonic or tonic; 39% showed decrementing discharge during the hold period, a pattern that was the reverse of the ramp CM cells. Five percent were only phasic and showed showed no sustained discharge during the hold period. All four types of motor units were facilitated by microstimuli at a given cortical output site.

Encoding of Movement Parameters by CM Cell Output

The encoding of movement parameters by motor cortex output has attracted much attention since Evarts (1967) reported that pyramidal tract neuron discharge is related to active force. However, in experiments in which the aim is to characterize the output signal transmitted from motor cortex to motoneurons, it is essential to know that a

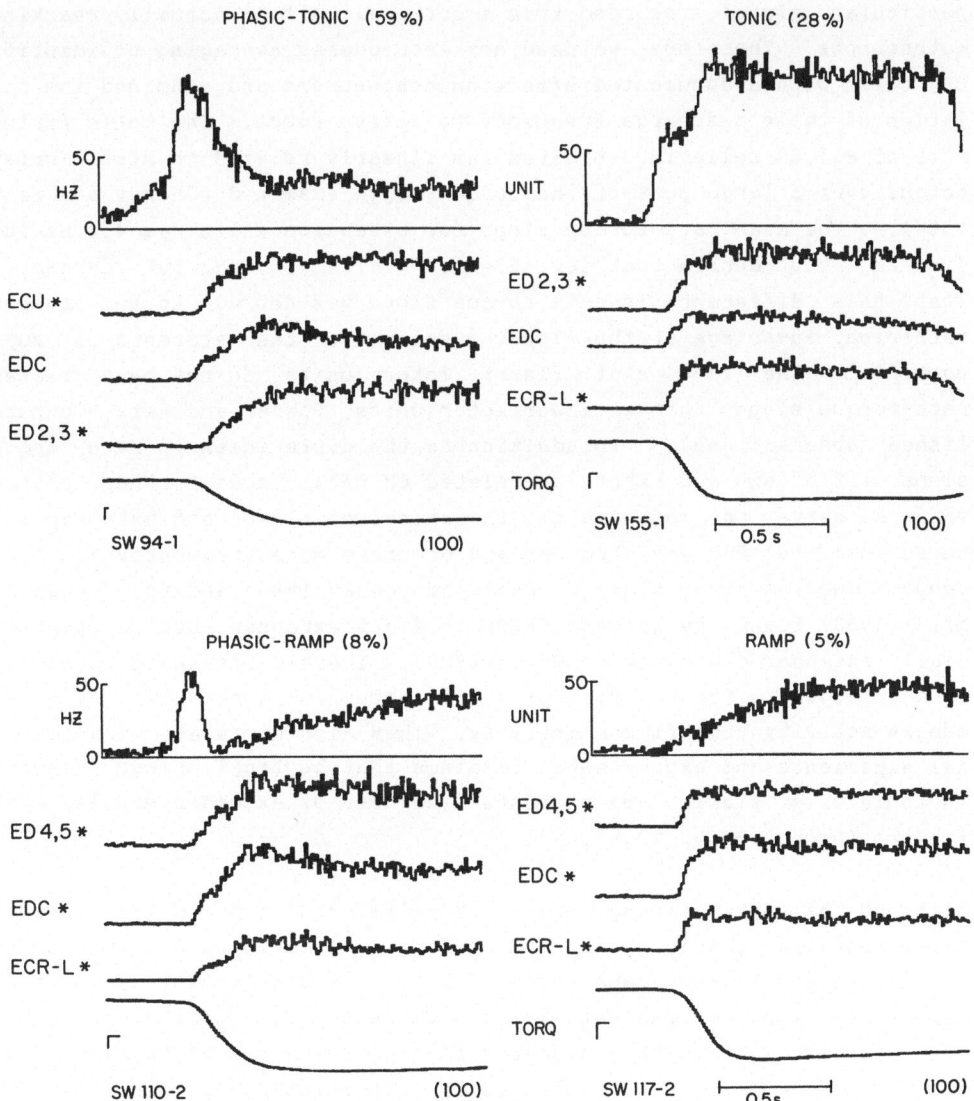

Fig. 5. Four response patterns charcteristic of CM cells during isometric ramp-and-hold wrist responses. Responses during ramp-and-hold wrist displacements were qualitatively the same. Average rectified muscle activity is also shown; *asterisks* indicate the cell's target muscles. (Cheney and Fetz 1980)

particular signal recorded from a cortical cell is actually reaching motoneurons. Therefore, we used spike-triggered averaging to identify CM cells with a documented effect on motoneurons and examined the relation of their discharge frequency to active force. The tonic firing rate of all CM cells investigated was linearly related to static wrist torque over a large part of the torque range examined (Cheney and Fetz 1980). The mean rate-torque slope for extension cells was 4.8 Hz/10^5 dyne-cm, about double that for flexion cells (2.5 Hz/10^5 dyne-cm). That this difference in rate-torque slope was not due to any obvious mechanical advantage of the flexor muscles over the extensors is supported by the fact that flexor motor units do not have greater rate-torque slopes than extensor motor units (Sawyer and Fetz, unpublished observations). In addition to the differences in rate-torque slope of flexion- and extension-related CM cells, some further observations serve to contrast the CM control of flexor and extensor motoneurons: (a) PSF was stronger and occurred more frequently in extensor muscles than flexors (Fetz and Cheney 1980) and (b) Clough et al. (1968) found the largest EPSPs in digit extensor muscles, particularly extensor digitorum communis (EDC). These differences taken together suggest a greater role of motor cortex in generating extensor muscle activity than flexor activity. They also correlate with clinical experience and experimental findings that cortical damage results in tonic wrist flexion and a greater weakness of extensor muscles than flexors (Denney-Brown 1966).

Although the static firing rate of CM cells encodes the active muscle force required to hold a steady position, it also shows smaller variations consistent with compensation for the length-tension properties of muscle (Cheney and Fetz 1980). For example, a particular active force is associated with a higher discharge frequency if the length of the target muscles is decreased by appropriate joint displacement. Therefore, the discharge frequency of CM cells encodes the force of movement through a largely linear rate-torque relation which is shifted appropriately to compensate for the length-tension properties of muscle. In view of this, it may be most accurate to regard CM cell output as encoding a particular level of motor unit discharge or muscle activity. The force associated with this activity will then depend upon the muscle's length-tension property.

Dissociation of CM Cell and Target Muscle Activity

The activity of alpha motoneurons is rigidly linked to the activity of
the muscles they innervate and is predictable, based on the principle
of orderly recruitment. Since CM cells are premotor neurons, we were
interested in whether a similar rigid linkage would apply to their activity or whether the linkage might show greater flexibility. To
answer this question we trained monkeys to perform two different motor
tasks - alternating wrist movements and power grip - which differed in
the temporal pattern of agonist and antagonist muscle activity (Kasser
and Cheney 1983). Alternating wrist movements involved a reciprocal
pattern of wrist flexor and extensor muscle contraction, whereas power
grip required the monkey to squeeze a pair of nylon bars and involved
co-contraction of flexor and extensor muscles to stabilize the wrist.
Some CM cells (4 of 12) increased their firing rate during both power
grip and either the flexion or extension phase of alternating movements. However, the remaining cells (8 of 12) increased their activity only during alternating movements and were unrelated to power grip
despite the fact that power grip involved activation of the cell's
target muscles. Figure 6 illustrates one such cell whose target muscles were extensors. Its activity co-varied reliably with the extension phase of alternating wrist movements, but during power grip,
which involved coactivation of the cell's target muscles and their antagonists, its firing rate decreased sharply. This functional uncou-
pling of activity in a CM cell from that of its target muscles may be
related to the fact that it reciprocally suppressed the antagonist
muscles (Fig. 6c). Since such suppression would interfere with
co-contraction, the neural mechanisms producing co-contraction may exclude reciprocal CM cells.

Another situation in which we observed dissociation of the activity of
a CM cell and its target muscles was during ballistic wrist movements
(Cheney and Fetz 1980). Ballistic movements were rapid, uncontrolled
oscillations between flexion and extension position zones which one of
our monkeys periodically produced after becoming frustrated with the
task requirements. The two cells in Fig. 7 increased their discharge
consistently during the extension phase of ramp-and-hold movements,
but failed to show a consistent relation to ballistic movements involving much greater target muscle activity. Both CM cells began dis-
charging during ramp-and-hold movements well in advance of target muscle EMG activity; hence, the results are not explained by inadequate

SPIKE AND STIMULUS-TRIGGERED AVERAGES

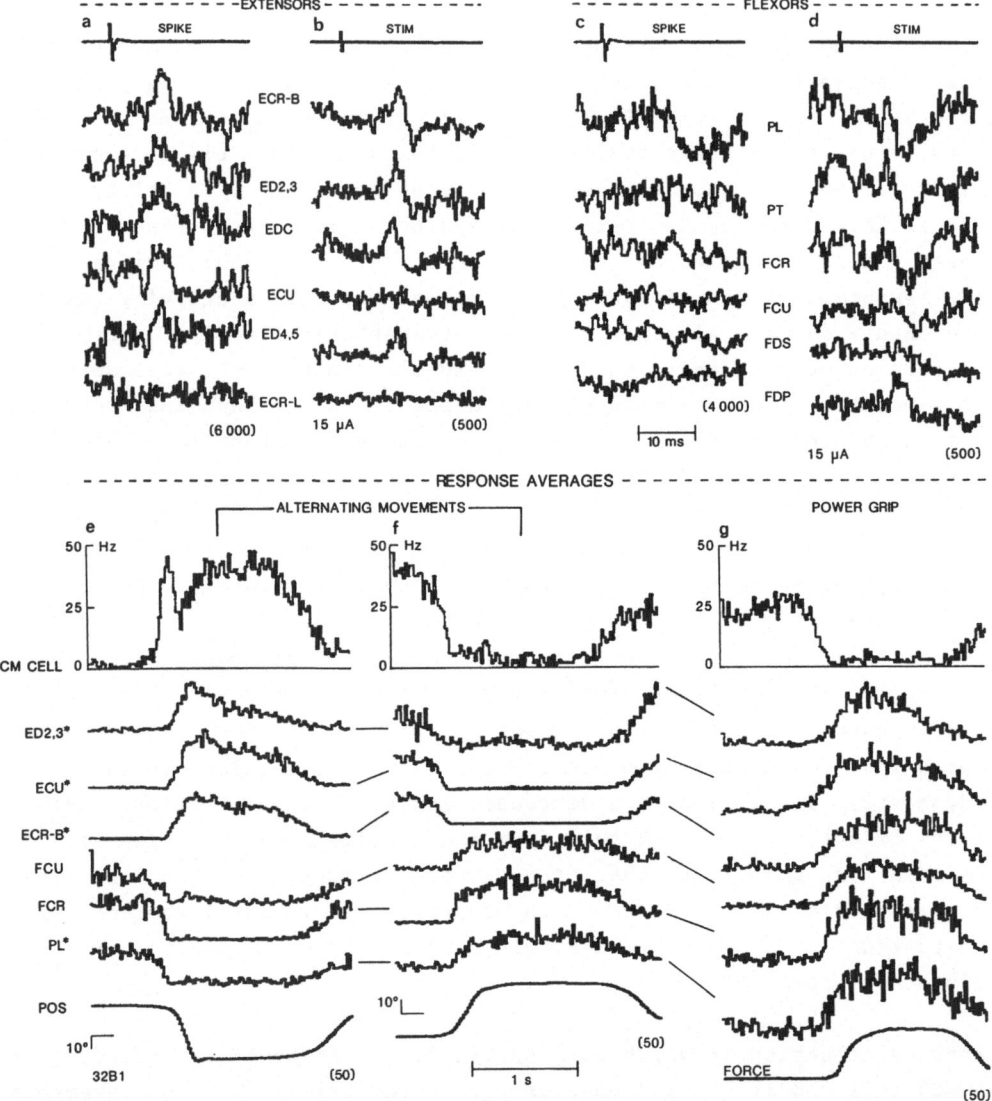

Fig.6a–g. Average activity of a reciprocal CM cell and its target muscles during alternating movements and power grip. Note the opposite relation between cell and target muscle activity during power grip (**g**), compared with the extension phase of alternating movements (**e**), even though both involve activation of cell's target muscles. Spike-triggered averages (**a, c**) computed during ramp-and-hold movements identify this cell's reciprocal output effects, i.e., postspike facilitation of agonists and postspike suppression of antagonists. These output effects were confirmed in stimulus-triggered averages (**b, d**) computed from stimuli applied to the site of CM cell recording during ramp-and-hold movements. *Asterisks* indicate muscles facilitated or suppressed by the cell

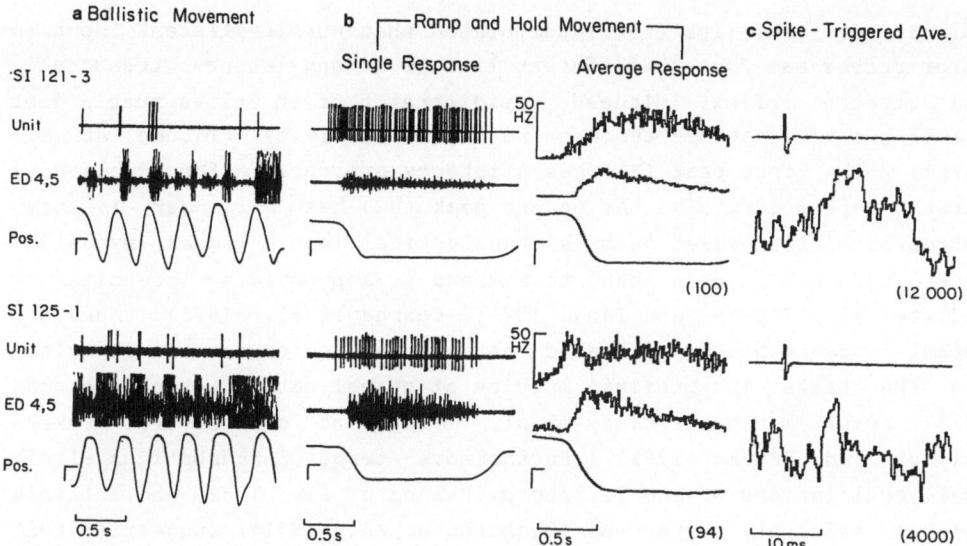

Fig. 7a-c. Activity of two CM cells during controlled ramp-and-hold wrist movements and ballistic movements. Note the intense activity during ramp-and-hold movements **(b)** and inactivity during ballistic movements **(a)** despite greater activation of the cell's target muscles. Spike-triggered averages **(c)** identify each as a CM cell. Position calibration bar is 10°. (Cheney and Fetz 1980)

time for cell activation. Some CM cells, therefore, appear to have a preferential role in accurate movements and these cells are excluded by motor programs for ballistic movements. This view is consistent with the finding of Fromm and Evarts (1977) that motor cortex neurons discharge more intensely during small, accurate movements than during ballistic movements. Similar findings have also been reported by Muir and Lemon (1983) who found greater CM cell activity related to precision grip than to a power grip.

Sensory Properties of CM Cells

Role of CM Cells in Transcortical Stretch-Evoked Muscle Responses

Motor cortex cells, including corticospinal neurons, are known from work in anesthetized animals to respond to afferent signals from both cutaneous and muscle receptors (Wiesendanger 1973; Hore et al. 1976; for additional references see Phillips and Porter 1977). Although the existence of these inputs is now generally accepted, their functional role is not. Phillips (1969) postulated that muscle afferent input to motor cortex may form the afferent limb of a long-latency transcortical stretch reflex. Indeed, rapid stretch of an active muscle does elicit two or sometimes three peaks of EMG activity (Tatton et al. 1975). The first peak (M1) has a latency appropriate for a segmental stretch reflex (Fig. 8), the second peak (M2) has a longer latency, appropriate for mediation by a transcortical loop. Indeed, pyramidal tract neurons have been shown to respond at appropriate latencies to mediate M2 (Evarts and Tanji 1976; Conrad et al. 1975). However, recent evidence has demonstrated that muscle stretch can elicit multiple EMG peaks in proximal muscles of spinal cats (Ghez and Shinoda 1978), spinal monkeys (Tracey et al. 1980), and decerebrate monkeys (Miller and Brooks 1981). Furthermore, torque perturbations elicit small oscillations of muscle length (Eklund et al. 1982) and multiple spindle afferent responses (Hagbarth et al. 1981), suggesting that the multiple EMG peaks may simply represent sequential spinal stretch reflexes.

Since CM cells have a documented effect on EMG activity, their response to torque perturbations is an important test of the role of motor cortex in long-latency stretch-evoked muscle responses. Therefore, we examined the responses of the CM cells to perturbations which stretched the cell's target muscles. Of 21 cells, 20 responded to target muscle lengthening torque perturbations; only one cell was unresponsive to the torque perturbations applied. Figure 8 illustrates the average torque pulse response of one CM cell and a target muscle. Torque perturpations which stretched the target muscles elicited both M1 and M2 EMG peaks and a brisk CM cell discharge, whose onset preceded M2 onset. In these experiments, transcortical loop time was measured as the sum of its afferent component (the onset latency of the stretch evoked CM cell discharge) and its efferent component (the onset latency of postspike facilitation). The mean PSF onset latency of 7.0 ms sums with the mean CM cell onset latency of 23.4 ms to yield a total mean transcortical loop time of 30.4 ms, which is comparable to the mean M2 onset latency of 27.9 ms. The duration of a CM cell's response to torque perturbations provides an additional measure of the extent of its potential contribution to the M2 muscle response. In all cases but two, the CM cell response, delayed by PSF onset time, overlapped with some part of the M2 EMG response.

Spike-triggered averages revealing PSF from our cells have been computed from spikes occurring during the static hold period of active

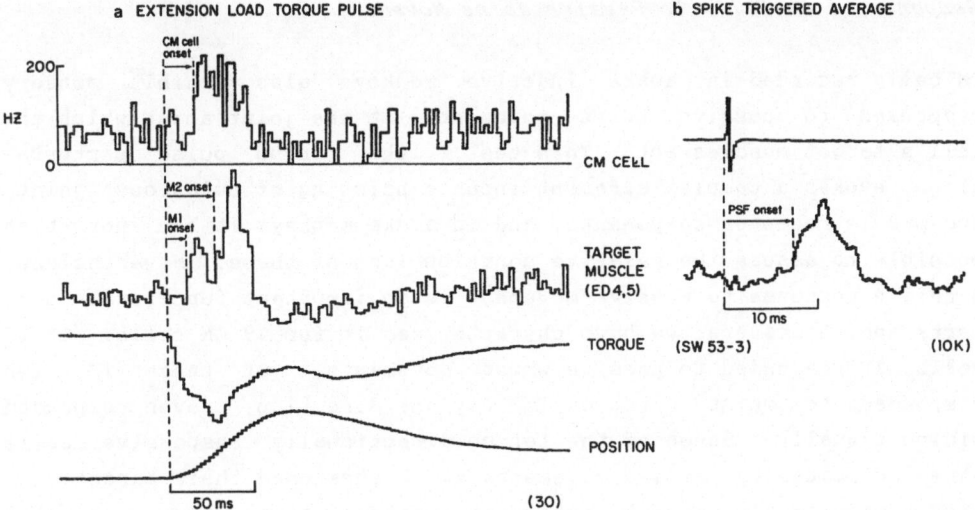

a EXTENSION LOAD TORQUE PULSE

b SPIKE TRIGGERED AVERAGE

Fig. 8a,b. Average response of a CM cell **(a)** and its target muscle evoked by transient muscle lengthening torque perturbations applied during active wrist extensions. The spike-triggered average **(b)** showing postspike facilitation identifies this as one of the cell's target muscles. (Cheney and Fetz 1983)

movements. But do the spikes evoked by the torque pulse also facilitate muscle activity during the M2 response? Thorough testing of this issue with existing data has been limited by the relatively small number of torque-pulse-evoked spikes available for spike-triggered averaging for a given cell. Nevertheless, torque-pulse-evoked spikes from one CM cell produced potent facilitation with only 685 triggers. Based on these findings, we conclude that CM cells do contribute to the long-latency M2 response evoked by target muscle lengthening perturbations.

In addition to responding to perturbations which stretched their target muscles, 8 of 18 CM cells tested also responded at short latency (22.0 ± 7.4 ms) to perturbations which shortened their target muscles. Although these responses seem paradoxical, they are consistent with the fact that the cell's target muscles also exhibited a shortening response which had an onset latency similar to M2 in lengthened muscles (33.9 ms compared with 27.9 ms for M2). These results suggest that the transcortical CM cell loop, and the muscle responses to which it contributes, may constitute a mechanism for stiffening the joint through co-contraction of flexor and extensor muscles.

Response of CM Cells to Passive Joint Movement

CM cells recorded in awake inactive monkeys also exhibit sensory responses to passive wrist movements at the joint about which the cell's target muscles act. This test, like torque pulse perturbations, evokes a complex afferent input consisting of cutaneous, joint, and muscle receptor components, and in awake monkeys it has not been possible to assess the relative contributions of these. Nevertheless, a cell's response to passive movement is an important functional property and, therefore, we have characterized it for 19 CM cells. Of 19 cells, 17 responded to passive wrist movements. Of these 17, ten responded to wrist rotation in only one direction; seven responded bidirectionally. Seven of the ten unidirectionally responsive cells were activated by passive movements which stretched their target muscles (opposite direction to active movements); three were activated by passive movements and active movements in the same direction. All evoked responses were phasic.

None of four wrist-related CM cell's we tested were activated by natural stimulation of the glabrous or hairy skin of the hand. However, Lemon and Muir (1983) reported that four of seven CM cells they tested could be activated by brushing the glabrous skin of the hand. These cells all produced PSF in small hand muscles and were highly active during exploration movements of the fingers.

Summary and Conclusions

Since functional properties may vary widely with a cell's axonal projection, identification of these projections is particularly important in establishing the cell's functional role in movement. Spike-triggered averaging of rectified EMG activity has emerged as a useful means of identifying CM cells in awake monkeys, where functional relations between cell activity and movement can also be investigated. This method is capable of revealing both postspike facilitation and postspike suppression as well as their distribution across different muscles. Glutamate iontophoresis can be combined with spike-triggered averaging to increase cell activity and enable adequate testing for suppression of antagonist muscles.

Using spike-triggered averaging in awake monkeys we have established the following properties of the primate corticomotoneuronal system.

1. CM cell output organization is of three basic types: *pure facili-tation* - these cells facilitate agonist muscles, but have no effect on antagonists; *pure suppression* - these cells suppress antagonist muscles, but have no effect on agonists; *reciprocal* - these cells simultaneously facilitate the agonists and suppress the antagon-ists. Postspike suppression is weaker and about 3 ms longer in la-tency than in facilitation. We conclude that it is probably medi-ated by a spinal inhibitory interneuron. These three cell types constitute fundamemtal organizational units involved in direct motor cortex control of forearm muscles.

2. Half of the wrist-movement-related CM cells produced PSF in two or more synergist muscles. We conclude that many CM cells make exci-tatory synaptic connections with motoneurons of multiple agonist muscles.

3. Neighboring CM cells produce the same profile of PSF across muscles and therefore appear to have similar synaptic connections with mo-toneurons of agonist muscles. Such neighboring cells with similar-ly organized output effects may form clusters in layer V of motor cortex.

4. CM cells can be divided into four types, based on their discharge during ramp-and-hold wrist movements: phasic-tonic, tonic, phasic-ramp, and ramp. The net output effect of a CM cell on tar-get muscle activity during active movement is a function of both its PSF and its firing pattern during movement.

5. CM cell discharge encodes relatively simple parameters of active movement. Under static conditions the discharge rate of CM cells encodes the force of active movement.

6. CM cells are the efferent limb of a transcortical reflex loop ac-tivated by external perturbations. The consequence of this long-latency reflex is to increase joint stiffness by coactivating flexor and extensor muscles.

7. CM cell discharge, unlike that of alpha motoneurons, is not invari-ably linked to the activity of its facilitated target muscles. Rather, CM cells exhibit more complex movement relations in which activation depends not only on muscles the cell facilitates but also on those which it suppresses.

230

References

CHENEY PD, FETZ EE (1980) Functional classes of primate corticomotoneuronal cells and their relation to active force. J Neurophysiol 44: 773-791

CHENEY PD, FETZ EE (1983) Primate corticomotoneuronal cells contribute to long latency stretch reflexes. J Physiol (Lond) (to be published)

CLOUGH JFM, KERNELL D, PHILLIPS CG (1968) The distribution of monosynaptic excitation from the pyramidal tract and from primary spindle afferents to motoneurones of the baboon's hand and forearm. J Physiol (Lond) 198: 145-166

CONRAD B, MEYER-LOHMANN J, MATSUNAMI K, BROOKS VB (1975) Precentral unit activity following torque pulse injections into elbow movements. Brain Res 94: 219-236

DENNY-BROWN D (1966) The cerebral control of movement. Liverpool University Press, Liverpool

ECKLUND G, HAGBARTH K-E, HAGGLUND JV, WALLIN EU (1982) Mechanical oscillations contributing to the segmentation of the reflex electromyographic response to stretching human muscles. J Physiol (Lond) 326: 65-77

EVARTS EV (1966) Pyramidal tract activity associated with a conditioned hand movement in the monkey. J Neurophysiol 29: 1011-1027

EVARTS EV (1967) Representation of movements and muscles by pyramidal tract neurons of the precentral motor cortex. In: YAHR MD, PURPURA DP (eds) Neurophysiological basis of normal and abnormal motor activities. Raven, New York, pp 215-253

EVARTS EV, TANJI J (1976) Reflex and intended responses in motor cortex pyramidal tract neurons of monkey. J Neurophysiol 39: 1069-1080

FETZ EE, CHENEY PD (1980) Postspike facilitation of forelimb muscle activity by primate corticomotoneuronal cells. J Neurophysiol 44: 751-772

FROMM C, EVARTS EV (1977) Relation of motor cortex neurons to precisely controlled and ballistic movements. Neurosci Lett 5: 259-266

GHEZ C, SHINODA Y (1978) Spinal mechanisms of the stretch reflex. Exp Brain Res 32: 55-68

HAGBARTH K-E, HAGGLUND JV, WALLIN EU, YOUNG KK (1981) Grouped spindle and electromyographic responses to abrupt wrist extension movements in man. J Physiol (Lond) 312: 81-96

HORE J, PRESTON JB, DURKOVIC RG, CHENEY PD (1976) Responses of cortical neurons (areas 3a and 4) to ramp stretch of hindlimb muscles in the baboon. J Neurophysiol 39: 484-500

JANKOWSKA E, TANAKA K (1974) Neuronal mechanisms of the disynaptic inhibition evoked in primate spinal motoneurons from the corticospinal tract. Brain Res 75: 163-166

JONES EG, WISE SP (1977) Size, laminar and columnar distribution of efferent cells in the sensory motor cortex of primates. J Comp Neurol 175: 391-438

KASSER RJ, CHENEY PD (1982) Double-barreled electrode for simultaneous iontophoresis and single unit recording during movement in awake monkeys. J Neurosci Methods 7: 235-242

KASSER RJ, CHENEY PD (1983) Motor cortical mechanisms mediating reciprocal activation and co-contraction of forearm flexor and extensor muscles in the primate. Soc Neurosci Abstr 9: (to be published)

LEMON RN, MUIR RB (1983) Cortical addresses of distal muscles: a study in the conscious monkey using a spike-triggered averaging technique. Exp Brain Res [Suppl] 7: 230-238

MILLER AD, BROOKS VB (1981) Late muscular responses to arm perturbations persist during supraspinal dysfunction in monkeys. Exp Brain Res 41: 146-158

MUIR RB, LEMON RN (1983) Corticospinal neurons with a special role in precision grip. Brain Res 261: 312-316

PHILLIPS CG (1969) Motor apparatus of the baboon's hand. Proc R Soc Lond [Biol] 173: 141-174

PHILLIPS CG, PORTER R (1977) Corticospinal neurones. Their role in movement. Academic, New York

RANK JB (1975) Which elements are excited in electrical stimulation of the mammalian central nervous system: a review. Brain Res 98: 417-440

SAWYER S, FETZ EE (1981) Effects of single intracortical microstimuli on activity of individual motor units in behaving monkeys. Soc Neurosci Abstr 7: 564

SAWYER S, CHENEY PD, MARTIN RF, FETZ EE (1979) Facilitation and suppression of forelimb muscle activity from single intracortical microstimuli in behaving monkeys. Soc Neurosci Abstr 5: 385

TRACEY DJ, WALMSLEY B, BRINKMAN J (1980) Long-loop reflexes can be obtained in spinal monkeys. Neurosci Lett 18: 59-65

TATTON WG, FORNER SD, GERSTEIN GL, CHAMBERS WW, LIU CN (1975) The effect of postcentral cortical lesions on motor responses to sudden limb displacements in monkeys. Brain Res 96: 108-113

WIESENDANGER M (1973) Input from muscle and cutaneous nerves of the hand and forearm to neurones of the precentral gyrus of baboons and monkeys. J Physiol (Lond) 228: 203-219

Effects of Lesions of the Corpus Callosum on Tactile Cross-Localization

G. Geffen[1], J. Nilsson[1], K. Quinn[1], and E. L. Teng[2]

1 Psychology Discipline, Centre for Neuroscience, The Flinders University of South Australia,
  5042 Australia
2 Division of Biology, California Institute of Technology

Introduction

If a finger on one hand is lightly touched, and vision excluded, a person with an intact brain can localize the correct finger either by touching it with the tip of the thumb of the same hand or by apposing the corresponding finger and thumb of the other hand. Patients who have had a forebrain commissurotomy to relieve intractable epilepsy can indicate which finger was touched using the same hand as was stimulated, but cannot do so with the opposite hand (Sperry et al. 1969). Previous studies of tactile disconnection in humans with complete or partial sections of the corpus callosum have been directed toward establishing whether a cross-localization deficit for various types of task was present, rather than quantifying its extent on a single standardized task. Further, individual results of finger localization, either with the same or the opposite hand responding, have not been correlated with the extent of commissurotomy and extracallosal cortical damage. This information is of interest for two reasons. Firstly, accuracy of localization within each hand could be affected by extracallosal cortical damage. Secondly, in the absence of preexisting cortical damage and with complete commissurotomy, the extent of cross-localization compared with within-hand performance could indicate the functional significance of ipsilateral sensory projections and subcortical interhemispheric connections.

In primates the distal portion of the limbs (hands and feet) are the only regions which do not have direct callosal connections of either sensory or motor areas (Ghez 1981, p. 283). It is proposed that information reaching the primary somatosensory area (S1) in the postcentral gyrus from the contralateral fingers has to be relayed to the posterior parietal association in order to be transferred across the corpus callosum to the opposite hemisphere.

The aims of the present study were (a) to compare the tactile localization performance of subjects with complete and partial section of the corpus callosum to determine whether the extent of the section is correlated with the extent of the deficit in cross-localization of fingers and (b) to determine whether the locus of the partial lesion of the corpus callosum differentially affects cross-localization ability. We investigated left- vs right-hand performance when responding with the same hand, as well as performance when responding with the opposite hand. Task complexity was manipulated by stimulating one vs two or three fingers and crossing of the arms. The stimulated hand could be in the palm-up, or palm-down position. The response hand in the between-hand conditions was either held in the same or the opposite orientation. With the same orientation (both palms up or both down), finger positions are spatially mirror imaged, whereas with opposite orientations (one palm up, the other down) the left-to-right finger positions are similar between the hands.

Methods

Subjects

Four patients who had undergone single-stage surgical division of the forebrain commissures, including the entire corpus callosum and the anterior commissure, as a part of the treatment for intractable epilepsy, were tested at the Psychology Laboratory, California Institute of Technology (C.I.T.). Their case histories have been fully documented (Milner and Taylor 1972). Relevant details are summarized in Table 1.

Five subjects with partial surgical section of the corpus callosum and one subject with a circumscribed lesion of the posterior trunk of the corpus callosum due to a closed head injury caused by a car accident were tested. One subject's (N.F.) case history has been described (Gordon et al. 1971). Case reports for W.F., G.O., and S.A. have previously been published (Jeeves et al. 1979; Geffen et al. 1980).

234

Table 1. Brief case histories of the 4 patients with complete commissurotomy

Patient	Sex	Age when tested	Operation
A.A.	M	30	At age 13 (October 18, 1964) complete section of forebrain commissures with retraction of the right hemisphere, including anterior commissure, corpus callosum, and hippocampal commissure. Massa intermedia probably absent; operation difficult; to relieve epileptic seizures. Residual lesions in the left cerebral arm area and right cerebral leg area. Two-point threshold and position sense of right hand impaired when tested 3 years postoperatively.
R.Y	M	57	At age 43 (March 7, 1966) operation as above. Seizures due to car accident at age 13. Slight spasticity of left hand with motor control of left hand poor.
L.B.	M	29	At age 13 (April 1, 1965) operation as above. Massa intermedia not visualized. No radiological or neurological signs of localiced brain damage preoperatively. Recovery rapid.
N.G.	F	49	At age 30 (September 5, 1963) operation as above, including massa intermedia. Radiological evidence of 1-cm calcified lesion beneath right cortex and EEG signs of abnormality in left posterior temporal region preoperatively. 1-year postoperative EEG normal and no neurological signs of localized cortical damage.

Relevant details for these subjects, as well as W.Y.   and   M.C.,   are summarized in Table 2.

Figure 1 shows the approximate location and extent of the partial lesions of the corpus callosum for all six subjects. Eight control subjects were tested; four 13-year-old males, and four adults (one male, three females, age range 32 to 42 years). Localization of three- and four-finger sequences was tested in 16 intact adults (eight males, eight females, aged between 19 and 29 years).

Procedure

The finger localization task involved light stimulation to the tip of one finger by touching and depressing the skin by about 2 mm for roughly 0.25 s using Lafayette esthesiometer. Movement of the stimulated finger, either by the experimenter or the subject, was avoided. The subject wore a sleep mask and indicated which finger was stimulated by touching it with the thumb of the same hand in the same-hand condition, or by touching the corresponding finger on the opposite hand with the thumb of that hand (opposite-hand condition). With sequences, two or three fingers were touched before a response was required. The interstimulus interval was about 0.25 s. Subjects had to touch the correct fingers in the same order as stimulated. The subject was told which hand was to be stimulated, how many fingers would be touched, and which hand was to respond. The test was demonstrated without the sleep mask until it was understood. No feedback concerning the accuracy of response was given in any trials.

Twenty-four random one-finger and 24 random two- or three-finger sequences were generated is sets of 16 trial sequences for each condition, with the constraint that each finger had to be represented within each sequence of 16 trails the same number of times, and within each two- or three-finger trial each finger was stimulated only once. Each condition was counterbalanced in an ABBA design to control for fatigue and practice effects and the test given in two sessions, 1 week apart. By reversing the first order of presentation for the second session, counterbalancing of all the factors was maintained. Results were summed across palm orientation of the stimulated hand (up/down) to yield a total of 32 trials per condition. There were 768 trials altogether, 384 with one finger stimulated per trial and 384

Table 2. Brief case histories of patients with partial lesions of corpus callosum

Patient	Sex	Age when tested	Operation
N.F.	F	38	At age 26, section of anterior commissure, anterior 5 cm of corpus callosum, and underlying hippocampal commissure; retraction of right hemisphere. Section included all of the genu, body, and anterior part of the splenium of corpus callosum; to relieve epilepsy.
W.F	F	25	At age 14, 3-cm section of trunk of corpus callosum posterior to the genu to remove IIIv cyst. Pillars of the fornix transected.
G.O.	M	72	At age 63, 3-cm section of trunk of corpus callosum to remove IIIv cyst. Right fornix transected.
W.Y.	M	61	At age 61, section of genu of corpus callosum to remove IIIv cyst. Tested pre- and post-operatively.
S.A.	M	18	At age 14, 2-cm section of anterior splenium of corpus callosum to remove AVM.
M.C.	M	20	No operation. Closed head injury sustained in car accident at age 17. CAT scan showed circumscribed hematoma in caudal half of body of corpus callosum.

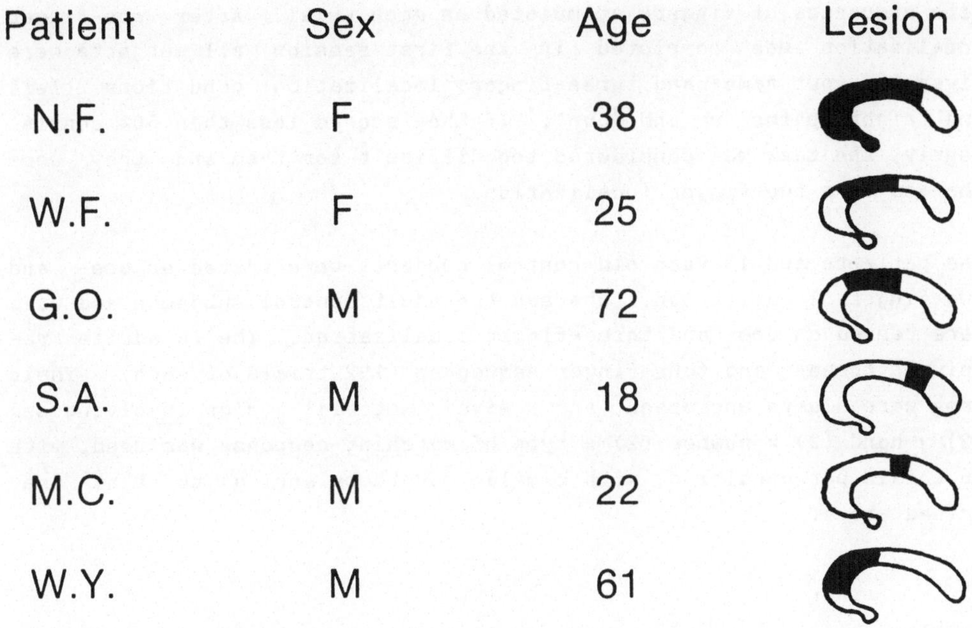

Patient	Sex	Age	Lesion
N.F.	F	38	
W.F.	F	25	
G.O.	M	72	
S.A.	M	18	
M.C.	M	22	
W.Y.	M	61	

Fig. 1. Approximate location and extent of partial lesions of the corpus callosum in 6 patients assessed from CAT scan (M.C.), operation notes (N.F.), operation notes and angiography (G.O., S.A.), and neurosurgeons' indications on diagrams (W.F., G.O., W.Y.)

with sequences of fingers stimulated on each trial. After one-finger localization was completed in the first session, all subjects were given the four same-hand three-finger localization conditions (left and right palms up and down). If they scored less than 50% consistently, the task was considered too difficult for them and they were changed to a two-finger localization.

The patients and 13-year-old control subjects were tested on one- and two-finger localization, whereas the adult control subjects and W.Y. were tested on one- and three-finger localization. The 16 adults received three- and four-finger sequences (192 trials of each). Their arms were always uncrossed, and a mixed factorial design involving sex (2) x hand (2) x number (2) x type of matching response was used, with 16 trials per condition, with similar counterbalancing to that described above.

Results

The finger localization task yielded two measures of accuracy and two types of order errors. The accuracy measures were: trials correct, including item (correct finger) and sequence information (correct order), and items correct, regardless of order of stimulation. Position errors were scored when a finger indicated in response had occurred in the stimulus sequence, but was not in the same position in the response sequence. Sequence errors were scored only if all the items were correct, but the pair relationships (first and second, first and third, second and third pairs) were reversed or disordered.

Between-group analyses were conducted to examine the effects of extent of corpus callosum section (complete vs trunk vs none/intact) on cross-localization. The groups consisted of the four subjects with forebrain commissurotomy and the four with lesions involving only the trunk of the corpus callosum. The adult control data was used for the analysis of trials correct. For the other dependent measures, the younger control group was a more appropriate comparison because they had the same number of fingers stimulated as the lesioned groups.

Five-way analysis of variance [group (3) x type of matching response (3) x number of fingers (2) x hand (2) x arm position (2)] were conducted on trials and items correct. On trials correct there was a significant main effect of extent: $F(2,9) = 80.13$, $p < 0.001$. However, a significant three-way interaction, extent x number x type of match [$F(4,18) = 19.27$, $p < 0.001$] modified this as well as the significant two-way interactions of extent x number, and number x type of match. Mean percentages are shown in Figs. 2a and b. With a single-finger stimulus and same-hand response, performance was close to perfect in all three groups. However, with a response by the other hand, the accuracy of the subjects with commissure sections declined by an average of 61%. The group with partial sections showed a 21% decrease, and the controls a 5% decrease. With two fingers stimulated (Fig. 2b), all three subject groups performed significantly less accurately, except for the control group when responding with the same hand.

Figure 2c shows the percentage of fingers correctly indicated, regardless of the order in which they were stimulated. The subjects with complete sections performed much less accurately than the control and partially sectioned groups (who did not differ significantly), particularly during cross-localization with two fingers stimulated. Comparison of Figs. 2b and c indicated that the lesioned subjects showed a limited ability to reproduce the correct stimulus sequence, the deficit being most apparent in the cross-localization of sequences for the group with complete sections.

Since the subjects with complete sections performed at chance level when responding with the other hand and especially with stimulus sequences, the order error measures dependent upon correct finger responses would be depressed in those conditions. Thus, the results of the two groups with lesions were separately analyzed.

Cross-localization of sequences was performed very poorly (see Table 3). However, the 14% correct sequences scored by L.B. when responding with the other hand was comprised of 19% correct left-to-right (L $\rightarrow$ R) transfer and 8.6% in the opposite direction. Since the chance correct rate for a two-finger sequence is 1/12 or 8.3%, application of the binomial test yields a L $\rightarrow$ R z score of 4.45, $p = 0.00003$. Similarly, N.G. scored 14% correct transferring tactile sequence information L $\rightarrow$ R ($z = 2.2$, $p = 0.0139$) and 4% R $\rightarrow$ L. Transfer in either direction was at chance levels for A.A. and R.Y. The transfer deficiency appears smaller for sequential than for single-finger localization because sequence scores with the same hand responding were also depressed. Thus, a deficit of tactile

240

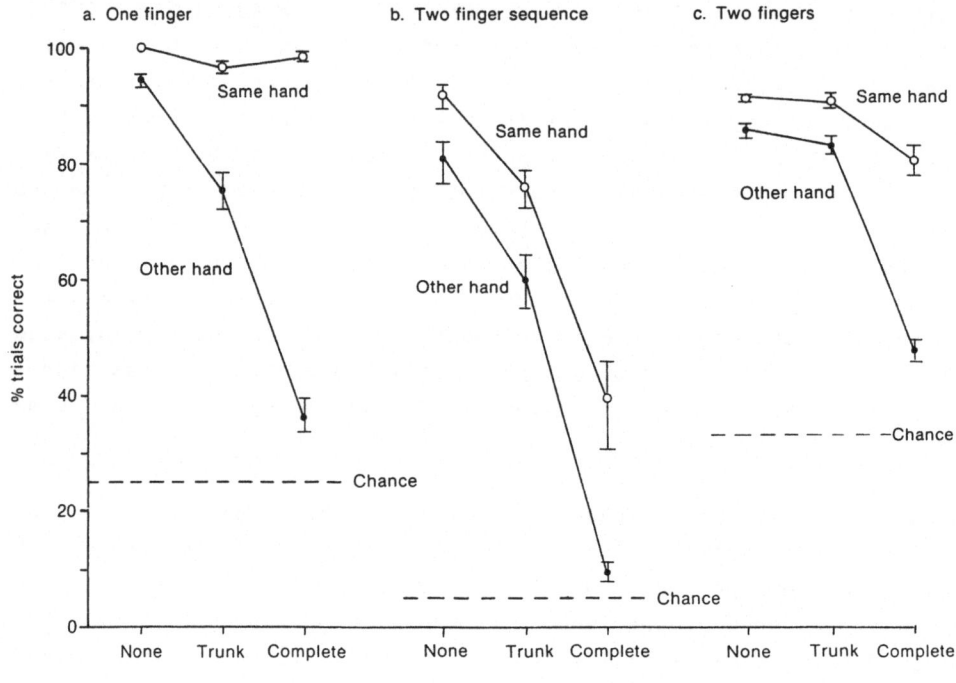

Fig. 2. Mean percent trials correct for one **(a)** and two **(b)** fingers stimulated for *control* subjects and subjects with partial lesions *(trunk)* and *complete* commissurotomies ($N = 4$ in each group) when the same hand (0-0) and the other hand (●-●) responded. **c** Mean percent items correct with 2 fingers stimulated, regardless of the order of fingers indicated in response. Scores were averaged across palm orientation of stimulated hand, left and right hands, and crossed vs uncrossed arms. For the "other-hand" data points, results were averaged across orientation of responding hand as well. Thus, for the same-hand condition there were 256 trails per subject per data point, and for the other-hand condition there were 512 trials per subject per data point. The *vertical bars* indicate standard errors

Table 3. Mean percent items and trials correct for subjects with complete commissurotomies

		Response hand		
Subject	Number of fingers stimulated	Same	Other	Difference
A.A.	1	99	28	71
	2*	16	8	8
R.Y.	1	97	30	67
	2	54	8	46
L.B.	1	97	52	45
	2	46	14	32
N.G.	1	98	36	62
	2	38	9	29
$\bar{X}$ (SD)	1	98 (1)	37 (11)	61 (11)
	2	38 (16)	10 (3)	28 (16)
Control	1	100 (0)	95 (6)	5 (6)
$\bar{X}$ (SD)	2	92 (5)	81 (11)	11 (6)

* Correct fingers in same order as stimulated

Scores were summed and averaged across the factors of hand, palm orientation of stimulated hand, and arm position, yielding 128 observations per subject mean. The other-hand responses are summed across same and opposite palm orientations as well, yielding 256 observations per subject mean cross-localization is more reliably assessed by one-finger stimulation. Chance correct performance is 1/4 or 25%. While cross-localization of single fingers by L.B. and N.G. was significantly worse than their same-hand performance, their between-hand scores were significantly better than chance (L.B. $z = 0.03$, $p = 0.00003$; N.G. $z = 4.26$, $p = 0.00003$) in either direction. A.A. obtained higher scores transferring L → R (32%, $z = 1.74$, $p = 0.0409$) than vice versa (24%). Similarly, R.Y.'s L → R transfer (33%, $z = 1.94$, $p = 0.0262$) was superior to his R → L score (27%). In summary, tranfer of the correct stimulus sequence produced the worst performance. However, limited transfer of information about which fingers were stimulated was possible, even with complete forebrain commissurotomy.

Partial Section of the Corpus Callosum

Four-way analysis of variance position and sequence errors revealed significant main effects of type or match: position $F_{(2,6)} = 11.79$, $p < 0.01$; sequence $F_{(2,6)} = 12.24$, $p < 0.01$. The means of the position errors are shown in Fig. 3. Responding with the other hand in the same orientation as the stimulated hand (mirror imaged) produced more order errors than either of the other conditions. Cross-localization improved when the left-right or right-left sequence was the same in the response and the stimulated hand. The intact subjects showed this result on all measures (see Fig. 4, for trials correct), while the position and sequence error data was more sensitive to this effect in subjects with partial sections.

Discussion

The difference between same- and other-hand scores provides an index of cross-localization deficit. Complete forebrain commissure section was found to produce a mean deficit of 61% with a single finger stimulated. Partial section involving the trunk of the corpus callosum produced a mean deficit of 21%. With only the splenium spared (N.F.), the deficit (23%) did not increase. After genu section (W.Y.), the 3% difference did not differ from that shown by the adult control subjects (5%) or W.Y.'s preoperative scores. These results implicate the

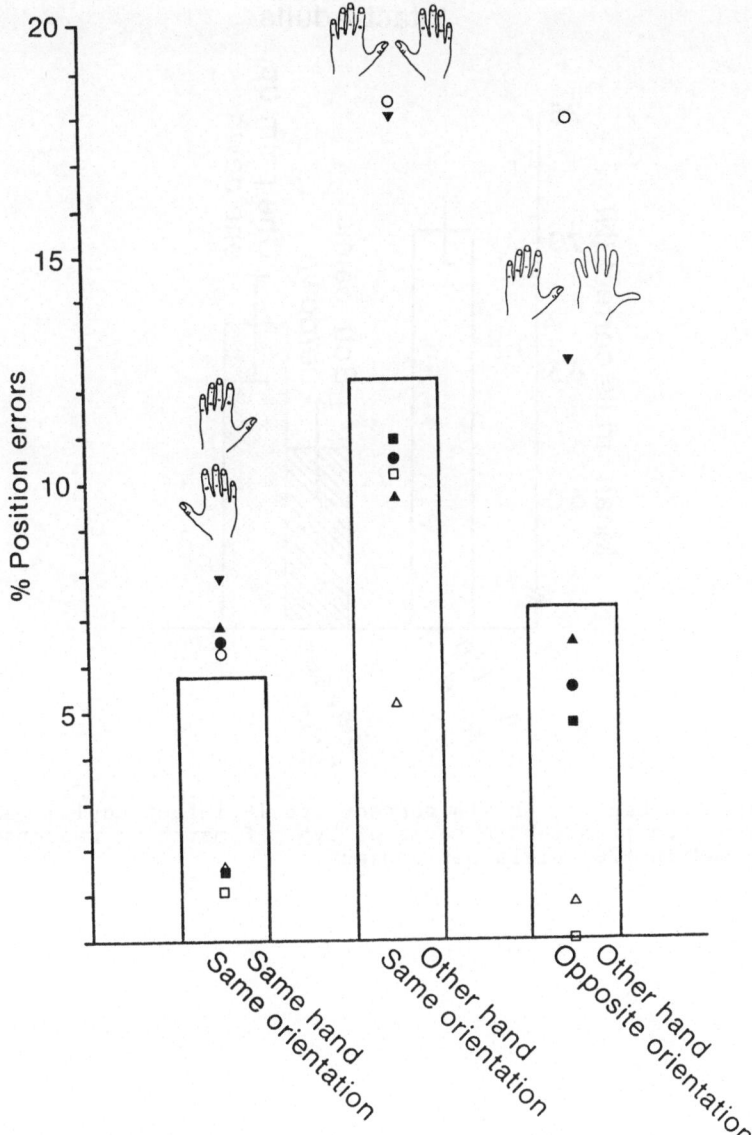

Fig. 3. Mean percent position errors with two-finger localization in subjects with partial section of the corpus callosum in terms of type of matching response, 256 trials per subject contributed to each mean. ● W.F., ■ G.O., ▼ S.A., ▲ M.C., O N.F., □ W.Y. pre-, △ W.Y. postoperation

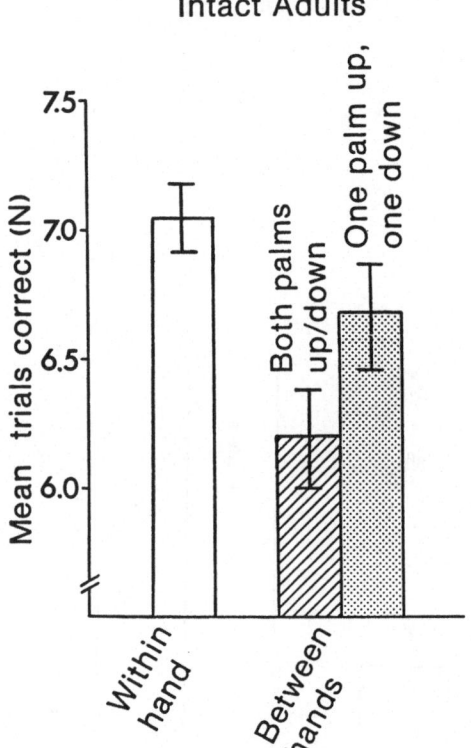

Fig. 4. Mean numbers of trails correct for 16 intact adults performing three-finger localization in terms of type of matching response. Each mean is based on 128 trials per subject

functional specificity of the trunk of the corpus callosum for the interhemispheric transfer of simple tactile information, quantifying
previous clinical reports (Ettlinger 1965; Gazzangia et al. 1975)
and extending the experimental findings on monkeys (Myers and Ebner
1976) to humans. However, the functional specificity is not absolute,
since a considerably reduced deficit (by 40%) was found with only the
splenium spared. Thus, the splenium of the corpus callosum appears to
be responsible for the extra 40% capability of intermanual transfer.
Schwartz and Goldman-Rakic (1982) have described two types of
cross-callosal projections in rhesus monkeys: (a) homotopic callosal
neurons, which project an axon across the callosum to the same cytoarchitectonic cortex in the contralateral hemisphere and (b) heterotopic
callosal neurons, with the target in the contralateral hemisphere
being an area different from the site of origin. If similar cortical
neurons exist in human brains, this would account for the spared portion of the corpus callosum, viz, the splenium, carrying out the interhemispheric transfer of tactile stimuli.

The subject with only the splenium spared, N.F., differed from the
subjects with partial sections whose sections did not extend anteriorly because she showed a large transfer deficit with two-finger sequences: 42% compared with the mean of 16% shown by the other four
subjects. It is possible that information about sequences of stimuli
could be transferred between the frontal lobes in the four subjects
with the more rostral portion of the corpus callosum spared. Since
genu section on its own did not result in an increase in the transfer
deficit (W.Y.), the rostral portion of the corpus callosum would appear to be used only when the trunk of the corpus callosum cannot be
used.

Evidence for limited interhemispheric transfer of tactile information
that could not have been transmitted across the corpus callosum was
found. It is proposed that ipsilateral sensory information available
from the left hand in the left hemisphere sometimes enabled the correct right-hand responses. While the right hemisphere could use similar ipsilateral sensory information from the right hand to control
correct left-hand, single-finger localization, it could not translate
sequential information from the right hand into a correct left-hand
response.

The orientation of the response hand produced no differences in performance in the subjects with complete sections. However, the intact

subjects showed worst performance with a mirror image transfer of the finger sequence. With one palm up and the other down, the direction of the stimulus sequence was maintained in the response and facilitated performance. Since all the subjects with partial sections also showed the most order errors when a mirror image transfer was required, it appears that the splenium of the corpus callosum relays hand position to the opposite hemisphere.

The absence of any effects due to crossing of the arms suggests that the cortical representation of the limbs remains constant whether they are in the same or in the opposite halves of space.

Acknowledgements. This work was supported by grants from the Australian Research Grants Scheme to Gina Geffen.

We are grateful to Dr. R.W. Sperry for access to L.B., N.G., R.Y., N.G., and N.F., and laboratory facilities during May, 1981. Mr. D.A. Simpson referred W.F., G.O., and S.A. Dr. J. Willoughby, Mr. D.A. Simpson, and Mr. P. Reilly referred W.Y. Mr. T.A.R. Dinning and Dr. M. Wood referred M.C. We are grateful for the cooperation of these doctors and to the subjects for their time.

References

ETTLINGER EG (1965) Functions of the corpus callosum. Churchill, London

GAZZANIGA M (1978) The integrated mind. Plenum, New York

GAZZANIGA M, RISSE GL, SPRINGER SP, CLARK E, WILSON DH (1975) Psychologic and neurologic consequences of partial and complete cerebral commissurotomy. Neurology 25: 10-15

GEFFEN G (1980) Fusion after partial section of the corpus callosum. Neuropsychologia 18: 613-620

GEFFEN G, WALSH A, SIMPSON D, JEEVES M (1980) Comparison of the effects of transcortical and transcallosal removal of intraventricular tumours. Brain 103: 773-788

GHEZ C (1981) Introduction to the motor system. In: KANDEL ER, SCHWARTZ JH (eds) Principles of neural science. Elsevier, North-Holland

GORDON HW, BOGEN JE, SPERRY RE (1971) Absence of deconnexion syndrome in two patients with partial section of the neocommissures. Brain 94: 327-336

JEEVES MA, SIMPSON DA, GEFFEN G (1979) Functional consequences of the transcallosal removal of intraventricular tumours. J Neurol, Neurosurg Psychiatry 42: 134-142

MILNER B, TAYLOR L (1972) Right hemisphere superiority in tactile recognition after cerebral commissurotomy: evidence of non-verbal memory. Neuropsychologia 10: 1-15

MYERS RE, EBNER FF (1976) Localization of function in corpus callosum: tactual information transmission in *Macaca mulatta*. Brain Res 103: 455-462

SCHWARTZ ML, GOLDMAN-RAKIC PS (1982) Single cortical neurones have axon collaterals to ipsilateral and contralateral cortex in foetal and adult primates. Nature 199: 154-155

SPERRY RW, GAZZANIGA MS, BOGEN JE (1969) Interhemispheric relationships: the neocortical commissures, syndromes of hemispheric disconnection. In: VINKEN PJ, BRUYN GW (eds) Handbook of clinical neurology, vol 4. Elsevier, Amsterdam

Activity of Dentate and Interpositus Neurons During Maintained Isometric Prehension

A. M. Smith[1], R. Wetts[2], and J. F. Kalaska[1]

1 Université de Montréal, Départment de physiologie, Faculté de médicine, Case postale 6208, Succursale A, Montréal, Québec, H3C 3J7, Canada
2 Developmental Genetics Laboratories, The Johns Hopkins Hospital, Baltimore, MD 21205, USA

Introduction

Smooth and coordinated movement requires the proper pattern of activation of both agonist and antagonist muscles. The incoordination resulting from inappropriate activation of antagonist muscles was called *asynergia* by Babinski and is symptomatic of cerebellar dysfunction (Babinski 1899, 1902, 1909; Tilney and Pike 1925; Thomas 1925; Rondot et al. 1979). From these early clinical studies it would appear that the cerebellum plays a key role in selecting whether antagonist muscles are contracted or inhibited simultaneously with the contraction of the agonists. However, because of the difficulty in inferring neurophysiological mechanisms on the basis of deficits following lesions, it was decided to apply single-unit recording methods for awake animals to the study of how cerebellar neurons behave when under different modes of muscular activation.

In this experiment, we studied the output of the cerebellum by recording the activity of single units in two cerebellar nuclei while a monkey performed a lateral, isometric pinch of the thumb and forefinger. This grasp elicits a coactivation of all forearm muscles; this observation has been substantiated by individual EMG recordings published previously (Smith 1981). Single-cell recordings made from the culmen-simplex area of the cerebellar cortex revealed that the majority of phasic Purkinje's cells decreased discharge during the cocontraction of the antagonist muscles associated with isometric pinching

Experimental Brain Research, Suppl. 10
© Springer-Verlag Berlin · Heidelberg 1985

(Smith and Bourbonnais 1981; Frysinger et al., to be published). Although Purkinje's cells are known to inhibit the neurons of the deep nuclei, some investigators have reported that in certain circumstances, the discharge frequency of Purkinje's cells and nuclear cells can increase simultaneously (Mortimer 1973, 1975; Thach 1970a, b). If Purkinje's and nuclear cell activity covaries, then one would predict that the majority of nuclear cells should decrease activity during prehension because the majority of Purkinje's cells decrease firing frequency. The alternative is that since the Purkinje's cells are inhibitory neurons (Eccles et al. 1967) and the major afferent to the cerebellar nuclei (Chan-Palay 1977), the discharge changes in the nuclear cells should be opposite in direction from the Purkinje's cell response. This latter alternative was observed in the present experiment, that is, during antagonist cocontraction, most cerebellar nuclear cells increased their discharge, which is opposite to the previously observed Purkinje's cell response.

Material and Methods

Two female monkeys *(Macaca fascicularis)* were trained to perform the sustained prehension task of pinching a transducer between the thumb and forefinger and maintaining a force level between a minimum and a maximum threshold for 1 s (Smith and Bourbonnais 1981). For purposes of analysis, the dynamic phase of the task was defined as the period from the initial increase of finger pressure until the minimum force threshold had been achieved. The static period was defined as the 500 ms of isometric force preceding the reward. At the completion of training, a microdrive receptacle and head-restraining bolts were implanted under general anesthesia, as described by Evarts (1965). Glass-coated tungsten microelectrodes were driven dorsoventrally through the cerebellum and the anterior lobe of the cerebellar cortex (identified by the peripheral receptive fields of the units) and into the deep nuclei (Fig. 1). Single-unit activity, prehensile force, and EMGs from the dorsal and ventral surfaces of the forearm were recorded on magnetic tape during performance of the task. These data were later analyzed with the aid of a laboratory computer.

250

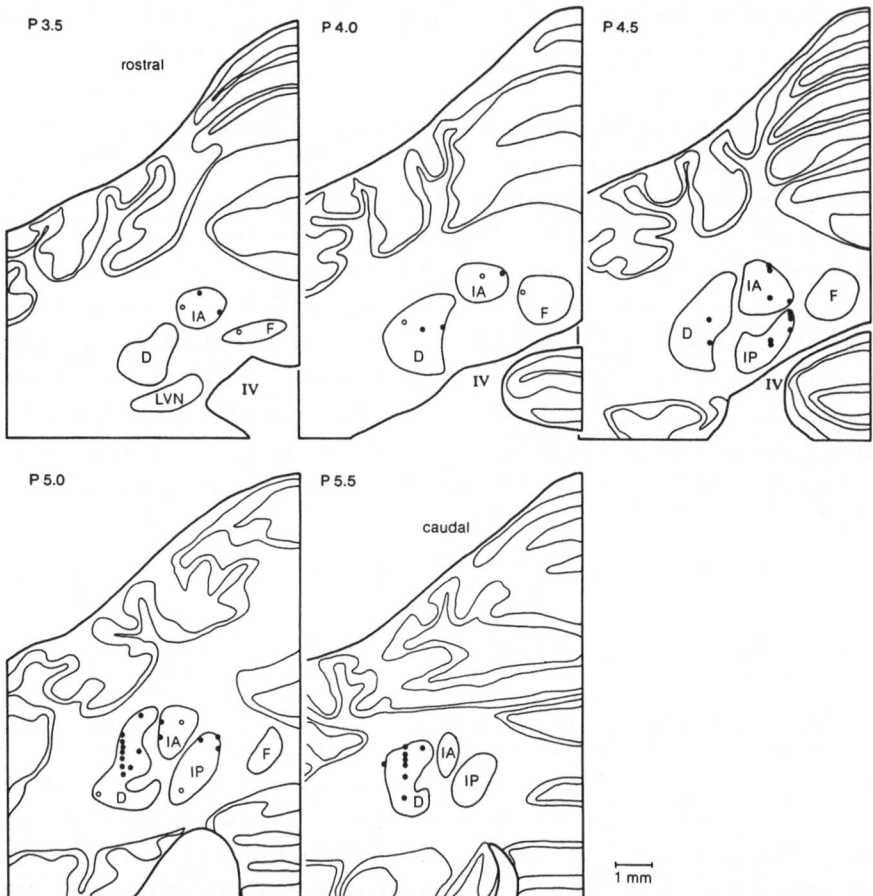

Fig. 1. Location of task-related cells in the cerebellar nuclei. One
of the two monkeys is shown here to illustrate that many of these
calls *(filled circles)* are located in the posterior third of the dentate, but that some task-related cells are located throughout the interpositus and dentate nuclei. The monkey was perfused with formalin,
and its brain was removed by dissection, embedded in paraffin, and
sectioned at 7 μm. Each drawing is a composite of 5 sections and
represents 500 μm. Electrode coordinates taken during the single-unit
recordings were aligned with the drawings, and the cell positions were
marked based on the coordinates. *Open circles* indicate those electrode trajectories which did not encounter any task-related units.
Seven units from this monkey were discarded because they were not located within the boundaries of the nuclei. *D,* dentate; *F,* fastigial;
NIA, interpositus anterior; *NIP,* interpositus posterior; *LVN,* lateral vestibular; *IV,* fourth ventricle

Results

To date, we have encountered 78 cells which displayed a change in firing frequency during the pinching task. Of these 78 cells, 23 were
located in the interpositus nucleus (either the anterior or the posterior subdivision), and 55 were located in the dentate nucleus
(Fig. 1). For both nuclei, the vast majority of cells (73) displayed
an increased discharge frequency during the application of the isometric precision grip (Table 1). Three cells displayed both increased
and decreased activity (at different times during the task), and only
two cells had decreased activity.

The excited cells were further divided into three groups. The cells
of the dynamic class displayed a peak of excitation during the dynamic
period, but had a discharge frequency similar to the resting frequency
during the static phase. In the static class, the increase in discharge frequency during the dynamic phase was maintained at about the
same level during the static period (Fig. 2). In the mixed
dynamic-static class, excitation again occurred throughout the duration of the task, but the increase in activity was greater during the
dynamic phase (Fig. 3). Decreases in discharge frequency below the
resting frequency were observed during either phase, but these cells
with diminished activity were so few in number that they could not be
further divided into subgroups.

Table 1. Change in discharge frequency during the application of the
isometric precision grip

	Interpositus		Dentate	
Increased activity				
- Dynamic	1	(4%)	8	(15%)
- Static	13	(57%)	20	(36%)
- Mixed	9	(39%)	22	(40%)
Increased-decreased activity	0	(0%)	3	(5%)
Decreased activity	0	(0%)	2	(4%)
Total	23	(100%)	55	(100%)

252

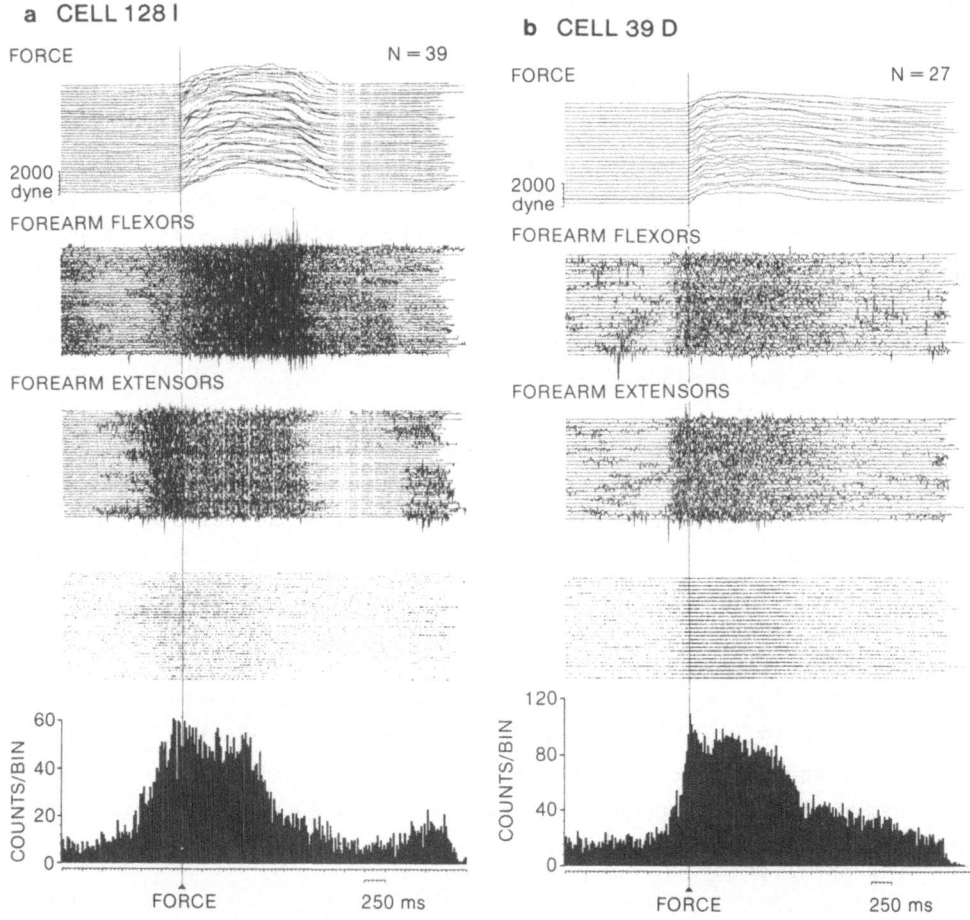

a CELL 128 I

b CELL 39 D

Fig. 2a,b. Nuclear cells with increased activity during the static phase. Cell 128 I **(a)** was located in the interpositus, and cell 39D **(b)** was located in the dentate. The histograms illustrate the pattern of cell activity during the task. It can be seen that the increase in cell discharge was similar for both dynamic and static periods. Also shown are the prehension forces, surface EMG traces, and the cell rasters for each trial. All are aligned on the force onset

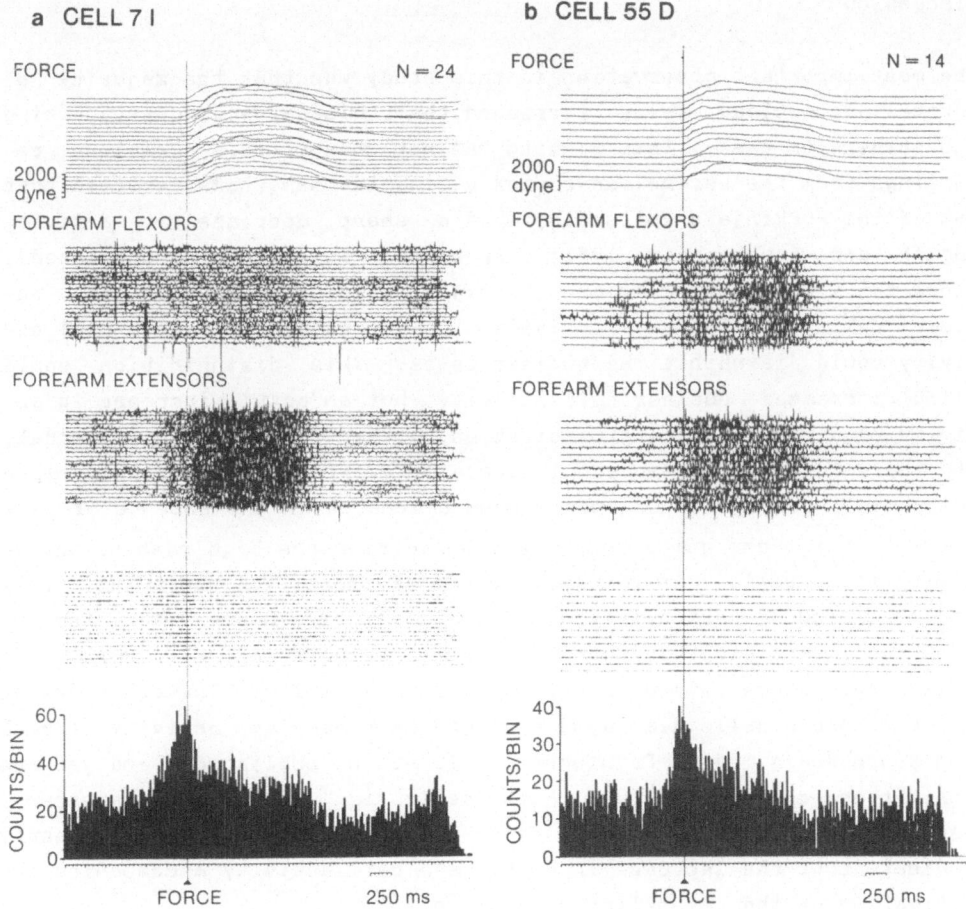

a CELL 7 I   **b** CELL 55 D

Fig. 3a,b. Nuclear cells possessing a dynamic peak of activity. Cell 7I **(a)** was located in the interpositus, and cell 55D **(b)** was located in the dentate. The arrangement is the same as in Fig. 2. In these cells, the increased discharge frequencies during the dynamic phase are noticeably greater than the increase during the static period

Discussion

The most important observation in this study was that the majority of
cerebellar nuclear cells increased their discharge frequency during
the prehension task. This finding was not unexpected. Previous recordings from the cerebellar cortex during the same task revealed that
66% of the Purkinje's cells displayed a sharp decrease in activity
(Smith and Bourbonnais 1981;   Frysinger et al. to be published).
Since the Purkinje's cells are the major inhibitory input to the nuclear cells (Eccles et al. 1967), the decreased Purkinje's cell activity could disinhibit the nuclear cells. This disinhibition would
allow increased nuclear cell activity, but an actual increase in activity requires a source of excitation, either internal or external.
If there existed an internal source of excitation, then the nuclear
cells would be capable of spontaneous discharge in the absence of any
input. This seems quite possible, considering the high resting activity of these cells. An external source of excitation, such as mossy
and climbing fiber collaterals, is also likely, since 94% of the nuclear cells are excited, wheras a smaller majority of the Purkinje's
cells (only about 66%) displayed decreased activity. The other 34% of
the Purkinje's cells displayed a slight increase in activity (Smith
and Bourbonnais 1981; Frysinger et al. to be published), and yet the
inhibition contributed by these cells is not very apparent.
Regardless of the source of excitation, the present results clearly
indicate that the decrease in Purkinje's cell activity accompanies the
activation of the cerebellar nuclear cells.

Phasic cells were located throughout the ipsilateral interpositus and
dentate nuclei (Fig. 1). Thach et al. (1982) localized hand-related
cells primarily to the middle third of both nuclei, but these authors
emphasized that some hand-related cells could also be found in all regions of the nuclei. It is possible that some of our cells were related to nonhand areas of the body which consistently moved during
each prehension. However, out of the 78 cells analyzed, the discharge
frequencies of 13 cells were significantly ($P < 0.05$) correlated with
the rate of force increase, the activities of five cells were correlated with static force, and the frequencies of four cells were correlated with both parameters (Fig. 4). In addition the responses to microstimulation at the recording sites frequently yielded movements of
the wrist and fingers. Together these observations make us confident
that most of the cells analyzed were related to movements of the hand.

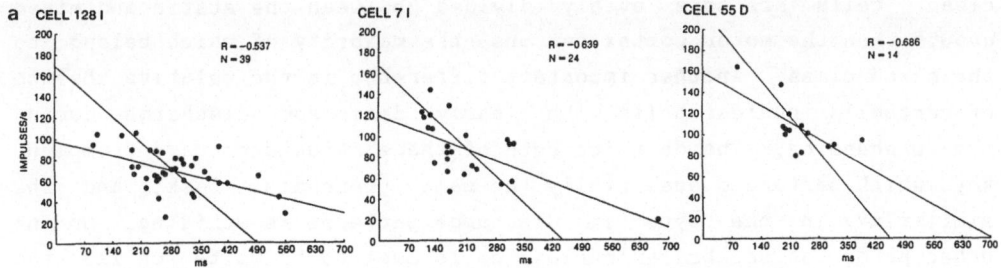

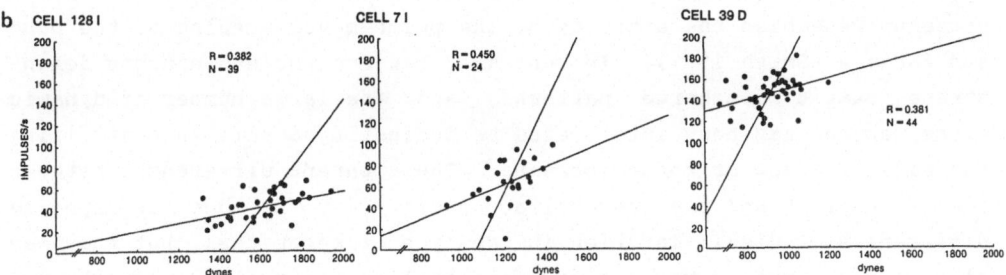

Fig. 4a,b. Correlations with task parameters. Each *dot* represents the average discharge frequency during the dynamic (a) or static (b) phases of a single trial. In a, cell discharge frequency during the dynamic period is graphed as a function of the duration of the dynamic period. Since the end of the dynamic period is defined by the minimum force threshold, the duration is inversely proportional to the rate of force increase. Thus, the discharge frequencies of these three cells are directly correlated with the rate of force increase. In b, the discharge frequencies of the three cells are directly correlated with the amount of force applied during the last 500 ms of the task

A comparison of the types of discharge patterns revealed no difference between interpositus and dentate neurons (Figs. 2 and 3). Although all five neurons with decreased discharge were located in the dentate (Table 1), this is probably due to greater sampling from the dentate, rather than reflecting any difference in function between the two nuclei. The classification of excited neurons into dynamic, static, and mixed classes has been previously used in the analysis of neuronal activity in the motor cortex (Smith et al. 1975) and the red nucleus (Ghez and Vicario 1978; Ghez 1981). Cerebellar nuclear cells were also grouped by these criteria, and this permitted the distribution of neurons into the three classes to be compared among the three structures. The cerebellar nuclear cells are similar to the cells of the motor cortex, in as much as both have relatively few neurons of the dynamic class. Cells in the mixed class also have a dynamic response, in addition to activity during the static phase. The cerebellar nuclear cells are more evenly divided between the static and mixed groups than the motor cortex neurons, the majority of which belong to the mixed class. Another important difference is the relative absence of cerebellar nuclear cells with clearly decreased discharge during the prehension. The data for both of these structures came from monkeys which performed essentially the same prehension task, and the similarity in the type of discharge patterns is striking. On the other hand, the cerebellar nuclear cells seem to be quite unlike the neurons of the red nucleus. Most of the cerebellar neurons are excited during the static phase (both static and mixed classes), and this pattern resembles the activity of the prime-mover muscles of the hand and forearm (Smith 1981). In contrast, few red nucleus neurons demonstrate static discharge patterns, and the large number of dynamic class neurons has been interpreted as indicating a role in controlling the rate of muscular force increase. The apparent differences between the red nucleus and the cerebellar nuclei may be due to species differences or dissimilarities in the type of motor task, but they may also reflect further processing of cerebellar afferents within the red nucleus itself (Tsukahara et al. 1968; Jeneskog and Padel 1983).

In summary, the overwhelming majority of cerebellar nuclear cells increase their discharge frequency during an isometric pinching task. This observation is consistent with the fact that most of the inhibitory Purkinje's cells sharply decrease their activity during the same task. These results provide further evidence that the cerebellum may play an important role in the voluntary cocontraction of antagonist and agonist muscles.

Acknowledgement. This research was supported by a grant to the MRC Group in Neurological Sciences at the University of Montreal.

References

BABINSKI J (1899) De l'asynergie cérébelleuse. Rev Neurol 7: 806-816

BABINSKI J (1902) Sur le rôle du cervelet dans les actes volitionnels nécessitant une succession rapide de mouvements (diadococinésie). Rev Neurol 10: 1013-1014

BABINSKI J (1909) Quelques documents réfletifs à l'histoire des fonctions de l'appareil cérébelleux et de leurs perturbations. Rev Interne 1: 113-129

CHAN-PALAY V (1977) Cerebellar dentate nucleus: organization, cytology and transmission. Springer, Berlin Heidelberg New York, p 548

ECCLES JC, ITO M, SZENTAGOTHAI J (1967) The cerebellum as a neuronal machine. Springer, Berlin Heidelberg New York

EVARTS EV (1965) Relation of discharge frequency to conduction velocity in pyramidal tract neurons. J Neurophysiol 28: 216-228

FRYSINGER RC, BOURBONNAIS D, KALASKA JF, SMITH AM (to be published) Cerebellar cortical activity during antagonist co-contraction and reciprocal inhibition of forearm muscles. J Neurophysiol

GHEZ C (1981) Cortical control of voluntary movement. In: KANDEL ER, SCHWARTZ JH (eds) Principles of neural science. Elsevier, New York, pp 323-333

GHEZ C, VICARIO D (1978) Discharge of red nucleus neurons during voluntary muscle contraction: activity patterns and correlations with isometric force. J Physiol (Paris) 74: 283-285

JENESKOG T, PADEL Y (1983) Cerebral cortical areas of origin of excitation and inhibition of rubrospinal cells in the cat. Exp Brain Res 50: 309-320

MORTIMER JA (1973) Temporal sequence of cerebellar Purkinje and nuclear activity in relation to the acoustic startle response. Brain Res 50: 457-462

MORTIMER JA (1975) Cerebellar response to teleceptive stimuli in alert monkeys. Brain Res 83: 369-390

RONDOT R, BATHIEN N, TOMA S (1979) Physiopathology of cerebellar movement. In: MASSION J, SASAKI K (eds) Cerebro-cerebellar interactions. Elsevier, New York, pp 203-230

SMITH AM (1981) The coactivation of antagonist muscles. Can J Physiol Pharmacol 59: 733-747

SMITH AM, BOURBONNAIS D (1981) Neuronal activity in cerebellar cortex related to control of prehensile force. J Neurophysiol 45: 286-303

SMITH AM, HEPP-REYMOND MC, WYSS UR (1975) Relation of activity in precentral cortical neurons to force and rate of force change during isometric contraction of finger muscles. Exp Brain Res 23: 315-332

THACH WT (1970a) Discharge of cerebellar neurons related to two maintained postures and two prompt movements. I. Nuclear cell output. J Neurophysiol 33: 527-536

THACH WT (1970b) Discharge of cerebellar neurons related to two maintained postures and two prompt movements. II. Purkinje cell output and input. J Neurophysiol 33: 537-547

THACH WT, PERRY G, SCHIEBER (1982) Cerebellar output: body map and muscle spindles. In: PALAY S, CHAN-PALAY V (eds) The cerebellum: new vistas. Springer, Berlin Heidelberg New York, pp 440-454 (Experimental brain research, suppl 6)

THOMAS A (1925) Pathologie du cervelet. In: ROGER GH, WIDAL F, TESSIR LJ (eds) Nouveau traité de médecine. Masson, Paris

TILNEY F, PIKE FH (1925) Muscular coordination experimentally studied in its relation to the cerebellum. Arch Neurol Psychiatr 13: 289-334

TSUKAHARA N, FULLER DRG, BROOKS VB (1968) Collateral pyramidal influences on the corticorubrospinal system. J Neurophysiol 31: 467-484

Topographical Aspects of Cerebral Cortex and Dentate Nucleus in the Control of Hand Movements

L. Rispal-Padel, F. Cicirata, and C. Pons

Départment de Neurophysiologie générale, INP, CNRS, B.P.71, 13277 Marseille Cedex 9, France

Introduction

Cortical control of movement is influenced by the cerebellum, which is involved together with the motor cortex in a central loop, the cerebellocorticocerebellar loop, which regulates the motor cortical activity (Evarts and Thach 1969; Massion 1973; Allen and Tsukahara 1974; Sasaki 1979). In addition, the disynaptic cerebellothalamocortical pathway contributes to the cortical output to the spinal centers.

Phylogenetic data suggest that the role of the cortex in the control of hand movements is particularly subject to the effects of the neocerebellum. In primates the considerable development conjointly affecting the neocerebellum and the neocortex occurs in parallel with the evolution of the hand and its increased utilization. Furthermore, it has been demonstrated that the motor cortical area receives significant projections from the dentate nucleus (neocerebellum) (Sasaki et al. 1976). There remain unresolved questions concerning both the spatial organization of the neocerebellar projections to the motor cortical hand region and their eventual participation in the command of hand muscles by way of the corticospinal pathways.

The first question is raised from observation of deficits caused either by cortical or by cerebellar lesions. Whereas the first type of lesion principally affects elementary movements (Napier 1962; Lawrence and Kuypers 1968; Hepp-Reymond and Wiesendanger 1972; Hepp-Reymond et al. 1974), cerebellar lesions mainly affect motor coordination (Babinski 1899; Holmes 1939). One is led to question whether these different types of deficit correspond to the loss of structures which are involved in the spatial organization of motor coordination. Are cerebellar motor incoordinations the result of the sum of deficits affecting the command of each of their elementary components (Holmes 1939; Brooks 1979)? Or are these cerebellar deficits due to the loss of a structure whose organization is responsible for motor synergies?

The second question concerns the role of dentate nucleus projections to the corticospinal system. The localization of dentatocortical projections onto the superficial layers of the motor cortex (Sasaki et al. 1976) and the impossibility of inducing hand movements through stimulation of the dentate nucleus in the anesthetized animal (Schultz et al. 1979) leads one to suppose that the connections with the corticospinal cells of the hand area are rather weak. The dentatocortical projections would be essentailly implied in the central loop associated with movement control. However, these data do not exclude the possibility of the dentatocortical participation in the control of descending pathways at the motor cortex level (Uno et al. 1970)

Thus, we tried to determine whether stimulation of the dentate nucleus could reveal the types of movement controlled by this nucleus, and whether or not the dentatothalamocortical pathway participates in the elaboration of these movements.

In order to compare this control with that exercised by the motor cortex on movements, we stimulated the two structures and analyzed the induced movements by means of electromyographic recordings. Cortical activation accompaying the movements induced from the dentate nucleus was also recorded. The eventual concomitancy of motor effects and cortical activation was studied in order to determine if a causal relation existed between the two responses.

Methods

The cortical and cerebellar effects on movement were examined in seven adult baboons (*Papio papio*). The awake animals were mildly restrained by a rigid collar which did not prevent them from moving the head or limbs. During preliminary surgery, under deep anesthesia (fluothane) and aseptic conditions, the cortical, pyramidal, and muscle electrodes were implanted, along with a base for a micromanipulator above the cerebellum. After 10 days of recovery the test sessions began. Myographic electrodes, consisting of two stainless steel wires insultated with Teflon, were implanted in the neck extensors, the infraspinatus, the biceps, the triceps, the extensor digitorum, the biceps femoris, the sartorius, and the gluteus maximus. The cortical and cerebellar electrodes were made of sharpened nickel-chrome wires which were insulated except at the tip.

Electrodes were positioned in pairs for recording from or stimulating the cortex: the distance between the two tips was equal to the cortcial thickness. The cerebellar nuclei were stimulated with monopolar nickel-chrome electrodes placed with the aid of the micromanipulator. Stimulation sites were located every millimeter, and the stimulus consisted of a 300-Hz train of eight shocks, each of 0.5 ms duration and an amplitude of 100 or 150 μA. We used the same pulse train for the transcortical bipolar stimulation.

The motor effects elicited by cerebellar or cortical stimulations were assessed by two observers and filmed on video tape. In subsequent sessions, electromyographic recordings were collected on a magnetic tape and photographed from an oscilloscope.

Each time a motor response was observed, the stimulus intensity was decreased in order to determine the threshold value (T). Secondly, in order to test the motor response latencies, the stimulus was adjusted to twice the threshold value (2T) and the train duration was diminished until the response again disappeared.

Cortical evoked potentials induced by dentate stimulation were recorded. At the end of each penetration, an anodal current of 100 μA was passed through the electrode for 20 s, so as to deposit iron particles (present as an impurity in the nickel-chrome alloy). The iron deposit was revealed by using the Prussian blue technique.

After the test sessions, each animal was perfused under deep anesthesia with 10% formol-saline solution. Parasagittal serial sections of the cerebellum were then prepared, on which the electrode trajectories were identified and the stimulated sites located.

Spatial Aspect of the Cortical Control of Hand Movements

Monopolar or transcortical stimulation of the whole precentral area revealed that hand movements were obtained from only a small region of the motor cortex (Fig. 1a).

These movements involved few joints, and most often only one. Figure 1b shows that the wrist was involved in a dorsiflexion. In the same way, a single phalangeal joint was involved either in flexion or extension. In two cases several joints were mobilized simultaneously: first, in a grip movement involving the thumb and the index finger; second, in the flexion or extension of four fingers conjointly.

Electromyographic recording revealed that only the muscle (Fig. 1c) implicated in the displacement of the appropriate segment was activated, whereas the more proximal muscles remained inactive. For example, during stimulation of motor cortical site 2, which induced a forelimb flexion (Fig. 1c-2), only the biceps was activated. The infraspinatus and the neck muscles showed no activity and the triceps activity was induced, only after a long latency, by a reflex effect. The same pattern was observed each time the motor cortical areas were stimulated. During a thigh extension, produced by the stimulation of the site 1, the two synergistic muscles were excited, but the antagonist (sartorius) remained inactive (Fig. 1c-1).

The hand movements induced by cortical stimulation seemed to conform *to a reciprocal pattern and to be limited to one or a few joints.* These observations fit with those already described by Asanuma (1973), Kwan et al. (1978), and Strick and Preston (1982). The finely organized movements obtained in this experiment with threshold stimulation appear to contradict the fact that each motor cortical site pojects to several spinal motor centers (Andersen et al. 1974; Jankowska et al. 1975b; Shinoda et al. 1981). It is possible that the movements induced by cortical stimulation are obtained by activation of the inputs to the motor cortex (Jankowska et al. 1975a). However, they could be revealing the predominant connections between each cortical site and a given part of the musculature. This suggestion is supported by the fact that there exist predominant projections to each motor spinal center from a "best motor cortical point," as shown by Jankowska et al. (1975b).

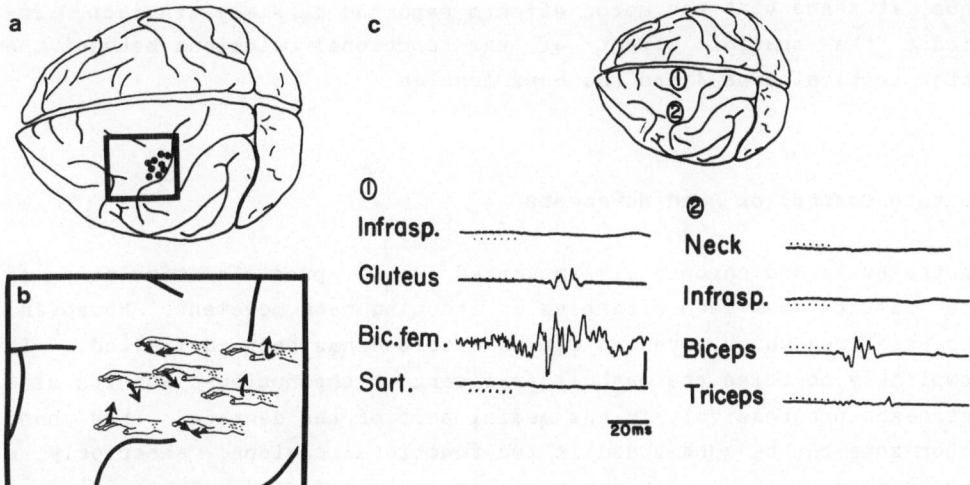

Fig. 1a–c. Spatial aspects of the cortical control of hand movement. **a** Cortical zone from which hand movements were obtained. The stimulated sites were located in and anterior to the precentral sulcus. **b** Illustration of the movements obtained by cortical stimulation. The *arrows* indicate the direction of the induced movement. **c** Patterns of muscular involvement. *1* and *2* are the stimulated sites in the motor cortex. The two muscles, biceps femoris and gluteus, were activated during stimulation at site 1. The infraspinatus and the sartorius remained inactive. In the same way, only the biceps was activated during forearm flexion induced by stimulation at site 2. Vertical scale: 0.4 mV

Thus, it seems that the motor effects reported here may, in fact, indicate the spatial aspect of the functional relations between the
motor cortical area 4 and the hand muscles.

Dentate Control of Hand Movements

In the awake and chronically implanted animal, punctate stimulation of
the dentate has been effective in inducing hand movement. Moreover,
the area from which movement can be elicited has been delimited. It
completely occupies the most lateral part of the nucleus, and its size
decreases progressively in the medial part of the dentate. This hand
motor zone can be subdivided in two functional regions: anteriorly, a
region from which complex movements are elicited, and posteriorly, a
region from which simple movements are elicited (Rispal-Padel et al.
1982).

Two kinds of hand movements were obtained by dentate stimulation:
simple and complex movements. *Simple movements* were characterized by
a brief displacement of a single segment. In general, only one joint
was involved in the movement, such as the wrist, either in a supination or a pronation (Fig. 2b). In the same way, the thumb or the
other fingers were often involved in a flexion or in an extension
(Fig. 2). Occasionally, grip movements or flexion of the fingers conjointly were also observed.

These movements were similar to those induced from the motor cortical
areas. However, the myographic recordings demonstrated some important
differences between these two types of movements. This point is discussed below.

The current intensities required to produce simple hand movements were
higher than those needed for inducing complex movements (Figs. 2 and
3). *Complex movements* resulted from the combination of several elementary movements. For example, the hand was involved simultaneously
with the shoulder or the head (Fig. 3). A brief head extension was
linked with either a hand supination or a hand flexion. Neither increasing nor decreasing the current intensity modified the pattern of
these complex movements. Thus, it was not possible to dissociate the
elementary components of these movements with threshold stimulation;
they occurred and disappeared in an all-or-none fashion.

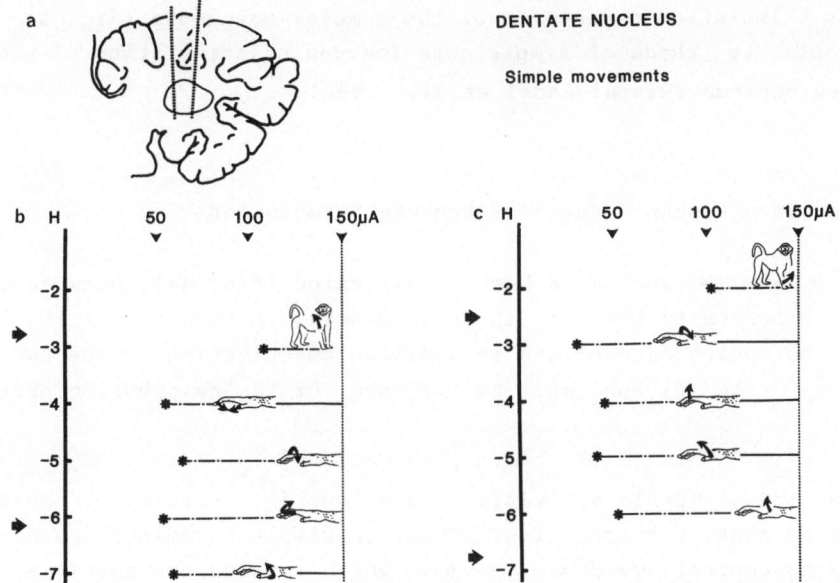

Fig. 2a-c. Dentate control of simple movements. **a** Parasagittal section of the neocerebellum. Two tracks were placed in the lateral part
of the dentate nucleus *(dotted lines)*. **b, c** Diagram showing the types
of movements in relation to the depths of the electrode tip (vertical
coordinate at left indicated in mm) and to the threshold value of
stimulation (horizontal coordinate). The boundaries of the nucleus
are shown by *arrows* along the vertical scale. The stimulation threshold of each movement is represented by a *star*. Movements involving a
single or few joints were induced from these dentate sites

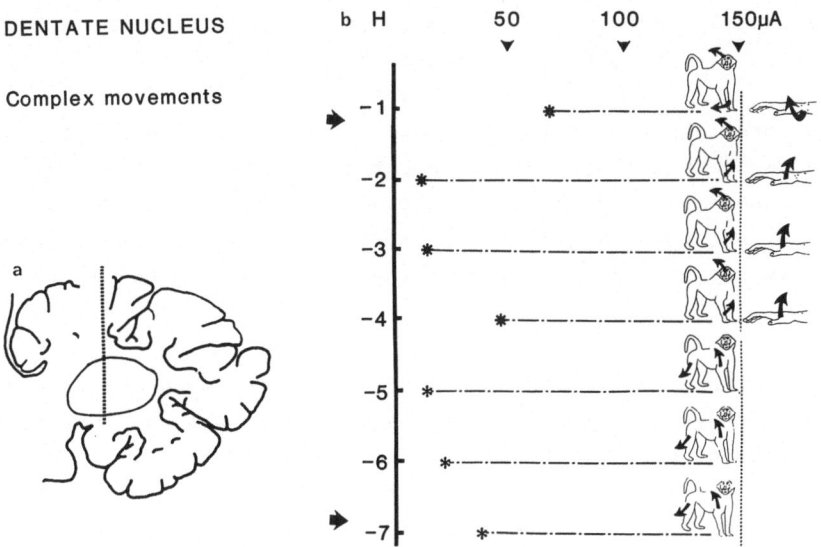

Fig. 3a,b. Dentate control of complex movements. **a** Parasagittal section of the neocerebellum from which complex (hand and head) movements
were induced. **b** Movements induced from sites situated along the electrode tract (depth in mm). Same format as in Fig. 2

The low stimulation thresholds of the complex movements (Fig. 3) were comparable to those of simple ones induced by stimulation of the interposed nucleus (Rispal-Padel et al. 1982).

Muscular Involvement Induced by Dentate Stimulation

Simple movements, such as a hand dorsiflexion (Fig. 4a), were concomitant with bursts of activity in three muscles, activation of the agonist of the moving segment and in addition coactivation of the two antagonists, the biceps and the triceps, of the neighboring proximal segment.

This pattern of muscle activation was a general feature of movement induced by dentate stimulation. Thus, in simple movements the dentate appears to control *muscular synergies* which involve the muscle of the moving segment and moreover ensure the cocontraction of the more proximal muscles, as observed in postural fixation mechanisms.

When a *complex movement* was elicited, a large number of muscles were activated at the same latency, and the activity disappeared in all the muscles according to an all-or-none principle. This excludes the possibility that some muscles were excited by a reflex effect. Moreover, as in simple movement, the antagonist muscles of the proximal segment were generally coactivated. This is in contrast with the effects obtained from stimulation of the motor cortex.

The characteristics of these complex movements (Rispal-Padel et al. 1982) suggest that they could be motor synergies initiated by a single central command. In contrast with the motor cortex, which appears to be organized for regulating the activity around a single joint, the cerebellum seems to play a role in the execution of the motor synergies.

This suggestion, made on the basis of the electromyographic analysis, is supported by clinical observations. Indeed, whereas lesions of the motor cortex or pyramidal tract affect manual skill, precise grip movements, or independent finger control (Tower 1940; Napier 1962; Lawrence and Kuypers 1968; Hepp-Reymomd and Wiesendanger 1972; Hepp-Reymond et al. 1974), neocerebellar lesions produce mainly motor incoordinations (Babinski 1899; Holmes 1917, 1939). In hand move-

ments, the loss of coordination between the fingers and the thumb or
the fingers and the wrist appears to be the most apparent disturbance
affecting writing or attempts to use simple or familiar tools (Holmes
1939). These deficits seem to be due to the loss of the cerebellar
command, which ensures the phasic activity on agonist muscles simultaneously with a postural fixation for the movement performance
(Fig. 4).

The defective programming of joint fixation could be mainly responsible for these deficits (Rondot et al. 1979). The specific role of
the dentate in the execution of the motor synergies is also suggested
by the fact that, unlike responses in the motor cortex, the dentate
unitary discharges accompanying a simple movement are rarely correlated with the activity of single muscles (Strick 1976; Thach 1978;
Massion and Sasaki 1979).

**Cortical Participation in the Simple and Complex Movements Controlled
by the Dentate Nucleus**

The dentate outputs reach the spinal centers via two principal pathways: the dentatothalamocortical (Uno et al. 1970) and the denta-
toreticulospinal pathway (Bantli and Bloedel 1975). It is possible to
obtain movements by dentate stimulation, even when the first of these
pathways is excluded (Schultz et al. 1979). The effects of the dentate outputs on the cortical cells are still unclear. However, con-
comitantly with the movements induced by dentate stimulation, evoked
potentials can be observed in the motor cortical area. The spatial
distribution of these potentials reveals that they could indicate excitation of the descending corticospinal pathways.

A topographical arrangement was observed in the projections from the
dentate regions to the motor cortical areas, but the relations were
not strictly point to point (Fig. 5). The head and hand dentate areas
(defined by the movements induced by punctate stimulation) have a
strongly effective projection to the face and to the hand area of the
motor cortex respectively. However, weakly effective projections were
also seen to other motor cortical regions. For example, in addition
to the strongly effective projections of the hand dentate region to
its homologous cortical zone, less effective projections reach the
face cortical zone and very feeble effects were seen in the hindlimb

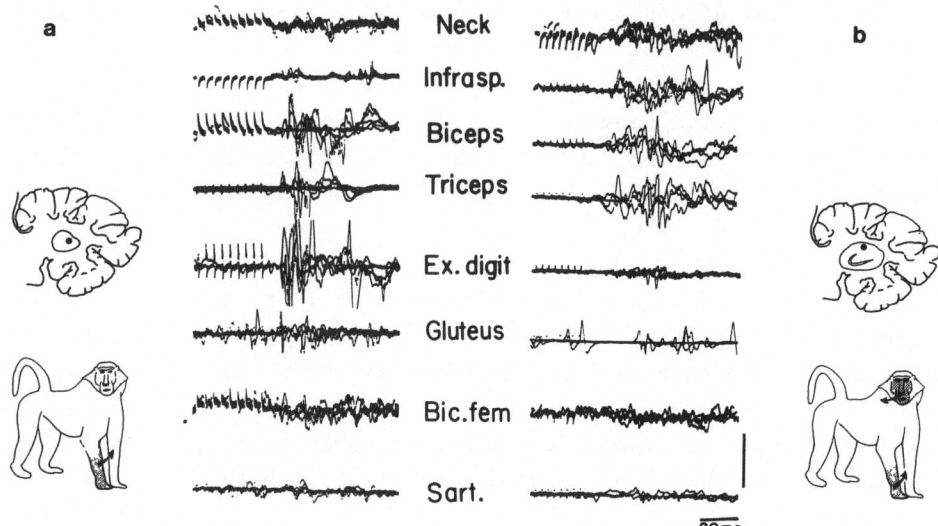

Fig. 4a,b. Muscular activation induced by dentate stimulation. **a** Dorsiflexion of the hand is represented in the monkey figurine. It was induced from the lateral part of the nucleus (parasagittal section). During this simple movement three forearm muscles were activated; the agonist of the movement (extensor of the digits) and two antagonists in the more proximal segment. **b** Activation of the muscles involved in a complex movement (figurine). All the muscles involved in this movement were simultaneously activated, excluding the possibility of excitation due to a reflex effect. Decreasing the stimulating current intensity did not allow dissociation of complex movements into their elementary components. Vertical scale: 0.5 mV. See methods for muscle details

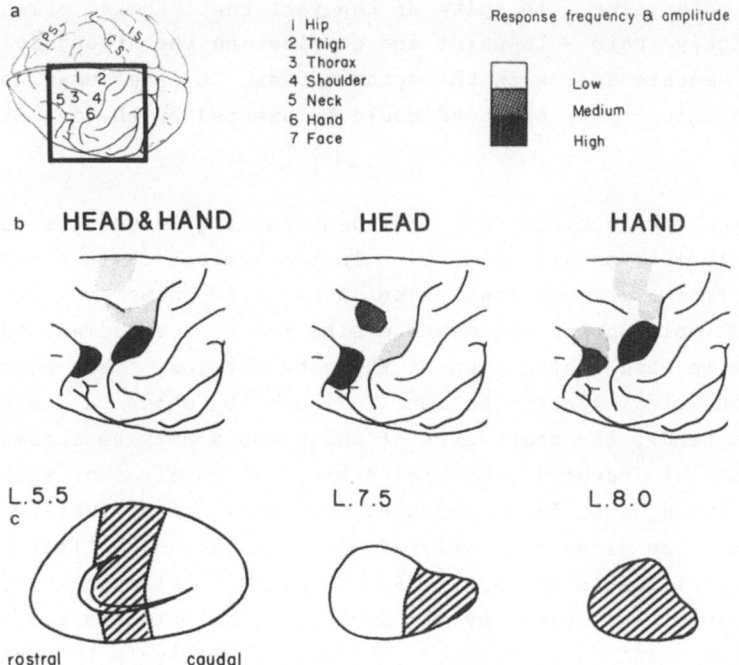

Fig. 5a–c. Cortical projections of the dentate nucleus. **a** Numbers on the cortical surface present different motor cortical zones and the associated body segments which move during stimulation of these zones. At the *right*, three intensities of projections were classified according to the amplitude and the frequency of the cortical evoked potentials induced by dentate stimulation. **b** Cortical projections of three dentate motor zones, which are shown **(c)** on parasagittal sections of the dentate nucleus. At the *left*, the denetate medial region *(hatched zone)*, originating complex hand movements, projected mainly to two cortical zones. Stimulation of these zones induced hand and head movements respectively. Feeble projections were seen in the hindlimb and shoulder motor cortical zones. In the *middle*, the head dentate motor region, as defined by the results of stimulation, projected strongly to the face motor area, but feebler projections reached the thorax motor cortical zone. At the *right*, the projections of the hand dentate region were strong on the hand motor cortical zone, but they also reached the face and the hindlimb motor cortical area

cortical motor area. In spite of the fact that these relations are not strictly point - to-point and considering the strong relations of the hand dentate area with the motor cortex, it is assumed that an efficient control of the hand could be exerted by the dentatocortical pathway.

The cortical projections from each dentate site inducing movement were examined in more detail (Fig. 6). In two typical cases the potentials produced from a given dentate site were also observed on cortical sites. Stimulation of the dentate site and of the corresponding cortical site induced contraction of the same muscles. This suggests dentatocorticospinal participation. Figure 6b shows a motor synergy (head and hand), the components of which could only be achieved by the activation of several cortical sites. Stimulation of each of these sites produced neck, face, and hand movements respectively. The movement was less closely correlated with the evoked cortical potentials in the case illustrated in Fig. 6a. Some cortical points not related to the muscles involved by the dentate stimulation were also excited (like sites 4 and 10). However, statistical analysis for stimulation at threshold showed that only the evoked potentials recorded at sites 8 and 11 were clearly concomitant with the movement.

Detailed analysis has shown that the movement produced by stimulation of a given site in the dentate nucleus is similar to the sum of movements induced by stimulation of each cortical point receiving excitatory projections from this particular dentate focus (Fig. 7).

The study of the dentatocortical relations reveals that the motor cortex is involved in the elaboration of the movements induced by the dentate stimulation. Several observations support this assumption. Firstly, there is concomitance between the cortical or pyramidal responses and the induced movements. Secondly, the timing of the cortical, pyramidal, and muscle response latencies shows that the message might be transmitted by this pathway (Rispal-Padel et al. 1981,

Fig. 7a,b. Comparison of the movements produced by dentate stimulation with those produced by cortical stimulation. **a** Dentate site inducing a hand dorsiflexion projects to cortical sites 8 and 11. Stimulation of these cortical sites induced shoulder and hand movements respectively. **b** In the same way, it was observed that the complex movement induced by dentate stimulation was similar to the sum of the movements induced from the cortical areas excited by this particular dentate site

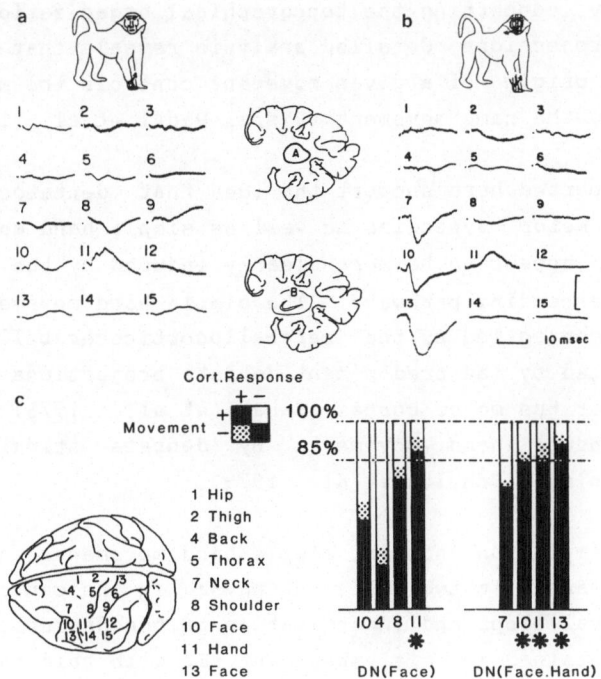

Fig. 6a–c. Detailed analysis of the dentatocortical relations. **a** Mapping of the evoked cortical potentials induced simultaneously with the hand dorsiflexion movement. **b** Evoked potentials concomitant with a complex movement (head and hand synergy illustrated by *arrows* on the figurine). The localizations of the dentate sites are indicated by *A* and *B* on the cerebellar parasagittal sections. **c** Statistical evaluation of the concomitance between the movement and the evoked cortical potentials. In the 2 x 2 array at left of the histograms, the *black* zones correspond to the concomitance between the presence (+) or the absence (−) of the two responses, and the *shadowed* and *white* zones correspond to the nonconcomitance between these responses. Histograms show 60 trials at the stimulation threshold for each cortical site identified by numbers below the histograms and on the cortical surface at left. The *stars* below the histograms indicate concomitance on more than 85% of the trials

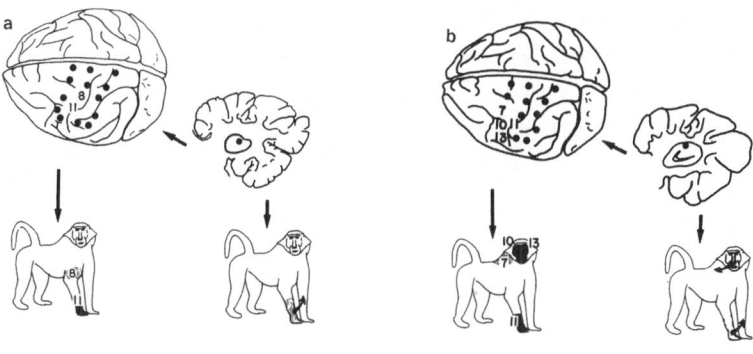

1983). Thirdly, concerning the topographical organization of the dentatocortical projections, detailed analysis reveals that the dentate site at the origin of a given movement controls the motor cortical sites producing the same movement (Rispal-Padel et al. 1981).

The results reported here support the idea that dentatocortical outputs control motor synergies as well as simple hand movements. The motor synergies appear to be more clearly induced by the dentate effects on the descending pathways. The simple hand movements could additionally be controlled by the cerebellocorticocerebellar circuits. This is suggested by the predominant dentate projections on the superficial layers of the motor cortex (Sasaki et al. 1976) and by the failure to induce hand movements by dentate stimulation in the anesthetized animal (Schultz et al. 1979).

The dentate stimulation inducing simple hand movements could implicate the cortical area in two different ways: in the control of central cerebellocerebral loops and in activation of the descending pathways. The results obtained in this experiment fit with this suggestion. It has been shown that in the dentate nucleus, stimulation thresholds necessary to induce simple hand movements were higher than those needed to induce other movements (Rispal-Padel et al. 1982). This fact indicates a more elaborate circuitry between the dentate and the muscles for the former movements. Pyramidal responses could also be observed simultaneously with the simple movements induced by dentate stimulation, clearly indicating the involvement of cortical output in these movements (Rispal-Padel et al. 1981).

The dentatothalamocortical circuit has been analyzed at the cellular level in the cat; its organization exhibits divergence and convergence throughout the pathway. It appears not to be responsible for a diffuse control of motor activity, but rather underlies localized synergistic effects on the muscles (Rispal-Padel and Grangetto 1977). The observation of motor synergies shown in simple and complex movements could be due, in part, to this spatial organization of the dentatothalamocortical pathway.

References

ALLEN GI, TSUKAHARA N (1974) Cerebro-cerebellar communication systems. Physiol Rev 54: 957-1006

ANDERSEN P, HAGAN PJ, PHILLIPS CG, POWELL TPS (1974) Mapping by microstimulation of overlapping projections from area 4 to motor units of the baboon's hand. Proc R Soc [Biol] 188: 31-60

ASANUMA H (1973) Cerebral cortical control of movement. Physiologist 16: 143-166

BABINSKI J (1899) De l'asynergie cérébelleuse. Rev Neurol 7: 806-816

BANTLI H, BLOEDEL J (1975) Monosynaptic activation of a direct reticulo-spinal pathway by the dentate nucleus. Pflugers Arch 357: 237-242

BROOKS VB (1979) Control of intended limb movements by the lateral and intermediate cerebellum. In: ASANUMA H, WILSON VJ (eds) Integration in the nervous system. Igaku-Shoin, Tokyo, pp 321-357

EVARTS EV, THACH WT (1969) Motor mechanisms of the CNS: cerebro-cerebellar interrelations. Annu Rev Physiol 31: 451-498

HEPP-REYMOND MC, WIESENDANGER M (1972) Unilateral pyramidotomy in monkeys: effect on force and speed of a conditioned precision grip. Brain Res 36: 117-131

HEPP-REYMOND MC, TROUCHE E, WIESENDANGER M (1974) Effects on unilateral and bilateral pyramidotomy on a conditioned rapid precision grip in monkeys *(Macaca fascicularis)*. Exp Brain Res 21: 519-527

HOLMES G (1917) The symptoms of acute cerebellar injuries due to gunshot injuries. Brain 40: 461-535

HOLMES G (1939) The cerebellum of the man. Brain 62: 1-29

JANKOWSKA E, PADEL Y, TANAKA R (1975a) The mode of activation of pyramidal cells by intracortical stimuli. J Physiol (Lond) 249: 617-636

JANKOWSKA E, PADEL Y, TANAKA R (1975b) Projections of pyramidal tract cells to alpha motoneurones innervating hind-limb muscles in monkey. J Physiol (Lond) 249: 637-667

KWAN HC, MACKAY WA, MURPHY JI, WONG YC (1978) Spatial organization of precentral cortex in awake primates. II. Motor outputs. J Neurophysiol 41: 1120-1131

LAWRENCE DG, KUYPERS HGJM (1968) The functional organization of the motor system in the monkey. I. The effects of bilateral pyramidal lesions. Brain 91: 1-14

MASSION J (1973) Intervention des voies cérébello-corticales et cortico-cérébelleuses dans l'organisation et la régulation du mouvement. J Physiol (Paris) 67: 117A-170A

MASSION J, SASAKI K (1979) Cerebro-cerebellar interaction: solved and unsolved problems. In: MASSION J, SASAKI K (eds) Cerebro-cerebellar interactions. Elsevier, Amsterdam, pp 261-285

NAPIER J (1962) The evolution of the hand. Sci Am 207: 56-62

RISPAL-PADEL L, GRANGETTO A (1977) The cerebello-thalamo-cortical pathway. Topographical investigation at the unitary level in the cat. Exp Brain Res 28: 101-123

RISPAL-PADEL L, CICIRATA F, PONS C (1981) Contribution of the dentato-thalamo-cortical system to control of motor synergy. Neurosci Lett 22: 137-144

RISPAL-PADEL L, CICIRATA F, PONS C (1982) Cerebellar nuclear topography of simple and synergistic movements in the alert baboon (*Papio papio*). Exp Brain Res 47: 365-380

RISPAL-PADEL L, CICIRATA F, PONS C (1983) Neocerebellar synergies. In: MASSION J, PAILLARD J, SCHULTZ W, WIESENDANGER M (eds) Neural coding of motor performance. Springer, Berlin Heidelberg New York, pp 213-223 (Experimental brain research, suppl 7)

RONDOT P, BATHIEN N, TOMA S (1979) Physiopathology of cerebellar movement. In: MASSION J, SASAKI K (eds) Cerebro-cerebellar interactions. Elsevier, Amsterdam, pp 203-230

SASAKI K (1979) Cerebro-cerebellar interconnections in cats and monkeys. In: MASSION J, SASAKI K (eds) Cerebro-cerebellar interactions. Elsevier, Amsterdam, pp 105-124

SASAKI K, KAWAGUCHI S, OKA H, SAKAI M, MIZUNO N (1976) Electrophysiological studies on the cerebello-cerebral projections in monkeys. Exp Brain Res 24: 495-509

SCHULTZ W, MONTGOMERY EB, MARINI R (1979) Proximal limb movements in response to microstimulation of primate dentate and interpositus nuclei mediated by brain-stem structures. Brain 102: 127-146

SHINODA Y, YOKOTA JI, FUTAMI T (1981) Divergent projection of individual corticospinal axons to motoneurons of multiple muscles in the monkey. Neurosci Lett 23: 7-13

STRICK PL (1976) Cerebellar neuron response to imposed limb displacement: dependence of short latency dentate activity on intended movement. Soc Neurosci Abstr 2: 876

STRICK PL, PRESTON JB (1982) Two representations of the hand in area 4 of a primate. I. Motor output organization: J Neurophysiol 48: 139-150

THACH WT (1978) Correlation of neural discharge with pattern and force of muscular activity, joint position and direction of the intended movement in motor cortex and cerebellum. J Neurophysiol 41: 654-676

TOWER S (1940) Pyramid lesion in the monkey. Brain 63: 36-90

UNO M, YOSHIDA M, HIROTA I (1970) The mode of cerebello-thalamic relay transmission investigated with intracellular recording from cells of the ventrolateral nucleus of cat's thalamus. Exp Brain Res 10: 121-139

Compensatory Motor Function of the Somatosensory Cortex in the Monkey Following Cooling of the Motor Cortex and Cerebellectomy

K. Sasaki and H. Gemba

Department of Physiology, Institute for Brain Research, Faculty of Medicine, Kyoto University, 606 Kyoto, Japan

Introduction

Cortical field potentials occurring prior to visually initiated hand movements have been recorded with chronically implanted electrodes in various cortical areas of the monkey (Gemba et al. 1981; Sasaki et al. 1982; Sasaki and Gemba 1982). In the forelimb motor cortex contralateral to the moving hand, early surface-positive, depth-negative (s-P, d-N) potentials at a latency of about 40 ms are followed by late s-N, d-P potentials, and both precede the hand movement (Gemba et al. 1981). The early potentials are more intimately related to the visual input than are the late potentials, whereas the late potentials are considered to be directly involved in the execution of the movement (Gemba et al. 1981). It was demonstrated that the late potentials are induced by impulses conveyed through the neocerebellum and the superficial thalamocortical (T-C) projection (Sasaki et al. 1982). Thus it may be assumed that the motor command, which presumably originated in some association and premotor cortices, is processed at least in part through the cerebrocerebellar interconnections and activates the motor cortex to execute the movement (Sasaki 1979; Sasaki et al. 1982).

To examine the role of the motor cortex as the cortical output station for the motor command, temporary inactivation of the forelimb motor area was attempted by local cooling, and changes of the premovement field potentials and of the motor performance were observed during such cooling in monkeys. Dysfunction of the motor cortex was also elicited by cerebellar hemispherectomy. It will be shown in the present report that the primary somatosensory cortex becomes predominant in motor function under these conditions and compensates, although incompletely, for the motor cortex dysfunction.

Methods

Eight adult monkeys *(Macaca fuscata)* were used in this study. Brass chambers with the undersides shaped to fit the respective cortical areas were insulated by a thin resin film and placed on the dura mater of the cortical areas and fixed to the bone with dental cement (Fig. 1a). Input and output pipes of the chamber were connected to a circulating pump with silicon rubber tubes. Such chambers were placed bilaterally on the forelimb areas of motor (FM) and primary somatosensory (FS) cortices, premotor (PM), prefrontal (PF), and parietal association (PA) cortices, etc. The chamber for cooling was perfused with cold water (about $1°C$, $0.6° - 1.2°C$) by the pump, but the other chambers were perfused with warm water ($38° - 39°C$) to prevent cooling of the surrounding areas.

The temperature of the brain was measured with a needle probe thermometer in some experiments (Figs. 1b and 2a). After switching from the warm to the cold water, the temperature of the cortex under the chamber at a depth of about 2 mm gradually fell, reaching a steady state within 5 min (Fig. 2a). At a depth of about $0.5 - 1.0$ mm, the steady state temperature was $20° - 27°C$ and at a depth of $2.5 - 3.0$ mm it was $24° - 29°C$. At a distance of 3 mm from the cooling chamber, the neocortex was cooled down to $27.8°C$ on the surface, $28.9°C$ at a depth of 0.5 mm, and $30.8°C$ at a depth of 2.5 mm, as shown in Fig. 1b. However, in the same area, the cortex was maintained at about $37°C$ at a distance of about 1 mm from the cooling chamber, provided it was under an adjacent warming chamber, as in Fig. 1b. Recovery from cooling was usually achieved by rewarming within $4 - 5$ min, as seen in Fig. 2a. Although cooling and rewarming effects occurred within a few minutes, we usually waited $5 - 10$ min, before measuring them in standard experiments.

Recording electrodes for cortical field potentials were implanted on the surface and at a depth of $2.5 - 3.0$ mm at each locus, as reported previously (Hashimoto et al. 1979). Two pairs of electrodes were placed under each chamber (Fig. 1c). The electrodes were made of insulated silver needles (0.25 mm in diameter), and their tips were

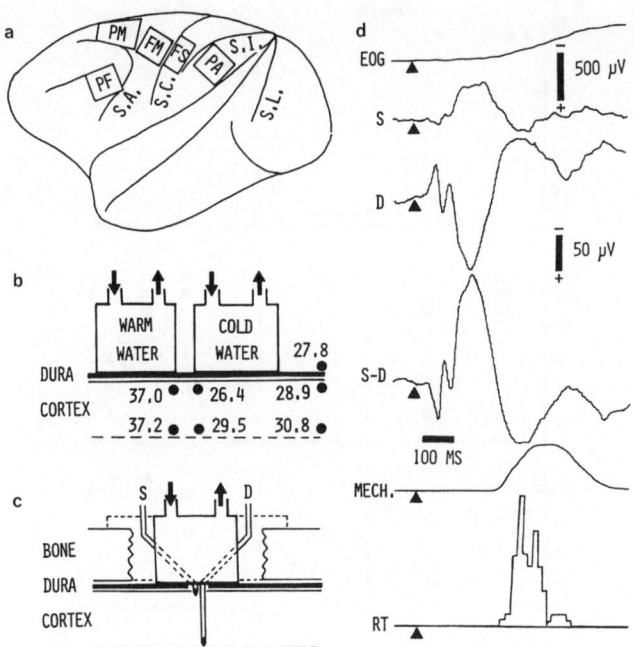

Fig. 1a–d. Arrangement of cooling of the cerebral cortex and recording of premovement cortical field potentials. **a** Outlines of cooling and warming chambers are drawn on the lateral view of the left hemisphere. *PF,* prefrontal cortex; *PM,* premotor cortex; *FM,* forelimb motor cortex; *FS,* forelimb somatosensory cortex; *PA,* parietal association cortex; *S.A.,* sulcus arcuatus; *S.C.,* sulcus centralis; *S.I.,* sulcus intraparietalis; *S.L.,* sulcus lunatus. **b** Two chambers for warming and cooling placed on the dura, and temperature (°C) in the cortex measured in the steady state after 10 min of circulation of warm and cold water. **c** Recording electrodes placed on the surface *(S)* and at a depth of 2.5 – 3.0 mm *(D)* under the chamber. The chamber and the electrodes were fixed to the bone by dental cement *(broken lines).* **d** Examples of *EOG,* and surface *(S),* depth *(D),* and surface minus depth *(S-D)* potentials in the forelimb motor cortex associated with the visually initiated hand movement. Muscular tension (mechanogram) was recorded through a transducer on an arbitrary scale *(MECH.).* These are all averages of 100 responses aligned on the onset of the visual stimulus *(triangle).* Reaction times of the same 100 movements are plotted in a histogram in 16-ms bins *(RT).* Calibration bars show 500 μV for EOG, 50 μV for cortical potentials, and 100 ms for all traces

278

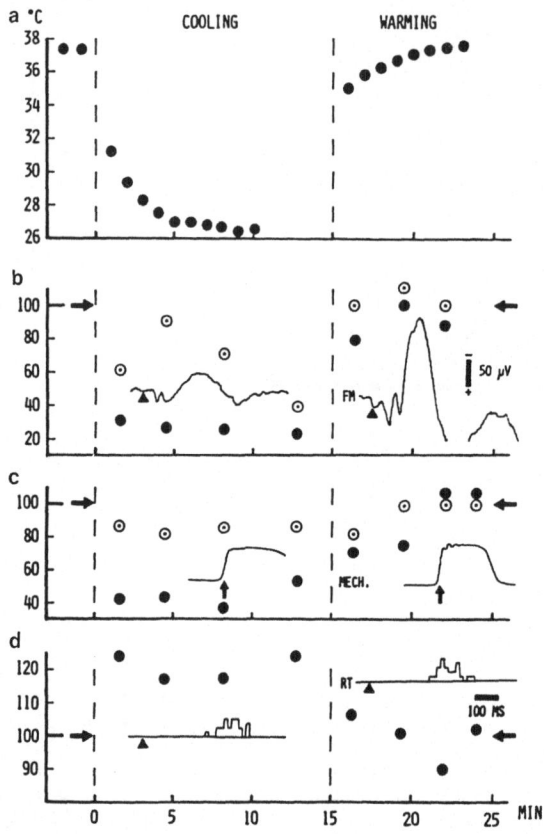

Fig. 2. Changes of cortical temperature **(a)**, premovement cortical potentials in the forelimb motor cortex **(b)**, muscular tension **(c)**, and reaction time **(d)** during cooling of the cortex in a monkey. **a** A thermometer needle probe was placed at a depth of about 2 mm from the cortical surface under the cooling chamber after the end of the experiments which gave the results in **b-d**. Before and after the cooling (between two *broken lines*), warm (38° - 39°C) water was perfused through the chamber. **b** Peak heights of early surface-positive, depth-negative (s-P, d-N) premovement potentials are plotted by *open circles* with *dot* in center, and those of late s-N, d-P ones by *filled circles*. S-D potentials illustrated for warming *(right)* and cooling *(left)* were averaged 25 times aligned on the stimulus onset *(triangle)*; calibration 50 µV. **c** Peak height of muscular tension is given by *open circle* with *dot* in center, and maximal rate of rise of the mechanogram by *filled circle*. Examples for warming *(right)* and cooling *(left)* were averaged 25 times aligned on the movement pulse *(arrow)*. **d** Mean reaction time. Specimen histograms of 25 trials are presented during warming *(right)* and cooling *(left)*. The stimulus onset is indicated by *triangle*. 100-ms time scale for all specimens in **b-d**. In **b-d**, ordinates show percentages of the control value *(arrows)*. All abscissae are in minutes after the start of cooling

pointed and exposed. Indifferent electrodes of silver wires were bu-
ried in the bone just behind the ears on both sides and were connected
together. A typical electrooculogram (EOG) measured at the rostrola-
teral edge of the frontal bone above the orbit is shown in Fig. 1d.
Cortical potentials and EOGs were measured with respect to the indifferent electrodes, amplified with ac amplifiers (2.0-s time constant),
and recorded on magnetic tape.

Monkeys were trained to lift a lever with wrist extension during the
light stimulus lasting 510 - 540 ms, which was delivered at random intervals of 2.5 - 6.0 s (Gemba et al. 1981). The cooling was continued for 10 - 30 min, during which the monkey performed more than 100
hand movements; the exception to this was, of course, the session in
which the wrist muscles were entirely paralyzed. Warming - cooling -
rewarming sequences were usually repeated two to four times in each
experiment. After repeated experiments for many weeks, the cerebellar
hemisphere ipsilateral to the moving hand was resected, and premovement cortical potentials and movements before and after the operation
were compared.

Cortical field potentials and EOGs recorded on tapes were averaged,
usually 100 times, aligned using the onset of either the light
stimulus or the movement (lever elevation) stored on the same tapes.
Potentials on the surface (S) and at a depth of 2.5 - 3.0 mm (D) and
the difference of these potentials (S-D) were averaged, aligned on the
stimulus onset, as shown in Fig. 1d. Simultaneously recorded EOGs and
mechanograms (MECH.) of the lever elevation were also averaged and
presented together with the reaction time histogram (RT) of the same
100 movements (Fig. 1d). For the mechanogram, a force transducer was
attached to the lever, but the transducer was nonlinear especially in
its higher range. Therefore, the maximal rate of rise in the tension
curve (MECH./T) expressed the muscle force better than the height of
tension (MECH.), as shown in Fig.2c. The mechanograms and their deri-
vatives are shown only on an arbitrary scale.

After experiments lasting usually for several months, the animal was
sacrificed and sites of recording electrodes and chambers were checked
morphologically. Extent of the cerebellar hemispherectomy was also
checked histologically.

Results

Effects of Cooling the Motor Cortex Upon Its Premovement Potentials and Visually Initiated Hand Movement

Cooling of the forelimb motor cortex suppressed its field potentials associated with visually initiated hand movements within a few minutes and made the movement slow and weak, as shown in Fig. 2. The temperature reached a steady low level of 26° - 27°C within 5 min, and the premovement field potentials (S-D) averaged 25 times before the cooling (right inset potential in Fig. 2b) were reduced within a few minutes (left inset potential in Fig. 2b). Surface-negative, depth-positive (s-N, d-P) potentials with a latency of about 100 ms were reduced in size immediately after the start of cooling (filled circle in Fig. 2b), but s-P, d-N potentials with a latency of about 40 ms decreased relatively slowly in this case (open circle with dot in center in Fig. 2b). This is consistent with the interpretation of the s-P, d-N potentials as resulting from EPSPs in the deep cortical layers, and of the s-N, d-P potentials resulting from EPSPs in the superficial layers (Sasaki 1979; Sasaki et al. 1981, 1982).

The height of the mechanogram (open circle with dot in center in Fig. 2c) decreased by 15% - 20% and the maximal rate of rise (filled circle) by nearly 60% immediately after the start of cooling. The small decrease in height would not necessarily represent the true loss in tension of the muscle (see Methods). The mean reaction time was prolonged by about 20% as shown in Fig. 2d, this effect also occurring immediately after the start of cooling. When the cold water was switched to the warm water, these changes quickly recovered within a few minutes.

Enhanced Activity of the Somatosensory Cortex During Motor Cortex Cooling

Simultaneous recording revealed that the primary somatosensory cortex increased its electrical activity during cooling of the motor cortex. Figure 3 presents examples of averaged cortical field potentials (S-D) aligned on the stimulus onset (V.S., upper half) or on the movement onset (L.E., lower half). In every column, 100 trials are averaged and plotted for periods before cooling (a), during the early (b) and late (c) part of cooling, and after rewarming (d). In the normal situation, the potentials in the motor cortex (FM) start earlier than the movement and precede it by more than the potentials in the somatosensory cortex (FS), as seen in the stimulus (V.S.) and movement (L.E.) aligned averages. The premovement field potentials in the forelimb motor cortex (FM rows) were strongly suppressed during the cooling (b and c), whereas those in the forelimb somatosensory cortex (FS rows) were enhanced. The enhancement is more marked in the area of s-N, d-P (upward) deflection before the movement (note area before the broken line in FS potentials on the L.E. average). Since reaction times were prologed during the cooling, as seen in RT row, the average aligned on the movement onset revealed that the potential in the somatosensory cortex preceded the movement when the cortex was cooled: this is analogous to the potentials in the motor cortex in the normal situation (compare FS records in column b and c with FM records in column a and d on L.E. averaging). These changes in the potentials were entirely reversible by rewarming the cortex. The data suggest that the somatosensory cortex compensates, although incompletely, for the motor cortex disabled by the cooling and controls the movement. No appreciable changes were noted in the premotor cortex.

Paralysis of Wrist Muscles by Simultaneous Cooling of the Motor and Somatosensory Cortices

Simultaneous cooling of both forelimb areas of the motor and somatosensory cortices resulted in the paralysis of wrist muscles, and the monkey stopped lifting the lever within a few minutes of cooling, as illustrated in Fig. 4. The premovement cortical potentials (S-D) in the forelimb motor (FM) and somatosensory (FS) areas were averaged 50 times aligned on the movement pulse, and revealed the enhanced activity of the somatosensory cortex during cooling of the motor cortex (Fig. 4b). Simultaneous cooling of the motor and somatosensory cortices stopped the movement within a few minutes, as shown in Fig. 4c, seven successful movements being noted in the reversed reaction time histogram (-RT). Fourteen movements, including weak ones which were insufficient to be rewarded, are plotted in the diagram for the distribution of maximal rate of rise in the mechanogram (MECH./T) in Fig. 4c. Cortical potentials in the motor and somatosensory cortices could be averaged over only seven successful movements and look smooth in this column (50-μV calibration is not applicable). After that only

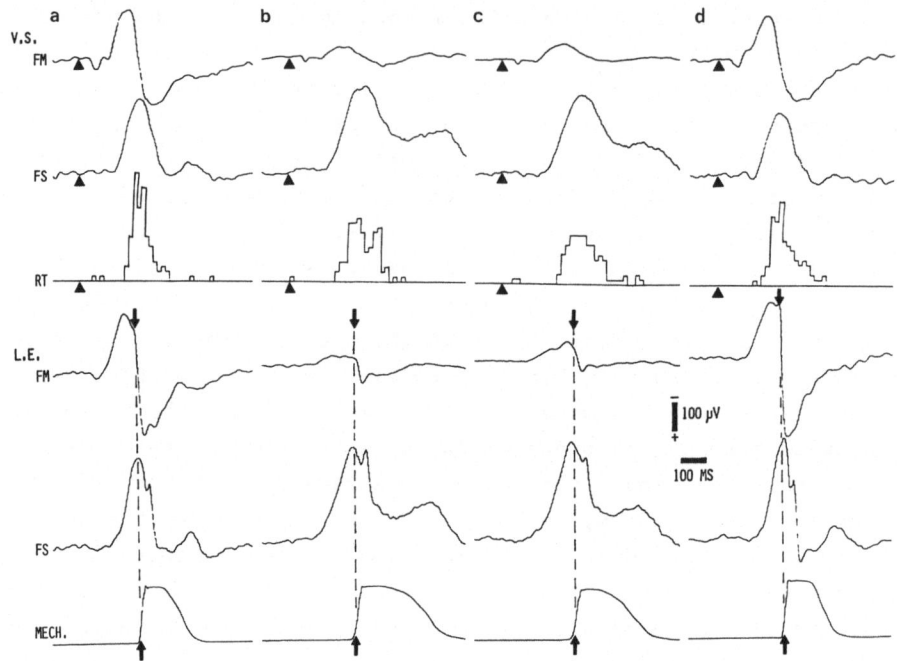

Fig. 3. Enhanced premovement potentials of the primary somatosensory cortex *(FS)* on cooling of the motor cortex *(FM)*. The same data from one monkey were averaged 100 times, aligned on the stimulus onset *(triangles* of upper half, V.S.) or on the onset of movement (upward and downward *arrows* with *broken lines* of lower half, L.E.). **a** 100 trials before cooling. **b,c** 100 trials during cooling. **d** 100 trials after rewarming. Calibration bars show 100 μV for all cortical potentials, 100 ms for all traces

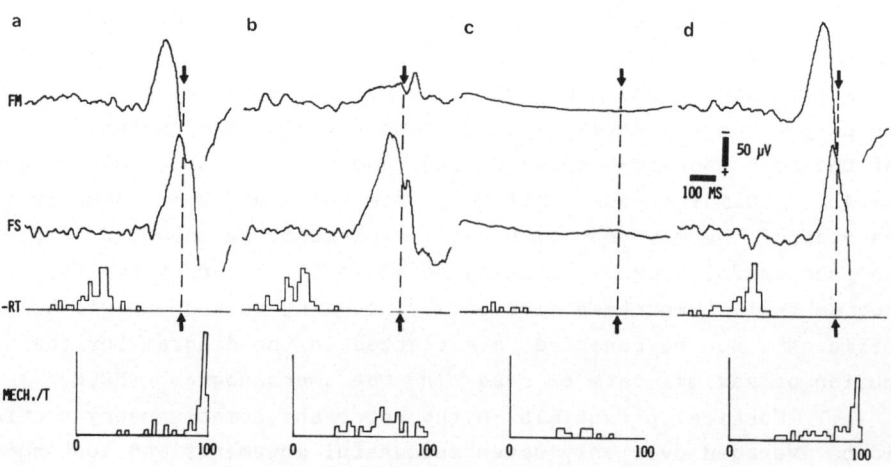

the somatosensory cortex was rewarmed and the movement, although slow and weak, just as in Fig. 4b, appeared again within a few minutes (not shown). Then the motor cortex was also rewarmed and complete recovery of the movement and of the potentials was observed, as shown in Fig. 4d.

It was confirmed that the additional cooling of the somatosensory cortex lowered the temperature of the precentral gyrus only a little (1° - 2° C) and that the effect of the somatosensory cooling was possibly due to the functional block of the somatosensory cortex itself, rather than due to the more complete dysfunction of the motor cortex. The cooling of the somatosensory cortex alone hardly affected the movement or the premovement potentials in the motor cortex.

Enhancement of Premovement Field Potentials in the Somatosensory Cortex After Cerebellar Hemispherectomy

Unilateral cerebellar hemispherectomy, ipsilateral to the moving hand and contralateral to the active motor cortex, eliminates s-N, d-P premovement field potentials in the motor cortex and delays the movement by 90 - 250 ms (Sasaki et al. 1982). Simultaneous recording from the primary somatosensory cortex showed a marked increase of s-N, d-P field potentials that preceded the delayed movement. In fact, the premovement area of the s-N, d-P potentials of the somatosensory cortex was increased to be about three times as much as the control (before operation). Such an enhancement was usually marked several days after the operation, but declined steeply within 30 - 60 days. Some other compensatory motor function replaces that of the somatosensory cortex within several weeks.

Fig. 4. Paralysis of wrist muscles due to simultaneous cooling of the forelimb areas of motor *(FM)* and somatosensory *(FS)* cortices in a monkey. **a** Control. **b** Cooling of the motor cortex. **c** Cooling of the motor and somatosensory cortices. **d** Rewarming of the cortices. All cortical potentials were averaged, aligned on the movement pulse *(arrows with broken line)* 50 times in **a**, **b**, and **d**, and 7 times in **c**. The 50-μV calibration is applicable to **a**, **b**, and **d**, but not to **c**. The onset time of the visual stimulus is plotted in histograms *(-RT)* preceding the movement in 16-ms bins. Maximal rate of rise in the mechanogram is plotted in histogram for the same 50 trials *(abscissa)* in percent (3%/bin) *(MECH./T of **a**, **b**, and **d**)*. In **c**, 14 trials, including 7 successful (to obtain reward) and 7 unsuccessful movements, are plotted in *MECH./T*. The cortical potentials and reaction time histogram in **c** are for the 7 successful movements. Calibration bar shows 100 ms for all cortical potentials and reaction time histograms

Discussion

The main advantage of the local cooling method is the capability of producing repeated, brief, local, reversible disruption of function of the brain tissue without the compensatory neural reorganization that may occur in the case of permanent destruction (see Brooks 1983). On the other hand, a grave problem of the method is the difficulty of delimiting the extent of inactivated tissue. In the present study, two devices were used to improve the method in this respect: firstly, several warming chambers were placed around the cooling chamber to prevent cooling effects from spreading to neighboring cortical areas (cf Reed and Miller 1978; Sherk 1978); secondly, recording electrodes were set not only in several cortical areas around the cooling area but also in the area under the cooling chamber itself in order to check the cooling effect and its extent. These procedures helped considerably to delimit the inactivated area and to estimate the degree of neuronal inactivation under the cooling chamber.

It was surprising to learn that cooling of the forelimb motor cortex results in only the paresis of wrist muscles. We had expected that the premotor cortex might have compensated for the motor cortex dysfunction, but no sign of increased activity in the premotor cortex was observed. Cooling of the premotor cortex produces quite different effects upon the visually initiated hand movement from those produced by cooling the motor cortex or the somatosensory cortex, which will be described elsewhere (Sasaki and Gemba, in preparation). When both the somatosensory cortex and the motor cortex were cooled enhanced electrical potentials were observed in the medial part of the premotor cortex, which presumably corresponds to the supplementary motor area. However, this enhanced activity was not able to move the hand, as was enhanced activity in the somatosensory cortex (Gemba and Sasaki, unpublished observation).

Another example of the compensatory motor function of the somatosensory cortex with its enhanced premovement potentials was observed in the case of cerebellar hemispherectomy. In this case, the enhanced potentials declined steeply to the preoperative level in several weeks. Some neural reorganization might occur in these weeks to replace progressively the enhanced compensatory activity of the somatosensory cortex (cf Goldberger 1974).

Morphological studies indicated that nearly half of all pyramidal tract neurons originate in the postcentral primary sensory cortex (areas 3, 1, and 2) and the parietal association cortex (areas 5 and 7) in primates, although they terminate mainly in the dorsal horn in the spinal cord (Kuypers 1960; Russel and DeMyer 1961; Jones and Wise 1977; Murray and Coulter 1981). Also, it was demonstrated that pyramidal tract neurons in the primary somatosensory cortex of the monkey exhibit similar responses to those of the primary motor cortex (Evarts 1974; Fromm and Evarts 1982).

In the normal animal, the motor cortex will be working as the main motor output station of the cerebral cortex, and the somatosensory cortex will contribute little to such motor function. When the motor cortex is directly blocked or indirectly disabled, the somatosensory cortex would assume a motor function as the result of some active central and/or peripheral feedback mechanism. The mechanism underlying the quick change in the working neural circuits, caused by the assumed active central and peripheral feedback mechanism, must be different from any compensatory neural reorganization caused by permanent destruction (see Goldberger and Murray 1978). The central nervous system may have various options for processing the information relating to the intended movement and may select from these as necessary. Even the motor output of the cerebral cortex reveals such possibilities, as shown in this study, and the associative cortical areas should have much more flexibility than the cortical output system. Nevertheless, the pyramidal tract neurons in the somatosensory cortex must normally have some other roles than such a compensatory motor function, as revealed in the present study (Evarts 1971; Bioulac and Lamarre 1979).

Acknowledgements. The authors thank Prof. N. Mizuno for his help and valuable advice on morphological examination of the hemispherectomized cerebelli. This work was supported by a grant-in-aid for scientific research from the Ministry of Education, Science, and Culture of Japan.

References

BIOULAC B, LAMARRE Y (1979) Activity of postcentral cortical neurons of the monkey during conditioned movements of a deafferented limb. Brain Res 192: 427-437

BROOKS VB (1983) Study of brain function by local, reversible cooling. Rev Physiol Biochem Pharmacol 95: 1-109

EVARTS EV (1971) Feedback and corollary discharge: a merging of the concepts. Neurosci Res Program Bull 9: 86-112

EVARTS EV (1974) Precentral and postcentral cortical activity in association with visually triggered movement. J Neurophysiol 37: 373-381

FROMM C, EVARTS EV (1982) Pyramidal tract neurons in somatosensory cortex: central and peripheral inputs during voluntary movement. Brain Res 238: 186-191

GEMBA H, HASHIMOTO S, SASAKI K (1981) Cortical field potentials preceding visually initiated hand movement in the monkey. Exp Brain Res 42: 435-441

GOLDBERGER ME (1974) Recovery of movement following lesions of the motor systems in monkeys. In: STEIN DG (ed) Recovery of function after brain damage. Academic, New York

GOLDBERGER ME, MURRAY M (1978) Recovery of movement and axonal sprouting may obey some of the same laws. In: COTMAN CW (ed) Neuronal plasticity. Raven, New York, pp 73-96

HASHIMOTO S, GEMBA H, SASAKI K (1979) Analysis of slow cortical potentials preceding self-paced hand movements in the monkey. Exp Neurol 65: 218-229

JONES EG, WISE SP (1977) Size, laminar and columnar distribution of efferent cells in the sensori-motor cortex of monkeys. J Comp Neurol 175: 391-438

KUYPERS HGJM (1960) Central cortical projections to motor and somatosensory cell groups. Brain 83: 161-184

MURRAY EA, COULTER JD (1981) Organization of corticospinal neurons in the monkey. J Comp Neurol 195: 339-365

REED DJ, MILLER AD (1978) Thermoelectric peltier device for local cortical cooling. Physiol Behav 20: 209-211

RUSSEL WR, DeMYER W (1961) The quantitative cortical origin of pyramidal axons of *Macaca rhesus*, with some remarks on the slow rate axolysis. Neurology 11: 96-108

SASAKI K (1979) Cerebro-cerebellar interconnections in cats and monkeys. In: MASSION J, SASAKI K (eds) Cerebro-cerebellar interactions. Elsevier, Amsterdam, pp 105-124

SASAKI K, GEMBA H (1982) Development and change of cortical field potentials during learning processes of visually initiated hand movements in the monkey. Exp Brain Res 48: 429-437

SASAKI K, GEMBA H, HASHIMOTO S (1981) Premovement slow cortical potentials on self-paced hand movements and thalamocortical and corticocortical responses in the monkey. Exp Neurol 72: 41-50

SASAKI K, GEMBA H, MIZUNO N (1982) Cortical field potentials preceding visually initiated hand movements and cerebellar actions in the monkey. Exp Brain Res 46: 29-36

SHERK H (1978) Area 18 responses in cat during reversible inactivation of area 17. J Neurophysiol 41: 204-215

Activity of Neurons in the Prefrontal Cortex During Visually and Pain-Induced Hand Movements in Monkeys

K. Kubota

Department of Neurophysiology, Primate Research Institute, Kyoto University, Inuyama City, Aichi 484, Japan

Introduction

Several single-neuron studies in monkeys performing a learned task have demonstrated neuronal activity in the prefrontal cortex prior to the execution of the necessary voluntary movement (cf. Kubota 1978). In these studies monkeys performed a manual lever press or lever release to obtain a reward. When there were two choices of the lever press (for example, left or right lever press), two kinds of movement-related neuronal activity were found: one occurred whenever the lever was pressed, regardless of the side of the lever, whereas the other occurred only when a particular lever - the left or right - was pressed. Since this movement-related activity in the prefrontal cortex appeared earlier than, or almost simultaneously with, activity of pyramidal tract neurons in the hand motor area, it has been assumed that prefrontal movement-related activity may influence neuronal activity of the motor system, so that the appropriate voluntary hand movement is executed (Kubota et al. 1980). Recently, comparable responses were also observed before and/or during flexion and extension movements at the wrist joint, and they were classified as bidirectional and undirectional movement-related activity respectively (Kubota and Funahashi 1982).

These movement-related responses were triggered by visual or, less frequently, auditory cues, since most studies used tasks with visual signals. In tasks such as delayed response and delayed matching-to-sample, some of the movement-related responses were activated by visual cues (Kubota et al. 1974, 1980). This prefrontal neuronal activity was thought to be in some way related to complex brain functions, such as short-term spatial memory, expectation, attention, etc. (cf. Kubota 1978). How prefrontal neurons behave in learned tasks involving other sensory modalities is not known. In multischeduled lever-release tasks with visual and auditory cues, it was shown that 11 out of 12 lever-release-related neurons were activated, regardless of the modality of the cue, and one neuron was activated only by a visual cue (Ito 1982). Whether or not there are movement-related neurons triggered by auditory stimuli in the prefrontal cortex is not known. Since the number of sampled neurons was so small, it was difficult to reach conclusions on the relations between the triggering sensory modalities and the movement-related responses.

In this presentation we approached the problem of sensory modalities vs movement-related activities, using visual and pain stimuli. Monkeys performed lever-release tasks on receiving visual and pain stimuli. Pain was applied to the forearm skin via an attached cylinder with a heating element through which heated water flowed. For comparison, visually triggered lever release was also performed.

Tasks

Two monkeys performed three kinds of task called the Visual GO Task, the Noxious GO Task, and the Nonnoxious NO-GO Task. These consisted of visual cues (red, green, and yellow lights), nonnoxious heat stimuli (warm, 43°C), and noxious heat stimuli (pain, 55°C), as defined below. Each trial was initiated by a "start" lamp (yellow). As soon as the monkey pressed the lever, the start lamp was turned off. Half a second later, either the green or the red lamp ("cue") came on for 1 s, after which its color changed to yellow, signaling the response period for GO. During this response period after the green cue, either a noxious or a nonnoxious heat stimulus was applied. After the noxious heat stimulus the monkey released the lever and then both the noxious stimulus and the yellow lamp went off. The monkey was not rewarded (Noxious GO Task). After the nonnoxious heat stimulus the monkey pressed the lever continuously for 5 s. Then he was rewarded and both the yellow lamp and the heat stimulus went off. After the reward he released the lever spontaneously (Nonnoxious NO-GO Task). During the response period after the red cue, the yellow lamp signal was not accompanied by a heat stimulus. The monkey released the lever, and then he was rewarded and the lamp extinguished (Visual GO Task). The succession of trials was presented pseudorandomly.

Responses During the Tasks

Neuronal responses related to the three tasks were recorded from the dorsolateral and dorsomedial prefrontal cortex in two monkeys. One hundred and seventeen neurons showed changes in discharge rate during the Noxious GO Task, 112 neurons during the Nonnoxious NO-GO Task, and 25 neurons during the Visual GO Task. About half of the neurons showed an increase in discharge rate, and the other half showed a decrease. Changes in the discharge rate were classified into different types according to onset latency (early or late), discharge pattern (transient or sustained), and their relation to the movement onset. Figure 1 illustrates schematically the time courses of the rate changes which would occur during the response periods of each of the three tasks.

During the nonnoxious NO-GO Task four types of responses were observed: (a) early transient, (b) early sustained, (c) late transient, and (d) late sustained. Early responses appeared 0.1-0.5 s after the GO signal onset. Therefore, these responses were induced by the visual GO signal. Late responses appeared after the skin temperature rose to more than 40° C. Therefore these responses were induced by nonnoxious heat.

With the Noxious GO Task, in addition to the four types of neuron response seen in the Nonnoxious NO-GO Task, there was a new type in which the cell discharged immediately before and during the lever release. Hence this response was classified as a lever-release type.

With the Visual GO Task, two types of neuron response were recognized. In one of these the cell was active in relation to lever release, whereas with the other the cell responded transiently immediately after the visual signal. Since the time interval between the Visual GO signal onset and the lever release was brief (of the order of several hundred milliseconds), it was often difficult to differentiate the two types of neuron response. With late transient and late sustained responses, most neurons were active during both the GO and the NO-GO tasks, but the magnitude of these changes were different for different tasks and for each neuron. If heat stimuli were applied naturally, the cell discharged, but its onset latency was slower and its magnitude was smaller. Thus, activated neurons are related to sensory GO signals, heat stimuli, or movement. It is unlikely that all induced activity is nonspecific to the sensory modality.

Lever-Release-Related Responses During the Two GO Tasks

Out of 62 neurons activated in association with the lever release, 30 neurons were activated with the Noxious GO Task and 17 neurons with the Visual GO Task. The remaining six neurons were activated during the performance of both tasks. These remaining neurons were also activated when the monkeys released the lever after the delivery of the reward in the Nonnoxious NO-G Task. Some of the neurons active in the Noxious GO Task were also activated during the Nonnoxious NO-GO Task, but neurons active in the Visual GO Task were not activated during the Nonnoxious NO-GO Task. These results indicate that some movement-related responses are specific to the sensory modality whereas others are nonspecific to the sensory modality.

It appears that these prefrontal neurons send commands for the execution of goal-directed behavior and that these commands vary according to the sensory modality of the triggering cue. Nonspecific movement-related responses may facilitate the execution of goal-directed behavior.

Summary

Figure 2 schematically illustrates the flow of impulses in the prefrontal cortex related to the initiation of the lever release. Information from sensory stimuli reaches the prefrontal cortex via separate channels and activates sensory-specific neurons transiently, which in turn activate sustained neurons. Sustained neurons activate movement-related neurons specifically and nonspecifically. Which of these movement-related neurons are more directly linked to the motor system is not known.

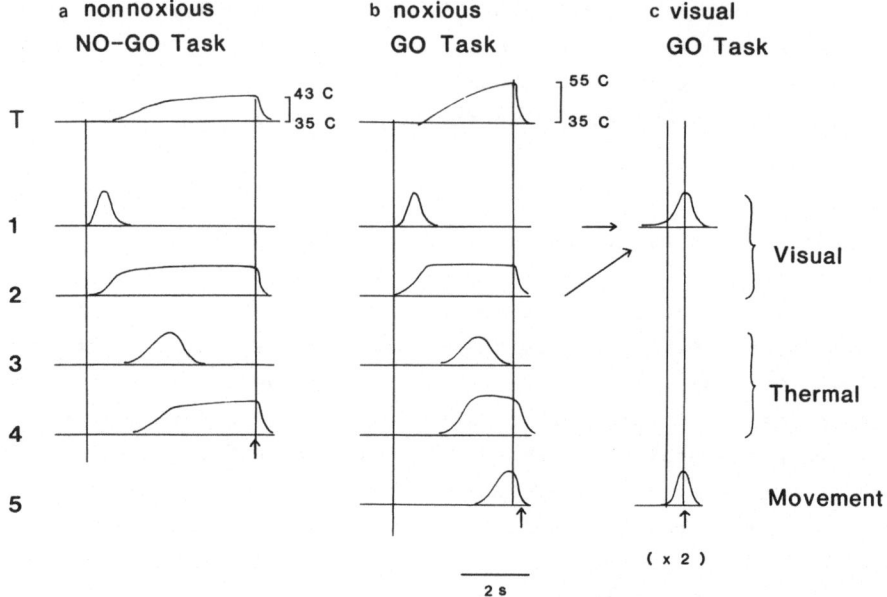

Fig. 1. Schematic illustration of changes of discharge rates in response periods of Nonnoxious No-GO Task **(a)**, Noxious GO Task **(b)**, and Visual GO Task **(c)**. Five different discharge patterns are shown related to the visual cue *(1, 2)*, to thermal stimuli *(3, 4)*, and to movement *(5)*. Top trace shows temperature change *(T)*

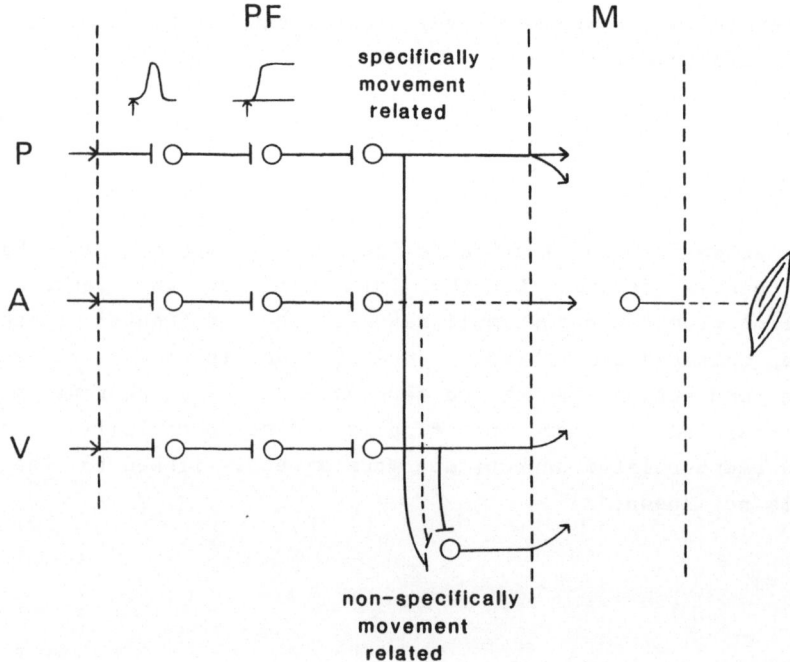

References

ITO S (1982) Prefrontal unit activity of macaque monkeys during auditory and visual reaction time tasks. Brain Res 247: 39-47

KUBOTA K (1978) Neuron activity in the dorsolateral prefrontal cortex of monkey and initiation of behavior. In: ITO M, TSUKAHARA N, KUBOTA K, YAGI K (eds) Integrative control functions of the brain, vol 1. Kodansha-Elsevier, Tokyo Amsterdam, pp 407-417

KUBOTA K, FUNAHASHI S (1982) Direction-specific activities of dorsolateral prefrontal and motor cortex pyramidal tract neurons during visual tracking. J Neurophysiol 47: 362-376

KUBOTA K, IWAMOTO T, SUZUKI H (1974) Visuokinetic activities of prefrontal neurons during delayed-response performance. J Neurophysiol 36: 1197-1212

KUBOTA K, TONOIKE M, MIKAMI A (1980) Neuronal activity in the monkey dorsolateral prefrontal cortex during a discrimination task with delay. Brain Res 183: 29-42

◁ **Fig. 2.** Schematic illustration of the flow of impulses in the prefrontal cortex. *P*, noxious input; *A*, auditory input; *V*, visual input; *PF*, prefrontal cortex; *M*, motor cortex

Thalamocortical Organization in the Raccoon: Comparison with the Primate

J. I. Johnson

Department of Anatomy, Michigan State University, East Lansing, MI 48824-1316, USA

Primates are not the only skilled hand users with neocortical governance. My involvement with raccoons began a quarter century ago when my professor, Kenneth Michels, just arrived with a fresh PhD from Harry Harlow's laboratory, sought to set up a monkey-experiment operation with "cheap monkeys." He needed a readily available subject who could work in the apparatus, and perform the behavioral tasks designed for monkeys. We used, successfully, raccoons (e.g., Johnson and Michels 1958).

At about the same time, Wally Welker, also from Harlow's laboratory and then working with Clinton Woolsey, was mapping the sensorimotor cerebral cortex of raccoons (Welker and Seidenstein 1959) and discovered an interesting parallel associated with the similar hand use of raccoons and monkeys. Unlike cats and dogs, raccoons possessed a fully developed central sulcus separating the sensory and motor regions of cortex (Fig. 1). The more striking finding was the vast expanse of sensory cortex devoted to receptive fields on the hand, with separate gyral crowns representing projections from each of the digits and the palm. In relative proportion, this is a greater hand representation than that seen in any primate, or any mammal studied thus far. When I later came to work with Welker, we found this expanded and segregated hand representation to have parallels in the lobulated architecture and projection pattern in the ventrobasal thalamus and cuneate-gracile complex (Welker and Johnson et al. 1965; Johnson 1968).

Primates show traces of these thalamic and cuneate-gracile lobules, furthering our hopes that raccoons can serve as a "giant" model of neural organization of information from the hand.

Experimental Brain Research, Suppl. 10
© Springer-Verlag Berlin · Heidelberg 1985

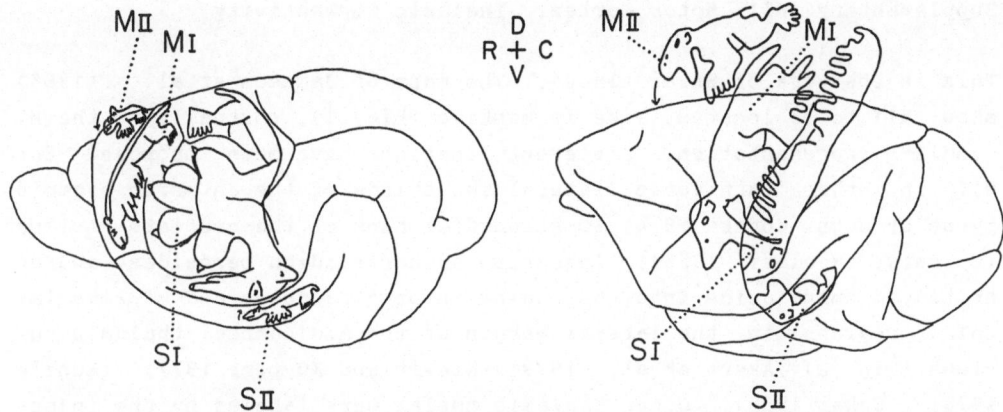

Fig. 1. Major subdivisions of sensorimotor cortex for monkey *(right)*, according to Woolsey (1958) and for raccoon *(left)* from the maps of supplementary motor cortex *(MII)* (Jameson et al. 1968), primary motor cortex *(MI)* (Hardin et al. 1968), primary somatic sensory cortex *(SI)* (Welker and Seidenstein 1959), and second somatic sensory cortex *(SII)* (Herron 1978). In this and all following figures: *D*, dorsal; *V*, ventral; *R*, rostral; *C*, caudal; *L*, lateral; *M*, medial

Later, in our laboratory, Herron (1978) showed the hand representation is enlarged in the second somatic sensory (SII) representation in the raccoon neocortex (Fig. 1). His SII map, along with the motor maps generated meanwhile in the Wisconsin laboratories (Hardin et al. 1968; Jameson et al. 1968), presents a picture of sensorimotor neo-cortical connections that closely parallels the older view of those regions in primates (Fig. 1). With the advent of newer ideas about the organization of the primary somatic sensory (SI cortex) and new methods of microelectrode recording and labeling of connections through axonal transport, we have in recent years been reexamining the thalamocortical connectivity of these four sensorimotor regions (MII, MI, SI, SII) in raccoons. The findings I will present here, in addition to my own studies, are from the PhD dissertations of Paul Herron (1980), Michael Ostapoff (1983), Sharleen Sakai (1980), and Sidney Wiener (1983). I shall consider corticothalamic connectivities of, in this order, MII, SII, MI, and finally SI (including the so-called area 3a subcutaneous projections).

Supplementary (MII) Motor Cortex: Thalamic Connectivity

This is the work of Sakai (1980). The maps of Jameson et al.   (1968)
show MII to be located, like in monkeys (Fig. 1), rostrally to the MI
hindlimb representation. Different locations have been proposed for
MII in other carnivores (lateral two-thirds of the anterior sigmoid
gyrus of dogs, Gorska 1974; rostromedial bank of the cruciate sulcus
in cats, Woolsey 1958). Injection of horseradish peroxidase and/or
tritiated amino acids into the region in both raccoons and monkeys labeled regions in the lateral margin of the mediodorsal thalamic nucleus (Fig. 2) (Akert et al. 1979; Kievit and Kuypers 1977;   Künzle
1978; Sakai 1980). Other thalamic nuclei were labeled by the injections in monkeys (including the ventroanterior, ventrolateral, centrolateral, and centromedian nuclei), but none of them was labeled as
consistently or as heavily as the mediodorsal nucleus. The injections
into raccoon MII also labeled a few loci in the ventroanterior, ventrolateral, centrolateral, and centromedian nuclei, but nowhere near
as many as in the lateral segment of the mediodorsal nucleus.

Second Somatic Sensory (SII) Cortex: Thalamic Connectivity

Herron (1980, 1983) injected horseradish peroxidase into physiologi-
cally identified representations of forepaw and trunk and hindlimb in
raccoon SII cortex. He found orderly sequences of cells labeled in
what he identified as the ventroposteroinferior (VPI) thalamic nucleus
(Fig. 2). This is in marked contrast with the results of injections
into the corresponding SI representations. Following the SI injec-

Fig. 2. Thalamic connections of hand-related regions of supplementary
motor *(MII)*, primary motor *(MI)*, primary somatic sensory *(SII)*, and
second somatic sensory *(SII)* cerebral cortex in raccoon and monkey.
MII data for raccoon from Sakai (1980, Fig. 20) and for monkey from
Künzle (1978, Fig. 7L). MI data for raccoon from Sakai (1980,
Fig. 24) and for monkey from Strick (1976, Fig. 1). SI data for raccoon from Herron (1980, Fig. 15) and for monkey from Nelson and Kaas
(1981, Fig. 11). SII data for raccoon from Herron (1979, Fig. 8) and
not available for monkey. Cortical regions *shaded* on brain outlines
showed connections with *dotted* regions indicated on diagrams of
transverse sections through the left thalamus in each case. *MD,* dorsomedial nucleus; *Vb,* ventrobasal complex; *VL,* ventrolateral
complex; *VLo,* ventrolateral nucleus oral part (connections were also
found in the caudal part and in the oral part of the ventroposterola-
teral nucleus); *VPI,* ventroposteroinferior nucleus; *VPLc,* ventropos-
terolateral nucleus, caudal part

RACCOON   MONKEY

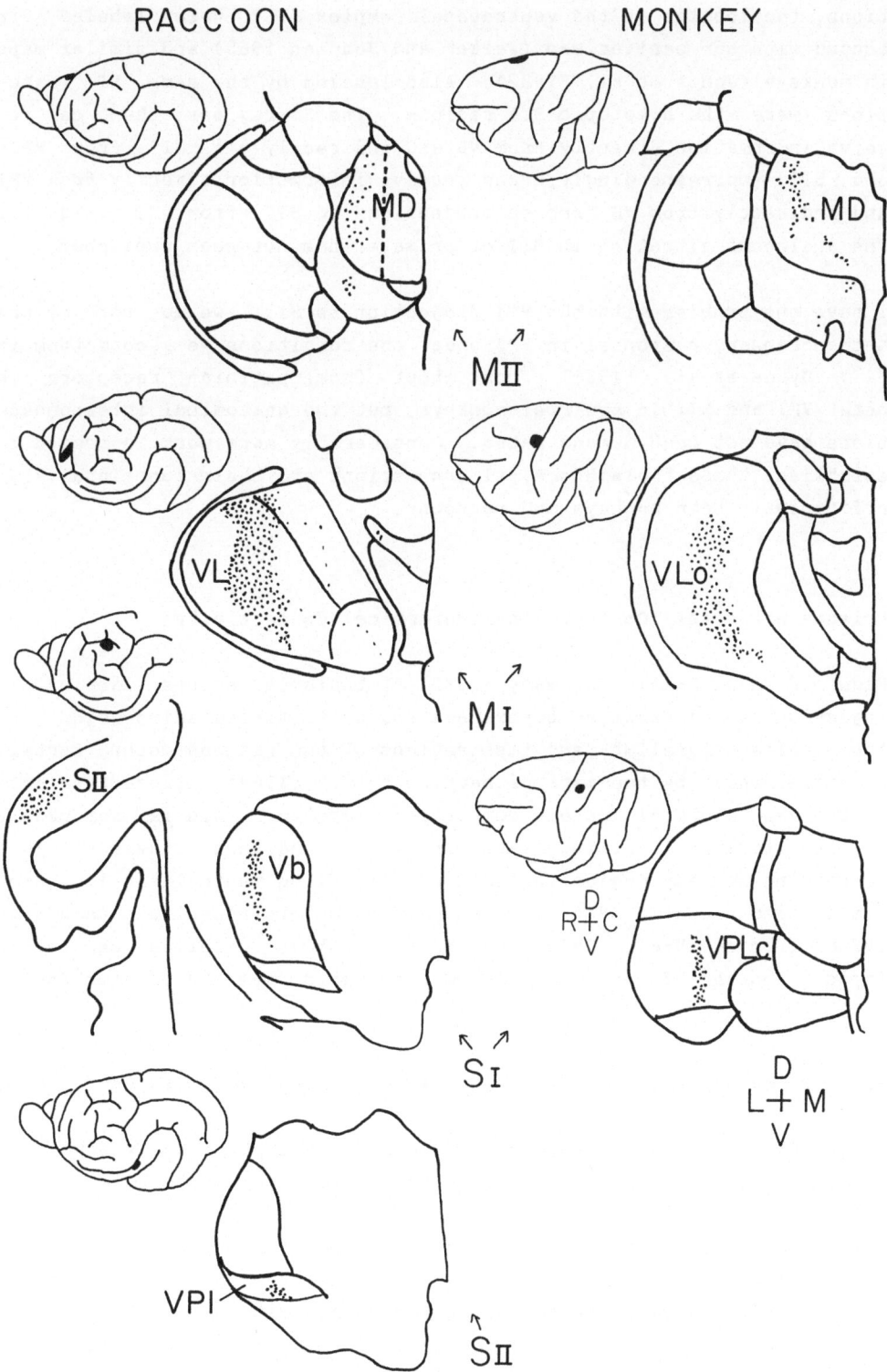

tions, the lobules of the ventrobasal complex (VB) were labeled, in accord with our earlier map (Welker and Johnson 1965) and similar maps in monkeys (Jones et al. 1982). Also labeled by the same SI injections were the homotopic SII regions. The SI regions, then, can receive information directly from VB and indirectly via SII from VPI, and SII, correspondingly, can receive information directly from VPI and indirectly from VB through projections to SII from SI (Fig. 2). The analogous situation in SII of primates has not been published.

I have one problem with the VPI projection to SII: we do not record mechanosensory responses in VPI under the conditions we record them in SII. Dykes et al. (1981) report input from pacinian receptors in both VPI and SII in squirrel monkeys, but the anatomical interconnections have not been demonstrated. Considerably more work is needed to establish these thalamocortical connections and their functional significance in both monkeys and raccoons.

Primary Motor (MI) Cortex: Thalamocortical Connectivity

Figure 2 shows Sakai's summary (1982) of thalamic regions labeled by injections of horseradish peroxidase and/or tritiated amino acids into the hindlimb, forelimb, and face regions of the raccoon motor cortex, as determined by the maps of Hardin et al. (1968). Successive arcs of the ventrolateral nuclear complex are labeled. This is remarkably similar to the labeling seen in monkey thalamus; there the corresponding nuclear region includes nuclei called ventrolateralis pars oralis and caudalis (VLo, VLc) and ventroposterolateralis pars oralis (VPLo) (Strick 1976). This region in monkeys receives cerebellar input; sources of input to VL have not been determined in raccoons.

Primary Sensory (SI) Cortex: Recently Determined Subdivisions

Recent fine-grain micromapping of SI cortex in monkeys has shown four subdivisions related to the long-recognized cytoarchitectonic fields of Brodmann (1905) (Fig. 3). The picture presented nowadays (e.g., Kaas et al. 1981) to those of us not working with primates is as follows. Separate, neighboring, somatotopically ordered representations of the cutaneous body surface are found corresponding to cytoarchitectonic areas 3b and 1. Similarly separate, but not neighboring, projections from deep subcutaneous tissues (possibly including muscles, tendons, fasciae, and joints) are found in the more outlying areas, 3a and 2. To my mind, this view is considerably modified by findings presented at this symposium concerning the characteristics of receptive fields of cells in area 2 of monkeys (see the chapters by Iwamura et al. and Darian-Smith et al. in this volume). These complex receptive fields, with multiple digits, large skin areas, and indeterminate surface versus deep origins, are similar to those we see in raccoons in the heterogenous anterior border zone of SI cortex, lying on the posterior bank of the central sulcus between the major simple-field cutaneous projection and the kinesthetic projections on the anterior bank of the central sulcus. This anterior border zone of SI in raccoons is reminiscent of the suggestion by McKenna et al. (1982) of a ring, receiving both cutaneous and deep input, surrounding SI in monkeys. We have evidence (Herron 1980) suggesting that the complexity of these receptive fields derives from corticocortical connections.

We have remapped portions of the raccoon sensory cortex in minute detail to see if corresponding subdivisions exist.

Rather than multiple representations of the hand, we find most of the raccoon hand projections to constitute a single representation described by Welker and Seidenstein (1959). The characteristics of this projection field, as well as the cortical cytoarchitectonics, are similar to those described for area 3b in monkey (e.g., Kaas et al. 1981). Allowing that the greatly expanded hand representation might have dislocated the area 1 and area 2 representations, we have examined in fine detail the less elaborated representation of the trunk in three raccoons thus far (Fig. 3). In each of the three, most of the trunk gyrus is occupied by a single somatotopic representation of the skin, with the ventral abdominal surface anterior and the dorsal midline projecting posteriorly. But continuing down the anterior bank of the ansate sulcus, we find another sequence of receptive fields, forming a tiny second representation of the trunk, adjacent to, and a distorted mirror image of, the larger representation. All receptive fields in this small representation are also cutaneous. This could conceivably correspond to the area 1 representation of primates, much reduced. We have found no evidence of a deep subcutaneous projection region, such as that in area 2 of primates, in the posterior sensory cortex of raccoons. In our animals, anesthetized with Dial-urethane,

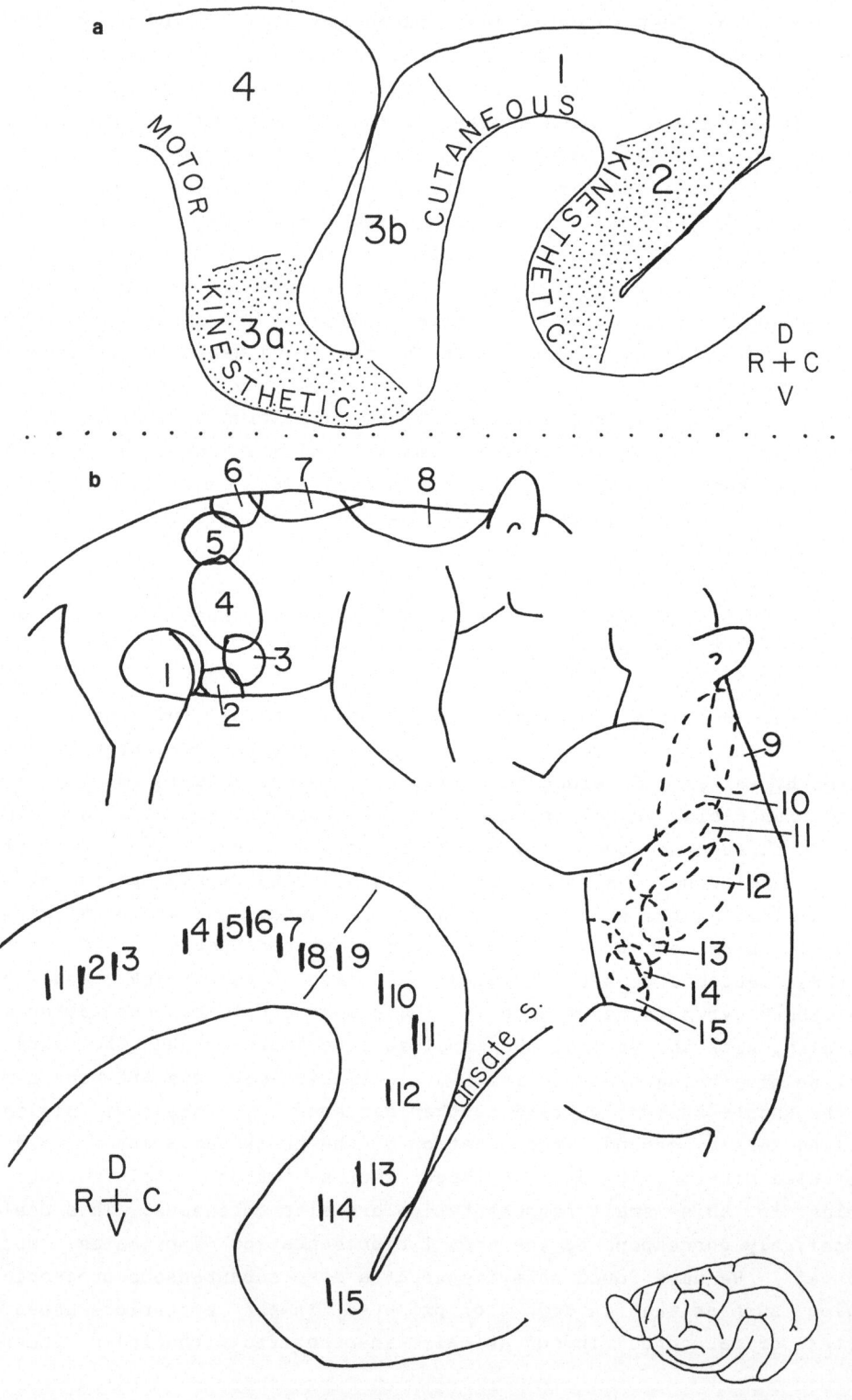

◁ **Fig. 3a.** Anteroposterior sequences of cytoarchitectonic fields in monkey SI cortex. A succession of roughly mirror image body representations, one for each of the cytoarchitectonic fields, 3a, 3b, 1, and 2,
has been reported (Kaas et al. 1981). Projections to 3a and 2 are
from subcutaneous kinesthetic receptive fields, and those to 3b and 1
are from cutaneous receptive fields.
b Anteroposterior sequence of somatic sensory projections in the trunk
representative of raccoon SI cortex (J.I. Johnson and S. Warach,
studies in progress). An anterior representation of subcutaneous
kinesthetic projections corresponds to the 3a projections in monkeys.
The major sequence of cutaneous projections (from ventral receptive
fields anteriorly to dorsal receptive field posteriorly) appears to
correspond to the 3b projections in monkeys. The tiny "reversed"
representation (shown by *dotted lines* around receptive fields) from
dorsal receptive fields anteriorly and ventral receptive fields posteriorly in the anterior bank of the ansate sulcus may be analogous to
the area 1 representation in monkeys. We have found no raccoon analogue to the subcutaneous kinesthetic projections to area 2 of monkeys

we find no mechanosensory responses posterior to the fundus of the ansate sulcus, dorsomedially to the hand representation. We have not as
yet mapped the more lateroposterior regions.

Anteriorly in SI, we find two border zones with characteristics different from the main body representation (Johnson et al. 1982).
These show clearly at the anterior edge of the representation of the
fourth and fifth digits of the hand. Going anteriorly in the main
hand representation, receptive fields are progressively more distal on
the volar glabrous surface of the hand. More anteriorly are projections from the claw, and anteriorly to these are what seem to be mixed
projections from claws, dorsal hairy surfaces of the digit, and of the
neighboring digits, and occasional projections from subcutaneous tissues. Still more anteriorly, near the depth of the posterior bank of
the central sulcus, are more projections from the volar surface of the
digit, and the sequence of receptive fields here is "reversed," the
most anterior ones from fields in the proximal digit and the palm. A
similar set of projections was reported in monkeys (Whitsel et al.
1971) in what is now regarded as the 3b region (Fig. 4).

Anteriorly to all of this, in the anterior bank of the central sulcus,
is a region of projections from deep, subcutaneous tissues, which corresponds to that known as the 3a region in primates and cats. I have
even more reservations than do the investigators using monkeys about
the cytoarchitectonic identity of the region (Jones et al. 1978;
Johnson et al. 1982); but there is striking similarity among rac-

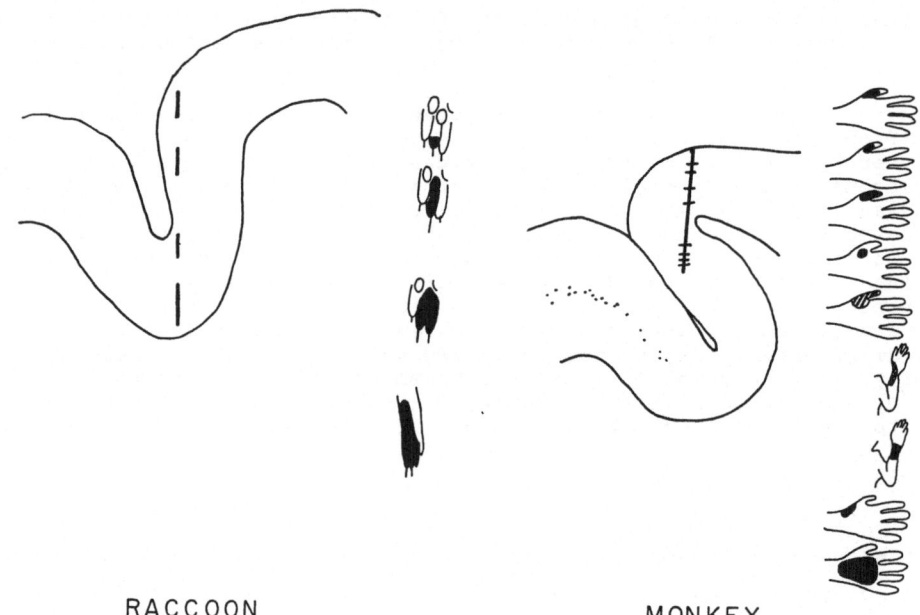

RACCOON   MONKEY

Fig. 4. In what has been termed "SI proper" in monkeys (area 3b, Kaas et al. 1981) and in raccoon SI, the somatotopic representation in general has the tips of the hand digits pointing rostrally. But in both raccoons (Johnson et al. 1982, Fig. 9) and monkeys (Whitsel et al. 1971, Fig. 7), there is a small region in the posterior bank of the central sulcus where this proximal-to-distal sequence is reversed

coons, primates, and cats in both the physiological character and the location of this field relative to the central sulcus (Fig. 5).

Deep Subcutaneous "3a" Projection Region: Thalamic Connectivity

Injections of horseradish peroxidase into either bank of the central sulcus of raccoons labeled cells along the dorsal and anterior borders of the ventrobasal complex and adjoining cells in the ventrolateral nucleus. A far more precise localization of the relevant thalamic region is obtained by fine-grain electrophysiological micromapping of projections to the ventrobasal thalamus from cutaneous vs deep subcutaneous receptive fields. This was the subject of the doctoral dissertation of Sidney Wiener, just completed this year. Examples of his, and some even more recent findings, are presented in Fig. 6. A layer, or "shell" of projections from deep subcutaneous receptive fields overlies the dorsorostral aspect of the mechanosensory region

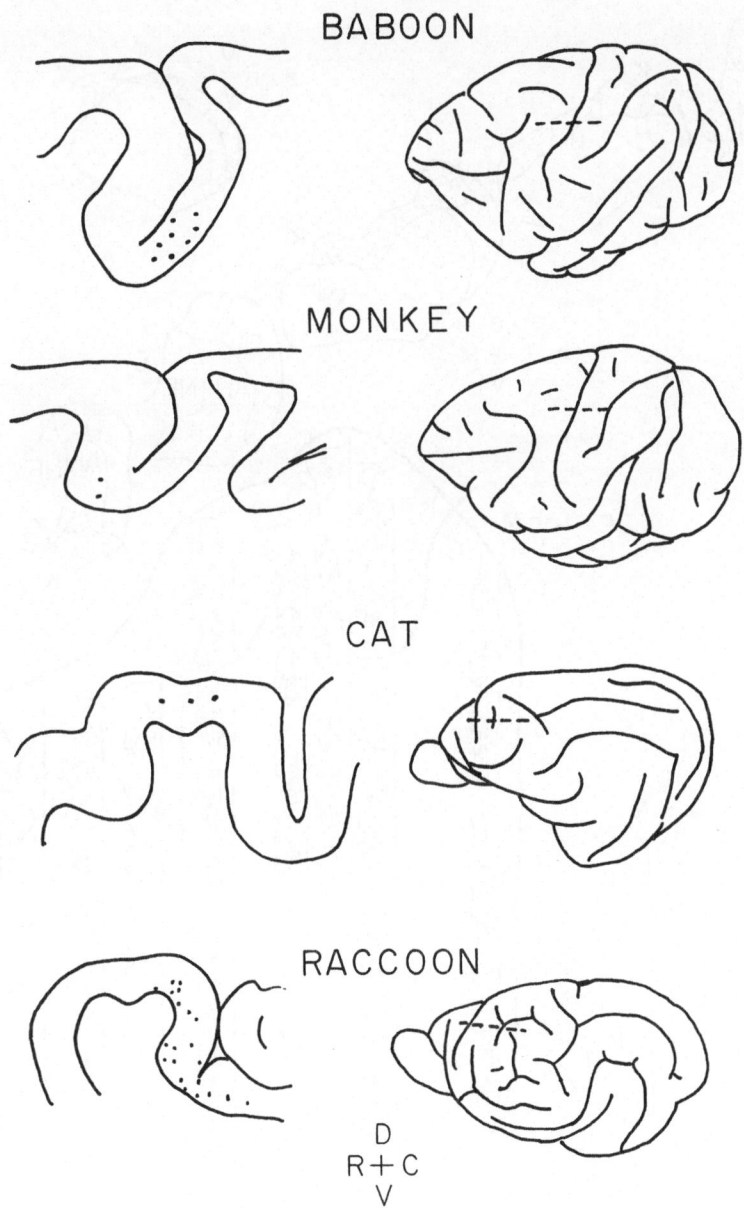

Fig. 5. Locations *(dots)* of projections from deep subcutaneous kinesthetic receptive fields to "area 3a" cerebral cortex, in relation to the central sulcus, as found in baboon (Phillips et al. 1971), monkey (Jones and Porter 1980), cat (Dykes et al. 1980 – central sulcus appears as the postcruciate dimple), and raccoon (Johnson et al. 1982). Drawings at *left* represent parasagittal sections through cerebral cortex at levels indicated by *dashed lines* on outline drawings of brains at *right*

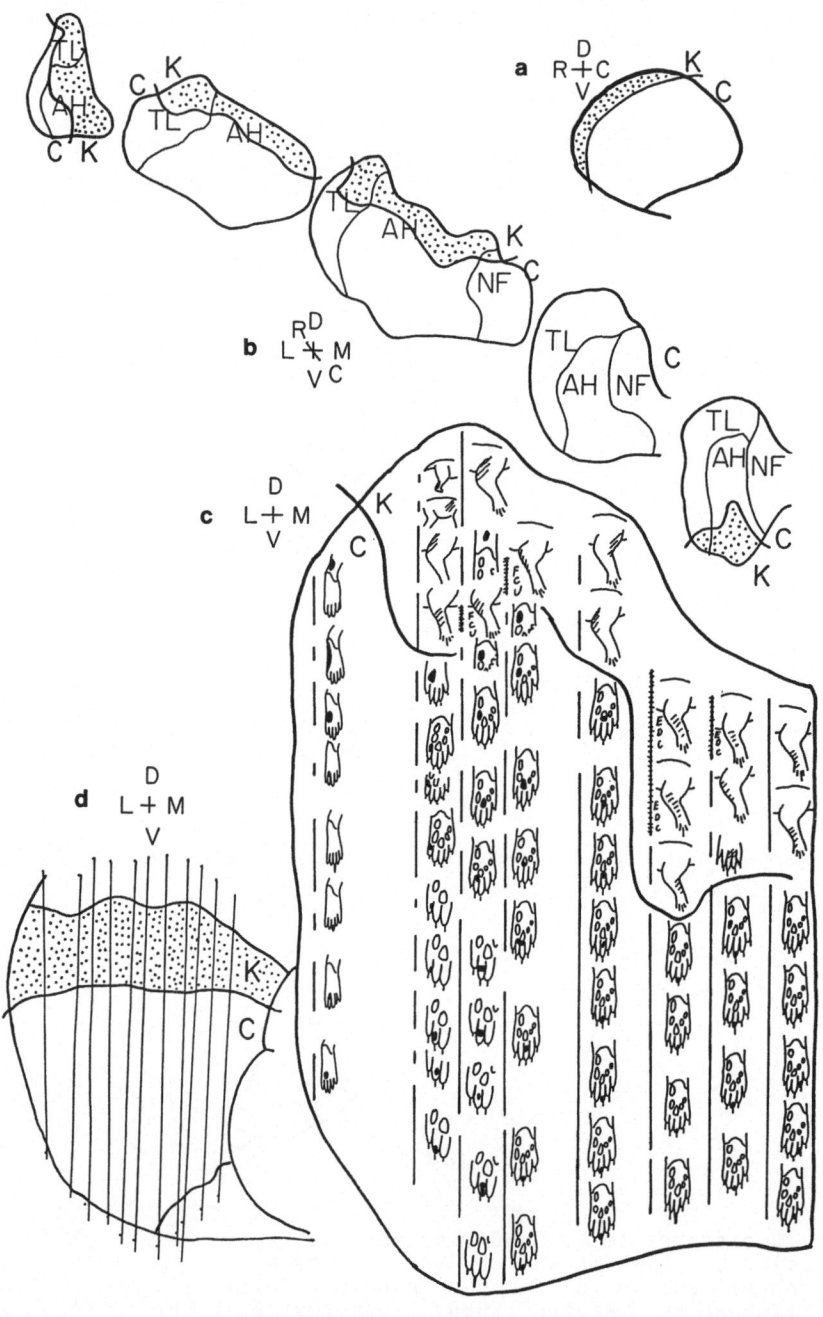

◁ **Fig. 6a.** Relative locations of deep subcutaneous kinesthetic *(K)* and cutaneous *(C)* projections to raccoon ventrobasal thalamus, as seen in a parasagittal section (Wiener 1983).
b Fields of kinesthetic *(K)* and cutaneous *(C)* projections in a set of 5 transverse sections through raccoon ventrobasal thalamus (Wiener 1983). Domains are indicated for projections: *AH,* arm and hand; *NF,* neck and face; and *TL,* trunk, tail, and hindlimb.
c Kinesthetic *(hatched)* and cutaneous *(solid black)* receptive fields activating loci along the tracks of a transverse row of dorsoventral microelectrode penetrations through raccoon ventrobasal thalamus. Data obtained in an experiment by C.M. Smith, J.I. Johnson, and S.I. Wiener, where overlying skin was reflected and the muscles flexor carpi ulnaris *(FCU)* and extensor digitorum communis *(EDU)* and their tendons were dissected free, but not detached, to accurately localize deep-lying kinesthetic receptive fields. *Solid bars* represent responsive regions along electrode tracks; those with *hatching* represent those responding to the exposed and freed muscles.
d Locations of deep kinesthetic and cutaneous projections to a transverse section of monkey somatic sensory thalamus (Jones and Friedman 1982, Fig. 7)

of the ventrobasal complex. Small additional regions of deep-tissue projections are found along the caudoventral nuclear boundary.

It is easy to discriminate deep from cutaneous fields on the trunk and limbs of raccoons; the skin hangs loosely and can be moved easily relative to underlying tissue. (This is not true of the face and extremities.) A large proportion of this "kinesthetic shell" of the ventrobasal complex is activated by mechanical displacement of the forelimb muscles responsible for hand movements. Carefully dissecting the skin away from over these muscles shows that some thalamic units respond to stretching of muscles and tendons; others, however, respond best to light stroking of the muscle surface and are unresponsive to stretch. Some of these are so delicately tuned that we were activating them by moving hairs on the overlying skin, before moving the skin away. This responsivity has also been reported for spinal interneurons (Cleland et al. 1982). We therefore must consider their suggestion that receptive free nerve endings in muscle, as well as the more frequently considered spindle and tendon organs, contribute to the sensory input to these spinal and forebrain regions.

There is one marked difference between the responses we obtain from the kinesthetic thalamus compared with those from 3a cortex. All responses in 3a cortex from deep receptive fields are fast adapting, whether the anesthetic used is a barbiturate (Dial-urethane), chloralose, or methoxyfluorane with nitrous oxide (Johnson et al. 1982).

All of our thalamic data are from animals anesthetized with barbiturates, and most, but not all, of the responses from deep tissues are slowly adapting. Frequently the clusters of units at a specific locus contain both fast- and slowly-adapting elements.

A similar rostrodorsal kinesthetic shell has been described in the mechanosensory thalamus of macaque (Maendly et al. 1981; Jones and Friedman 1982) and squirrel monkey (Dykes et al. 1981), (Fig. 6). In macaques it appears to supply the deep tissue cortical representations of both areas 3a and 2 (Jones and Friedman 1982).

The locus of the cell bodies of neurons responsible for delivering kinesthetic information to the ventrobasal thalamus is unknown for monkeys; suspicion has been directed to the thalamic projections from the external cuneate nuclei (Boivie and Bowman 1981).

Medullary Sources of Kinesthetic Projections to Raccoon Ventrobasal Thalamus

This is the doctoral dissertation work of Ostapoff (1983). Injection of small quantities of horseradish peroxidase from the tip of a microelectrode, while simultaneously recording responses from deep tissues in the kinesthetic shell of the ventrobasal thalamus, labeled cells in a group of medullary nuclei. All of these medullary regions respond themselves to kinesthetic stimulation; they include the external cuneate nucleus and its medial tongue, the basal cuneate region underlying the cutaneous receiving area of the cuneate-gracile complex, and cell groups z and x rostral to the cuneate-gracile and external cuneate nuclei (Fig. 7).

Parallels and Differences Between Raccoons and Primates

Figure 8 illustrates diagrammatically the current picture of thalamocortical organization of the sensorimotor regions of cerebral cortex of raccoons compared with those in primates. Parallels include thalamic mediodorsal nuclear connections with the supplementary motor area (MII, medial area 6); connections of the motor area (including area 4 (with the ventrolateral thalamic complex (including nucleus

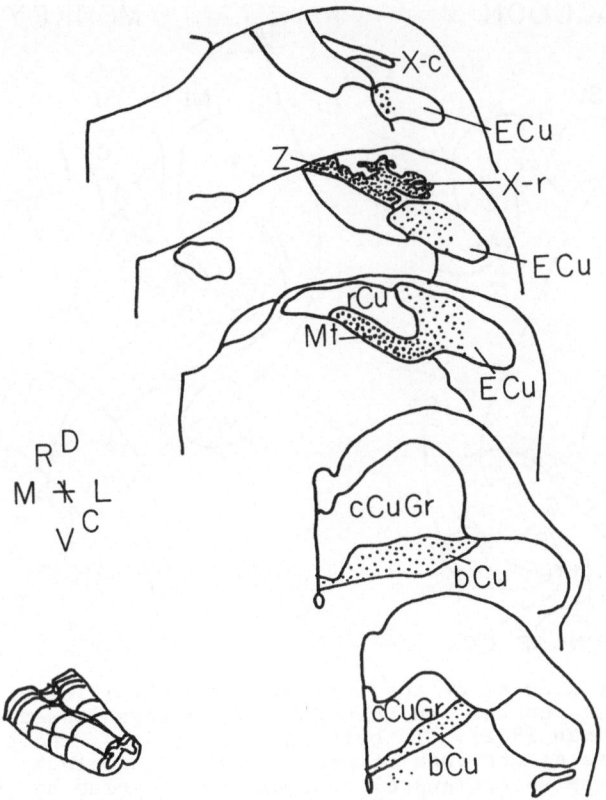

Fig. 7. Locations of cell bodies in dorsal medulla of raccoons labeled following small injections of horseradish peroxidase into the kinesthetic projection region in the rostrodorsal shell of the ventrobasal thalamus (Ostapoff 1983), shown in diagrams of the right sides of five transverse sections through the medulla at levels indicated in the sketch at low left. Labeled were virtually all cells in cell group z *(Z)* and the reticular portion of cell group x *(X-r)*, as well as many cells in the external cuneate nucleus *(ECu)* and its medial tongue *(Mt)*, and in the basal cuneate nucleus *(bCu)*. All of these regions of labeled cells show electrophysiological responses to mechanical stimulation of deep subcutaneous kinesthetic receptive fields. Labeled cells were not seen in the compact portion of cell group x *(X-c)* and the lateral third of ECu, which (like the rest of ECu and Mt) projects to cerebellum, or in the rostral cuneate nucleus *(rCu)* (which sends projections to cerebellum and to other regions of thalamus) and in the central and caudal cuneate-gracile complex *(cCuGr)*, which projects to cutaneous receptive fields in ventrobasal thalamus

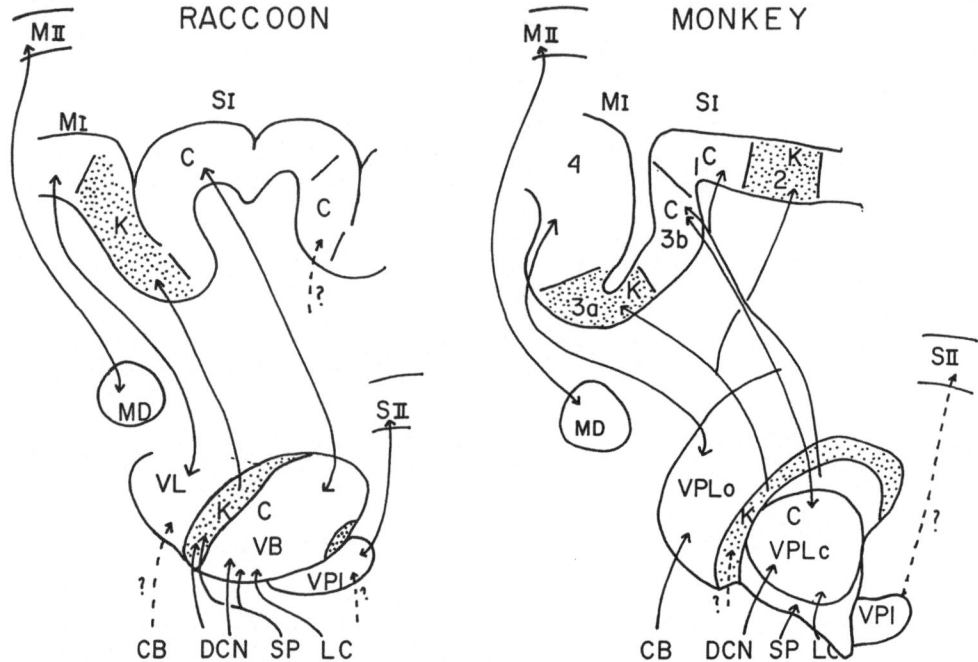

Fig. 8. Summary diagram of the thalamic connectivities of hand regions of sensorimotor cortex in raccoons and monkeys (based on Fig. 13 in Jones and Friedman 1982). In both species, MII is reciprocally connected with lateral portions of the dorsomedial nucleus *(MD)*, as is MI (area 4) with the ventrolateral complex *(VL)* (including nucleus ventroposterolateralis pars oralis *(VPLo)* in monkeys) and SI with the ventrobasal complex *(VB)* (including nucleus ventroposterolateralis caudalis *(VPLc)* in monkeys). Within VB in both animals, there is a rostrodorsal shell *(K, shaded)* projecting to kinesthetic cortical regions; reciprocal corticothalamic projections remain to be demonstrated unequivocally. The major difference between raccoons and monkeys in this regional organization is the absence of the posterior kinesthetic region in raccoons, which would correspond to area 2 in monkeys. The kinesthetic thalamic region in raccoons receives input from the dorsal column nuclei *(DCN)* and spinal regions *(SP)*, but not from the lateral cervical nucleus *(LC)* (Ostapoff 1983); sources of input in monkeys remain to be determined. All three sources - DCN, SP, and LC - project to cutaneous sensory regions *(C)* in both species. A separate outer cutaneous region in VPLc may project to area 1 and area 3b in monkeys (Jones and Friedman 1982); projections to the possible analogue of area 1 in raccoons have not been determined, nor have cerebellar *(CB)* projections to VL in raccoons, nor have projections from ventroposteroinferior nucleus *(VPI)* to SII in monkeys. Not shown in this diagram are connections of all these cortical regions, in both species, with intralaminar centrolateral or centromedian nuclei. *AH,* arm and hand; *NF,* neck and face; and *TL,* trunk, tail, and hindlimb

ventroposterolateralis pars oralis in monkeys); connections of the cutaneous receiving areas of the primary sensory cortex (areas 1 and 3b) with the "core" of the ventrobasal complex (ventroposterolateralis pars caudalis in monkeys); and connections of the kinesthetic receiving zone 3a with the rostrodorsal shell of the mechanosensory region.

The major difference is a kinesthetic projection, from the same rostrodorsal shell of mechanosensory thalamus, to a posterior kinesthetic area in sensory cortex (area 2) in monkeys; this is not apparent in raccoons. Yet to be determined are the medullary sources of kinesthetic input to thalamus in monkeys. Also needed is identification of cerebellar input to the ventrolateral complex in raccoons. Further questions of great interest are the identification of the peripheral receptors responsible for kinesthetic thalamic and cortical activation; the nature of the physiological processing of this information at spinal, medullary, thalamic, and cortical levels; and the relative routing of kinesthetic information, particularly that from the forelimb. Some goes from medullary nuclei through cerebellum and thalamus to motor cortex; some goes from medullary nuclei to thalamus to sensory cortex. Determining the role of all this information in hand function is a rich field for further investigation, and raccoons, as well as primates, offer promising opportunities for our enlightenment.

Acknowledgements. Raccoon studies were supported by NSF grants NB 78-00879 and NB 81-080731. Subjects were obtained with the cooperation of the Division of Wildlife, State of Michigan Department of Natural Resources and the State of Indiana Department of Fish and Game.

References

AKERT K, HARTMANN-VON MONAKOW K, KÜNZLE H (1979) Projection of precentral motor cortex upon nucleus medialis dorsalis thalami in the monkey. Neurosci Lett 11: 103-106

BOIVIE J, BOMAN K (1981) Termination of the separate (proprioceptive?) cuneothalamic tract from external cuneate nucleus in monkey. Brain Res 224: 235-246

BRODMANN K (1905) Beiträge zur histologischen Lokalisation der Grosshirnrinde. III. Die Rindenfelder der niederen Affen. J Physiol Neurol Lpz 4: 177-226

CLELAND CL, RYMER WZ, EDWARDS FR (1982) Force-sensitive interneurons in the spinal cord of the cat. Science 217: 652-655

DYKES RW, RASMUSSEN DD, HOELTZELL PB (1980) Organization of primary somatosensory cortex in the cat. J Neurophysiol 43: 1527-1546

DYKES RW, SUR M, MERZENICH MM, KAAS JH, NELSON RJ (1981) regional segregation of neurons responding to quickly adapting, slowly adapting, deep and pacinian receptors within thalamic ventroposterior lateral and ventroposterior inferior nuclei in the squirrel monkey (*Saimiri sciureus*). Neuroscience 6: 1687-1696

GORSKA T (1974) Functional organization of cortical motor area in adult dogs and puppies. Acta Neurobiol Exp 34: 171-203

HARDIN WB, ARUMUGASAMY N, JAMESON HD (1968) Pattern of localization in 'precentral' motor cortex of Raccoon. Brain Res 11: 611-627

HERRON P (1978) Somatotopic organization of mechanosensory projections to SII cerebral neocortex in the raccoon *(Procyon lotor)*. J Comp Neurol 181: 717-728

HERRON P (1980) Intrahemispheric and commissural connectioons between and within commissuraly and noncommissurally interconnected regions in the primary and secondary somatic sensory cerebral cortex: their possible role in interhemispheric tranfer of learning in the raccoon. PhD dissertation, Psychology and Neuroscience, Michigan State University, East Lansing

HERRON P (1983) The connections of cortical somatosensory areas I and II with separate nuclei in the ventroposterior thalamus in the raccoon. Neuroscience 8: 243-257

JAMESON HD, ARUMUGASAMY N, HARDIN WB (1968) The supplementary motor cortex of the raccoon. Brain Res 11: 628-637

JOHNSON JI, MICHELS KM (1958) Learning sets and object-size effects in visual discrimination learning by raccoons. J Comp Physiol Psychol 51: 376-379

JOHNSON JI, WELKER WI, PUBOLS BH (1968) Somatotopic organization of raccoon dorsal column nuclei. J Comp Neurol 132: 1-44

JOHNSON JI, OSTAPOFF E-M, WARACH S (1982) The anterior border zones of primary somatic sensory (SI) neocortex and their relation to cerebral convolutions, shown by micromapping of peripheral projections to the region of the fourth forepaw digit representation in the raccoon. Neuroscience 7: 915-936

JONES EG, FRIEDMAN DP (1982) Projection pattern of functional components of thalamic ventrobasal complex on monkey somatosensory cortex. J Neurophysiol 48: 521-544

JONES EG, PORTER R (1980) What is area 3a? Brain Res Rev 2: 1-43

JONES EG, COULTER JD, HENDRY SHC (1978) Intracortical connectivity of architectonic fields in somatosensory, motor and parietal cortex of monkeys. J Comp Neurol 181: 291-348

JONES EG, FRIEDMANN DP, HENDRY SHC (1982) Thalamic basis of place- and modality-specific columns in monkey somatosensory cortex: a correlative anatomical and physiological study. J Neurophysiol 48: 545-568

KAAS JH, NELSON RJ, SUR M, MERZENICH MM (1981) Organization of somato-sensory cortex in primates. In: SCHMITT FO, WORDEN FG, ADELMAN G, DENNIS SG (eds) The organization of the cerebral cortex. MIT Press, Cambridge, pp 237-261

KIEVIT J, KUYPERS HGJM (1977) Organization of the thalamo-cortical connexions to the frontal lobe in the rhesus monkey. Exp Brain Res 29: 299-322

KÜNZLE H (1978) An autoradiographic analysis of the efferent connec-tions from premotor and adjacent prefrontal regions (areas 6 and 9) in *Macaca fascicularis*. Brain Behav Evol 15: 185-234

MAENDLY R, RUEGG DG, WIESENDANGER M, WIESENDANGER R, LAGOWSKA J, HESS B (1981) Thalamic relay for group I muscle afferents of forelimb nerves in the monkey. J Neurophysiol 46: 901-917

McKENNA TM, WHITSEL BL, DREYER DA (1982) Anterior parietal cortical topographic organization in macaque monkey: a re-evaluation. J Neu-rophysiol 48: 289-314

NELSON RJ, KAAS JH (1981) Connections of the ventroposterior nucleus of the thalamus with the body surface representations in cortical areas 3b and 1 of the cynomolgus macaque, *(Macaca fascicularis)*. J Comp Neurol 199: 29-64

OSTAPOFF E-M (1983) The medullary sources of projections to the kinesthetic thalamus in raccoons. PhD dissertation, Psychology and Neuroscience, Michigan State University, East Lansing

PHILLIPS CG, POWELL TPS, WIESENDANGER M (1971) Projection from low threshold muscle afferents of hand and forearm to area 3a of baboon's cortex. J Physiol (Lond) 217: 419-446

SAKAI ST (1980) The thalamic connectivity of the primary motor (MI) and the supplementary motor (MII) cortices in the raccoon. PhD dissertation, Psychology and Neuroscience, Michigan State University, East Lansing

SAKAI ST (1982) The thalamic connectivity of the primary motor cortex (MI) in the raccoon. J Comp Neurol 204: 238-252

STRICK PL (1976) Anatomical analysis of ventrolateral thalamic input to primate motor cortex. J Neurophysiol 39: 1020-1031

WELKER WI, JOHNSON JI (1965) Correlations between nuclear morphology and somatotopic organization in ventrobasal complex of the raccoon's thalamus. J Anat 99: 761-790

WELKER WI, SEIDENSTEIN S (1959) Somatic sensory representation in the cerebral cortex of the raccoon *(Procyon lotor)*. J Comp Neurol 111: 469-501

WHITSEL BL, DREYER DA, ROPPOLO JR (1971) Determinants of body representation in postcentral gyrus of macaques. J Neurophysiol 34: 1018-1039

WIENER SI (1983) Kinesthetic and cutaneous mechanosensory regions of the ventrobasal thalamus of raccoons as determined by electrophysio-logical mapping of projections in relation to distributions of cyto-

chrome oxidase activity, acetylcholinesterase activity, and Nissl cytoarchitecture. PhD dissertation, Psychology and Neuroscience, Michigan State University, East Lansing

WOOLSEY CN (1958) Organization of somatic sensory and motor areas of
the cerebral cortex. In: HARLOW HF, WOOLSEY CN (eds) Biological and
biochemical bases of behavior. University of Wisconsin Press, Madi-
son, pp 63-81

Subject Index

arm movement   175-183, 196-210

blood flow, regional cerebral
   96-110

cerebellar nuclei   203, 248-258,
   259-274
cerebral cortex,
   neuron responses area 1   1-16,
   17-43, 59-76, 275-287, 294-312
--, -- area 2   1-16, 17-43,
   44-58, 59-76, 275-287, 294-312
--, -- area 3a   59-76, 294-312
--, -- area 3b   1-16, 17-43,
   59-76, 275-287, 294-312
--, -- area 4   17-43, 68,
   130-154, 155-174, 175-183,
   184-195, 196-210, 211-231,
   259-274, 275-287, 294-312
--, -- area 5   175-183
--, -- area 6   17-43
--, -- area 7   196-210
--, -- prefrontal cortex
   288-293
--, -- somatosensory area II
   14, 294-312
--, -- supplementary motor area
   184-195, 294-312
--, topography area 1   1-16,
   17-43, 294-312
--, - area 2   1-16, 17-43,
   44-58, 294-312
--, - area 3a   294-312
--, - area 3b   1-16, 17-43,
   294-312
--, - area 4   17-43, 259-274,
   294-312
--, - area 6   17-43
--, - somatosensory area II
   14, 294, 312

--, - supplementary motor area
   294-312
coactivation of forearm muscles
   248-258
cooling, local cortical
   275-287
cortical hand representation,
   motor   32, 123, 259-274
---, sensory   1-16, 23, 44-58
corticomotoneuronal cells
   142-154, 155-174, 211-231
corticospinal   155-174, 211-231

detection, effect of lesions
   98-110
-, tactile   93-98
-, vibratory   59-76
discrimination, effect of lesions
   8-16, 93-110
-, shape   104-110
-, size   8-16, 52
-, tactile   1-16, 17-43, 44-58,
   77-92, 93-110
-, texture   8-16, 17-43, 52,
   77-92

EMG   155-174, 211-221, 248-258,
   259-274
evoked potentials, cortical
   259-174, 275-287

field potentials, cortical
   275-287
finger localization   238-247
finger movements   155-174,
   248-258

grasping   111-129, 155-174,
   248-258
gratings   17-43, 77-92

Experimental Brain Research

Supplement 1:

Afferent and Intrinsic Organization of Laminated Structures in the Brain

(7th International Neurobiology Meeting)
Editor: **O. Creutzfeldt**
1976. I27 figures. XXIII, 579 pages. ISBN 3-540-07923-8

Supplement 2:

Hearing Mechanisms and Speech

EBBS-Workshop, Göttingen, April 26–28, 1979
Editors: **O. Creutzfeldt, H. Scheich, C. Schreiner**
1979. 85 figures, 12 tables. XXIII, 413 pages. ISBN 3-540-09655-8

Supplement 4:

The Renin Angiotensin System in the Brain

A Model for the Synthesis of Peptides in the Brain
Editors: **D. Ganten, M. Printz, M. I. Phillips, B. A. Schölkens**
1982. 108 figures, 46 tables. XVII, 385 pages. ISBN 3-540-11344-4

Supplement 5:

The Aging Brain

Physiological and Pathophysiological Aspects
Editor: **S. Hoyer**
1982. 52 figures, 66 tables. XIV, 281 pages. ISBN 3-540-11394-0

Supplement 7:

Neural Coding of Motor Performance

Editors: **J. Massion, J. Paillard, W. Schultz, M. Wiesendanger**
1983. 88 figures, 7 tables. XI, 348 pages. ISBN 3-540-12140-4

Supplement 8:

Sleep Mechanisms

Editors: **A. Borbély, J. L. Valatx**
1984. 53 figures. XIII, 315 pages. ISBN 3-540-13146-9

Supplement 9:

Sensory-Motor Integration in the Nervous System

Editors: **O. Creutzfeldt, R. F. Schmidt, W. D. Willis**
1984. 194 figures. Approx. 490 pages. ISBN 3-540-13680-0

Springer-Verlag
Berlin
Heidelberg
New York
Tokyo

Related titles

C. Porac, S. Coren
Lateral Preferences and Human Behavior
1981. 21 figures. XII, 283 pages. ISBN 3-540-90596-0

J. Hyvärinen
The Parietal Cortex of Monkey and Man
1982. 85 figures. XI, 202 pages. (Studies of Brain Function, Volume 8). ISBN 3-540-11652-4

E. Zrenner
Neurophysiological Aspects of Color Vision in Primates
Comparative Studies on Simian Retinal Ganglion Cells and the Human Visual System

1983. 71 figures. XVI, 218 pages. (Studies of Brain Function, Volume 9). ISBN 3-540-11653-2

G. A. Orban
Neuronal Operations in the Visual Cortex
1984. 188 figures. XV, 367 pages. (Studies of Brain Function, Volume 11). ISBN 3-540-11919-1

Neuroethology and Behavioral Physiology
Roots and Growing Points

Editors: F. Huber, H. Markl
1983. 183 figures. XVIII, 412 pages. ISBN 3-540-12644-9

Foundations of Sensory Science
Editors: W. W. Dawson, J. M. Enoch
1984. 190 figures. X, 577 pages. ISBN 3-540-12967-7

Cognition and Motor Processes
Editors: W. Prinz, A. F. Sanders
1984. 34 figures. X, 378 pages. ISBN 3-540-12855-7

Self-Regulation of the Brain and Behavior
Editors: T. Elbert, B. Rockstroh, W. Lutzenberger, N. Birbaumer
1984. 115 figures. XIV, 360 pages. ISBN 3-540-12854-9

Springer-Verlag
Berlin
Heidelberg
New York
Tokyo